W0245968

Communications and Control Engineering Series

Editors: B.W. Dickinson · A. Fettweis · J. L. Massey · J.W. Modestino
E. D. Sontag · M. Thoma

Torsten Bohlin

Interactive
System Identification:
Prospects and Pitfalls

With 53 Figures

Springer-Verlag
Berlin Heidelberg NewYork
London Paris Tokyo
Hong Kong Barcelona Budapest

Prof. TORSTEN BOHLIN
Automatic Control
Royal Institute of Technology
S-10044 Stockholm
Sweden

ISBN 978-3-642-48620-3   ISBN 978-3-642-48618-0 (eBook)
DOI 10.1007/978-3-642-48618-0

© Springer-Verlag Berlin, Heidelberg 1991
Softcover reprint of the hardcover 1st edition 1991

61/3020 543210 - Printed on acid-free paper

To Birgitta

Preface

The craft of designing mathematical models of dynamic objects offers a large number of methods to solve subproblems in the design, typically parameter estimation, order determination, validation, model reduction, analysis of identifiability, sensitivity and accuracy. There is also a substantial amount of process identification software available. A typical 'identification package' consists of program modules that implement selections of solution methods, coordinated by supervising programs, handling file administration, operator communication, and presentation of results. It is to be run 'interactively', typically on a designer's 'work station'.

However, it is generally not obvious how to do that. Using interactive identification packages necessarily leaves to the user to decide on quite a number of specifications, including which model structure to use, which subproblems to be solved in each particular case, and in what order. The designer is faced with the task of setting up cases on the work station, based on apriori knowledge about the actual physical object, the experiment conditions, and the purpose of the identification. In doing so, he/she will have to cope with two basic difficulties: 1) The computer will be unable to solve most of the tentative identification cases, so the latter will first have to be formulated in a way the computer can handle, and, worse, 2) even in cases where the computer can actually produce a model, the latter will not necessarily be valid for the intended purpose. The field of process identification is riddled with 'pitfalls', and recognizing that, the user must decide whether he/she can trust the result of the computer's calculations. Even with the best available program support identification is still a difficult task.

This means, as in other cases of Computer–Aided Design, that the quality of the result depends critically on how skillful the designer is in handling the tools. In order to get such skill the designer may basically consult text books or take courses in 'Process Identification' for support. But the text books contain mainly descriptions of methods to choose from, and theoretical properties of these methods, and the user will still have to decide whether the *assumptions* behind any particular method are valid in his/her particular case.

This book deals with the fundamental problems and possibilities of making a mathematical model of a physical object, and what one has to think of in order to do it correctly. It points out where the limits are, *i.e.* what can and cannot be inferred about the object under various circumstances. It introduces principles and general approaches to model design. It illuminates the 'pitfalls', *i.e.* approaches that may appear feasible, but which may easily go wrong. It uses abstract terms and general reasoning, in order to stress the generality of the problems, but leaves out technicalities. A basic assumption is that probabilistic ideas can be applied, in particular those of Bayes and of testing statistical hypotheses.

The purpose of the book is twofold:
1) To serve as a text book for the designer of stochastic dynamic models, in particular for the user of interactive identification software (present and coming). It should be possible to use also in graduate courses in process identification. In both cases it must be supplemented by another book treating the mathematical *techniques* of model design, but several excellent books on this topic have been published. This books stops where other text books start.
2) To lay a basis for the construction of operator's–guide functions for identification packages. The results of the analysis of the interactive model making process is presented in such detail that it should be possible to use as a basis for writing an operator's guide. (Chapters 7 – 10 contain general algorithms for modelling and identification in nonlinear stochastic cases.)

In this sense the book is both a text book and a monograph. It suggests a new and comprehensive approach to the model making problem, but also explains well–known ideas and concepts in a way that should be possible to understand intuitively. It is not an exhaustive treatise on modelling techniques and methods. It attempts to *unify* rather than survey.

The book stresses the intuitive interpretation of the theoretical assumptions behind identification methods, but does not linger on theoretical derivations. The reason is, again, that assumptions are the responsibility of the user of a method, while conclusions drawn from given assumptions are not; once they are known they can be programmed into a computer and hence arrived at automatically without interference from the user. However, the book will not be free of theory or derivations, since identification methods are in fact based on theory, sometimes quite deep, and an understanding of what the theory means in practice (and hence what the identification program will do) cannot reasonably be arrived at without reference to the same theory. The main tool will be applied mathematics with emphasis on *applied*. In particular, elementary probabilistic concepts will be used frequently and interpreted intuitively without reference to the usual 'probability space' $(\Omega, \mathcal{F}, P)$. The necessary elements in that space will just be assumed properly defined.

Hence, the book uses more *words* than is customary in a treatise on identification. The reason is that words are the tools we have to connect the real world with the abstract world of mathematics, and the latter is the role of the human designer in the interactive process. And, of course, mathematics will have to be used too, since this makes it possible to derive algorithms, which is the language computers understand.

Acknowledgement

The author has been influenced by long discussions with Dr S. Graebe on various technical and philosophical aspects on identification, in particular on the modelling of disturbances and the role of the elusive 'true model'. Dr Graebe also wrote the computer programs that make it possible to realize much of the prospects in this book.

Contents

1 Introduction

Design of mathematical models of physical objects involves a great number of conditions, aspects, and problems. Actual cases differ in purpose, access to the object for experiments, means for computing, and apriori information. Some are not very well defined in practice. The topic is therefore too diversified to be solved by means of a 'recipe' for what is correct model design under given circumstances. An engineer designing models must use 'art' as well as 'science'.

The fact that there is computer software available for 'process identification' only helps some. Even if identification software is no doubt a necessary tool in model design, and keeps developing in scope, power, and ease of use, it still requires *input* in the form of *specifications*. The specifications are not always obvious to the user. He/she has also to define the *problems* that the software is to solve. There is normally no program command that means "Make a model from this data". (If there would be one, and you would be tempted to use it, you should be very certain that you know how the program operates and what it produces, since *you* are the one who is responsible for the model, not the company that sold you the program.)

Ljung (1982b) formulated a sequence of questions that meet the potential user of identification methods:
– "Has System Identification anything to offer for my problem?"
– "How should I design the identification experiment?"
– "Within what set of models should I look for a suitable description of the system?"
– "What criterion should I use for selecting that model in the model set that 'best' describes the data?"
– "Is the obtained model 'good enough' for my problem?"

A satisfactory answer to each question raises the next one, and none can usually be answered by running some computer software. The answers depend on what one knows beforehand about the object, and what one wants to do with it, and the dependences do not lend themselves easily to analysis by computers, at least not by today's software. Identification 'tool boxes' can only give partial answers, and the user must answer what remains. To be able to do this the user must know about the problems involved in the modelling of reality, as well as the powers and limitations of the available techniques. These problems are the main topic in the present book.

The introductory text will review the basic conditions for model design. Generally, one has to take into account:

● The art, custom, and lore of this particular trade. They affect the terminology and the thinking.

● The software that will *use* the model. That will affect the model description, and hence the choice of method.

● The purpose of the modelling. That will affect the model description, the necessary accuracy, and the experimentation.

● The experiment facilities. They affect the accuracy and sometimes even the feasibility of designing models.

● The apriori information on the object. That affects profoundly the descriptions, methods, and achievable accuracy.

Sometimes one will also have to take into account the limitations of the software that is to be used for the model design. But they are not basic and therefore not a subject of this introduction — there may be better software in the future.

> **Remark:** Willems (1987) has questioned the common belief that 'science' must be supported by 'art' in system identification, at least for linear systems.

> **Remark:** A short history of the development of system identification is given by Eykhoff (1987). There are also several elementary surveys on its basic problems and techniques; see for instance (Åström, 1972, 1980; Strejc, 1981; Ljung, 1982b, 1987c; Eykhoff, 1988).

1.1   The terminology

The term 'process identification' or 'system identification' (applied to an 'object') has somewhat ambiguous meanings in the literature. Loosely, 'identification' is a tool for solving the following type of problems: One wants to know about important properties of the object's behaviour when stimulated in various ways. By observing the responses of the object when subject to different kinds of stimuli, one hopes to discover some regularity in the behaviour that can be formulated into a law or *mathematical model*. Literally, 'identification' means "finding out who the object is", "putting a label on it that points out the particular object among a set of possible objects", or, in analogy with criminology: "associating its fingerprints and other hard evidence with a particular individual". In process identification the 'fingerprints' are data records of stimuli and responses, and the 'individual' is a mathematical model, picked from one or more classes of models.

> **Remark:** In the usual application to dynamical systems the problem of 'identification' (find the dynamics, given input and output signals) may be regarded as an inverse problem to 'analysis' (find the output, given input and dynamics). This indicates that identification is, mathematically, a more difficult problem; there will obviously be some invertibility conditions for its solution.

> **Remark:** A third aspect is to regard identification as 'the (indirect) measuring of dynamics', usually for the purpose of providing data for design of control of the object.

Another term, '(mathematical) modelling' is closely related to 'identification', and would literally mean the same. However, it has come to mean in practice "setting up a mathematical model (for an object) based on other data than the object's response". In particular, for technical systems one can sometimes use socalled 'basic principles' of physics, chemistry, mechanics, and electricity. Hence, modelling does not necessarily involve any experimentation; one may be so sure of ones knowledge about the object, that one need not verify the model further.

Because of the more or less established interpretations of the two terms (and because one should not waste good words by letting several words mean the same), the following definitions will be used in this book (see also section 9.7 on "Terminology revisited"):

● *Identification* means finding a mathematical model of a physical object, given a class of tentative models and given a set of response data from experiments.

● *Modelling* means constructing a class of tentative models, based on any source of information (possibly including experiment data).

With these meanings the terms are complementary, and, in general, both tasks have to be carried out in order to provide a good model. This will be called 'model design' or 'model making'.

Often in practice 'modelling' involves more construction type of work, while 'identification' involves more computing. The designer often tends to concentrate his/her effort on one of the tasks, depending on skill and personal preference. One can generally distinguish between two approaches (Ljung, 1987a):

● **Emphasis on modelling:** Assemble equations (*e.g.* differential, algebraic, or integral) relating variables of interest and based on physical principles, empirical relations, or other apriori information. The set of equations (the 'model structure') normally contains a number of parameters, whose values are unknown apriori. The model is then constructed by the three steps of writing a program subroutine for the system of model equations, combining that with one or more library routines for solving the equations, and finally using another library routine for optimizing the unknown parameters to a best fit of model output to measured data. One aims at finding an 'internal' model, *i.e.* one whose structure approximates that of the object. Models are typically deterministic and nonlinear.

● **Emphasis on identification:** Assume that the object's behaviour can be described sufficiently well by some member of a sufficiently general standard class of models, at least for a given purpose. One may then construct the model more or less automatically by applying algorithms for statistical inference. This will provide estimates of parameters, the values of which point out one model in the given class. The parameters are either real–valued coefficients or integer–valued order numbers or structure indices. In socalled 'non–parametric' identification methods they are functions or sequence of values characterizing the model. One aims at finding an 'external' model, *i.e.* a model that is able to describe the input–output behaviour of the object (for the given purpose), but which does not necessarily have a structure that even resembles that of the object. Models are typically stochastic and linear.

The second case is generally called 'black box' identification, since there is no information of what is inside the object, neither required before, nor obtained after identification; only how its outputs respond to input stimuli is of interest. Contrasting this, the first case would correspond to a 'transparent box', since it is possible to peek inside and describe its structure, even if not to measure its parameters directly. However, for symmetry, and somewhat illogically, the latter case is usually called 'white box', and this term will be used in the sequel.

The modelling part is probably the most crucial step for the success of the model making (Gevers and Bastin, 1982). It is therefore unfortunate that it is also the task least supported by general theory, and consequently by computerized guidelines. Admittedly, there is modelling software available with a high degree of sophistication in the technical *specification* of the model (Jamshidi and Hergel, 1985; van den Boom, 1988; Nakamori, 1989). It is much more difficult to design software that also helps the user check whether the specifications are reasonable, considering the *purpose* of the modelling and the available *experiments*.

> **Remark:** In the case of parametric models the term 'parameter estimation' is sometimes used as a synonym for 'identification', since the latter is often done by estimating unknown parameters in a given model structure (Eykhoff, 1974). The International Federation of Automatic Control (IFAC) does distinguish between the two terms — its regular conferences on the subject bears the title "Identification and System Parameter Estimation". Caines (1986) defines 'system identification' as 'modelling' followed by 'parameter estimation'. The definitions of 'modelling' and 'identification' adopted here follow Ljung (1987a). However, he uses the term 'grey box' for the 'white box', which is not sufficient for the purpose of this book. The attribute 'grey' will be reserved for mixed 'black' and 'white' boxes, and the distinction is important for the sequel (see section 5.9 on "'black–box' and 'grey–box' models").

> **Remark:** The term 'model structure determination' is sometimes used as a synonym for 'modelling'. In practice determining a model structure means selecting a structure from an enumerable set of structures, where the members are distinguished by structure indices, or order numbers, or otherwise (Karny, 1983; Söderström and Stoica, 1989) — what else is there to do? A more adequate term would therefore be 'model structure selection' (Stoica *et al*, 1986; van Overbeek and Ljung, 1982). Determining integer parameters is also referred to as 'structure identification' (e.g. Natke and Samirovski, 1988). For definitions of these and related terms see chapter 9.

> **Remark:** The terms 'external' and 'internal' models are occasionally used in different meanings in the literature (see section 5.5). So are other general concepts, like 'validation' (see section 4.1).

1.2   The software

The approach to model making is also determined by how one intends to use the result. In particular, the form of the model must suit available analysis and synthesis software that is to use the model for input. Packages for control design (other than deterministic optimization) typically use 'black–box' models, since only an 'external' model is needed for that purpose, and available design methods work only for given and limited classes of models, mainly linear. Simulation and optimization packages are able to use 'white–box' models. In the first case the user of suitable identification software is asked for only little information about the object, in the second case to provide a full description of the model structure. The first case allows severely contaminated data, the second case does not. In the first case the feasible model structures are limited by available identification methods, usually statistical, and in the second case the identification methods are limited by the model structure, usually to those fitting model output to data. In both cases the structures are also limited to those suiting the purpose. For control design the 'black box' assumption is certainly appropriate in many cases; it has the very favourable effect of producing simple and useful feedback controllers. In that case it may not be relevant to ask whether the model structure is 'right' or not; the main requirement is that the model give a good output prediction.

In many cases, however, the user of an identification package will not have enough apriori information of the object to be able to specify the complete structure of a 'white–box' model. Assuming a 'black–box' model, on the other hand, will cause difficulties, if the partial information the user does have is not compatible with any of the model classes incorporated in the package. This means that he/she will have to ignore some of the apriori information. That has obvious drawbacks:

● It is a waste of information, logically resulting in less accurate models (Eykhoff, 1974).

● The structure (whether linear or not) belongs to a given class, prescribed by the designer of the package. Only exceptionally does that happen to agree with what the user would like to have, whether the latter stems from existing theory about the object, fits the user's purpose, or simply accords with what is the art of his/her particular trade.

● The parameters are often not possible to interpret physically or intuitively, and thus eludes further checking with other information sources, such as literature or other experiments.

● The 'black–box' model does not basically separate the object from the process of experimentation, and that makes it difficult to compare the results of differently designed experiments.

In practice, the amount of apriori information may range from a full specification of model equations to just a surmise that, for instance, a particular group of variables cannot reasonably affect another particular group. The cases in between full structure knowledge and the 'black box', will be called 'grey–box identification', since the object is analogous to a box where some structure can be seen through the casing, but not all. The approach of this book rests on the assumption that, after one has written down the *hypotheses* one may have, and in general all one knows about the object and the experiment condition, there generally remain as unknowns *i)* the values of some real or integer parameters, *ii)* variables whose variation can be described by stochastic time series, *iii)* unspecified but causal relations, and *iv)* an uncertainty as to whether particular parts of the collection of tentative equations, variables, and parameters are either adequate or relevant for the particular purpose of the model making. One may possibly be able to order the hypotheses after increasing degree of uncertainty. To this apriori information may or may not be added assumptions of apriori distribution or admissible ranges of some of the unknown parameters. If a particular, unknown variable is related to other variables in a completely unknown way (except causality), one must assume a 'black box' for that relation.

'Black–box' model design needs only identification, and the 'white box' needs only modelling, while the 'grey box' must involve both, since the result of the modelling is not certain and must be verified by identification.

1.3 The purpose

A need for identification may arise in practice when one wants to manipulate a physical object in such a way that its behaviour is more to ones liking, but its responses to the kind of control one is able to exert are not known sufficiently well. The purpose is then to use the model as a test case for investigating the effects of different control strategies (for instance by simulation), or as input data to general synthesis procedures, or simply to forecast what will happen in case of no control.

This means that one has implicitly defined a time segment, and divided it in two parts, *viz.* the *identification* and the *application phases*: During the first phase the object is stimulated in a particular way designed for identification, and the behaviour of the object is observed. At the end of the phase a model is constructed, that is, a concise description of the behaviour during this phase is formulated, usually in mathematics. After that, two things are usually required:

● *Representability*: The same description should be valid for its purpose, when it is applied during the second phase.

● *Accuracy*: The description should be sufficiently accurate for the purpose.

Representability means usually a sort of invariance property, or in case that does not hold, it should not matter for the purpose. Variables are obviously not invariant as a rule, and dynamics may or may not be invariant. But *something* must be invariant to make identification meaningful.

Three things limit the accuracy that can be attained: *i*) the length of the identification phase, *ii*) any constraints put on the stimulation of the object, and *iii*) the fact that the object characteristics sometimes change with time in an unpredictable manner. The first and second restrictions can often be relaxed at an increased cost (in terms of measurement cost, computing cost, storage cost, or production loss), but the third one may put a definite limit to the accuracy that can be achieved.

> **Remark:** Object characteristics that change unpredictably may also be regarded as a violation of the condition of Representability. That condition can be trusted to hold only for a limited time segment in practice (except for certain 'natural laws'). If the application is to last longer, the application phase must be followed by another identification phase, and so on. However, provided the identification phase can be made long enough to make the Accuracy condition a reasonable one, the need for repeated identification does not complicate the problem.

A number of cases may be distinguished here, depending on the general conditions for the identification:

● **Open–loop identification:** During the application phase the behaviour of the object is determined by its purpose. During the identification phase the same behaviour may not suit the purpose of identification, and, therefore, the very purpose of the object may constitute a constraint on its identification. If the problem allows that this constraint be relaxed during the identification phase, the object may be stimulated in a more efficient way. This is the ideal case for identification.

● **Closed–loop identification:** When the purpose of the object cannot be ignored for a sufficiently long time to identify it, the identification phase must overlap or coincide with the previous application phase. This means stimulating the object in a particular way, given by the purpose, and may therefore result in inferior models. It may also raise problems of a basic nature (identifiability), since there are simple and yet realistic cases where identification is not possible under closed loop (see Example 1.1). However, the Representability condition still holds, so that, if the object actually changes with time, it must do this so slowly that the changes over the interval it takes to identify it with sufficient accuracy (for the purpose) are negligible.

● **Real–time identification:** Logging input and output during the identification phase often yields a large amount of data, which has to be stored away for processing by the identifier. However, if the latter allows data to be processed, as time progresses, in such a way that the storage space needed does not grow with time (and *using* the measurement gauges does not cost much more than *having* them), the length of the identification phase (and hence the model accuracy) is limited only by the Representability condition. Real–time identification is treated by Ljung and Söderström (1983) and, more briefly, by Ljung (1987a).

● **Adaptive control (or stochastic control):** In some cases the interval where Representability holds may be so short that the data collected during the application phase will not allow identification with sufficient accuracy. In such cases separate identification phases usually cannot be allowed. Therefore the control strategy must take into consideration that the model is uncertain. One is led to a stochastic control problem. It has been shown that the object, when subject to the optimal control strategy, will behave as if the strategy had a dual purpose, *viz.* to control the object, and simultaneously to stimulate it in such a way that it can be identified (Fel'dbaum, 1965). The optimal strategy strikes a compromise between the two conflicting purposes of 'stimulating' and 'directing' (Eykhoff, 1974).

> **Remark:** The 'open' and 'closed–loop' cases are also called 'batch' identification, as opposed to 'real–time' identification, since a batch of data is available and can be stored for repeated processing if necessary, before the model will be needed. That is the case treated in this book.

─────────────── **Example 1.1** ───────────────

The example illustrates the four types of conditions for identification.

Let the object be a linear, discrete–time, first–order system with input noise (for instance a stirred–tank mixer with controlled input flow and measured concentration):

$$y(k) = a\, y(k-1) + b\, u(k-1) + \omega(k)$$

where

k = discrete time (= the sampling counter)
u = the control variable
y = the measured variable
ω = random disturbance
a,b = parameters.

Open–loop identification:

Let the parameters (a,b) be invariant and unknown. Let the input signal u be an independent sequence of numbers, for instance from a random–number generator. Then a and b can be estimated from sequences $\{u(k)\}$ and $\{y(k)\}$.

Closed–loop identification:

Let the parameters (a,b) be invariant and unknown. Let the input signal u be generated by the control law $u(k) = -y(k)$. This means that $y(k) = (a - b)\, y(k-1) + \omega(k)$. Since the dynamics of the closed system depend only on the difference $a - b$, and since the u–sequence does not carry more information than the y–sequence, the parameters a and b cannot be determined independently. If instead the controller is delayed one step, so that $u(k) = -c\, y(k-1)$, the closed system will be $y(k) = a\, y(k-1) - b\, y(k-2) + \omega(k)$, and both a and b can be estimated from the y–sequence.

Real–time identification:

Let the parameters (a,b) be invariant and let only b be unknown. Let the input signal u be generated by the control law $u(k) = -c\, y(k)$. Let the feedback gain c be weak, so that $c\, b \ll 1$. The latter condition implies that the data sequence $\{y(k)\}$ has to be very long to allow a sufficiently accurate estimation of b, and longer than the storage may hold. Data must therefore be processed using a statistic x of fixed length, *i.e.* by applying a 'recursive' algorithm of the type

For $k = 1,2,\dots$, repeat

$$x(k+1), \hat{b}(k) \leftarrow \textit{Real–time\_identifier}\,[y(k), x(k)]$$

such that $\hat{b}(k)$ tends to b as k tends to infinity. This means a constraint on feasible identification methods.

Adaptive control:

Let the a–parameter be invariant and known, but let the b–parameter be unknown and vary according to the random process

$b(k) = b(k-1) + \omega'(k)$.

The instantaneous values can still be estimated by an algorithm

$x(k+1), \hat{b}(k) \leftarrow Real-time\_identifier[y(k),x(k)]$

but the estimate $\hat{b}(k)$ will not converge, and will not equal $b(k)$. Let the input signal u be generated by an 'adaptive' control law, that is, one that uses the estimate $\hat{b}(k)$ and the fact that it is an uncertain estimate of the true $b(k)$

$u(k) \leftarrow Controller[y(k),x(k),\hat{b}(k)]$.

Generally, the two algorithms applied to the object make a nonlinear dynamic system, which is difficult to analyse, and even more difficult to design optimally (Goodwin and Sin, 1984; Åström and Wittenmark, 1989).

Forecasting and control are not the only purposes of identification. One may simply want to find out something about the object from available data, or, more ambitiously, to aim at finding 'truth' (like in Natural Sciences). Whether that purpose is meaningful or not will not be debated here, but the purpose of 'data compression' is a sound motif for wanting nothing more than to describe a segment of data. In the latter case there is no obvious 'application phase'.

Example 1.2

This is a case of modelling where the purpose is 'data compression'.

In the transmission of speech signals over a noisy channel it is advantageous to digitize the signal in an 'encoder', send the digital code, and then create a copy of the original signal in a 'decoder' at the receiving station. The reason is that digital signals are less susceptible than analogue signals to channel noise. A primitive way of encoding would be to sample the signal at a rate well above twice the signal bandwidth (the Nyquist frequency), convert the amplitudes to digital form, and transmit the resulting sequence of binary digital numbers. However, one can achieve the same result by a more clever encoding, resulting in fewer bits per second, and thus one can save channel capacity:

Model the speech signal (including the sampling process) by a process of the following type

$y(k) + a_1 y(k-1) + ... + a_n y(k-n) = \lambda \omega(k)$

where, $y(k)$ is the amplitude value at sample point #k, $\{\omega(k)\}$ is a sequence of uncorrelated numbers ('white noise'), and $a_1, \ldots, a_n$ are constants characterizing the speech.

Let the encoder compute the variables

$$e(k) = y(k) + a_1 \, y(k-1) + \ldots + a_n \, y(k-n)$$

and transmit the sequence $\{e(k)\}$ instead of $\{y(k)\}$. The receiver will then reconstruct the speech signal from

$$y(k) = - a_1 \, y(k-1) - \ldots - a_n \, y(k-n) + e(k)$$

In addition the transmitter will have to send the parameters $a_1, \ldots, a_n$, and possibly also the start values $y(1), \ldots, y(n)$. However, the latter has to be done with a much lower frequency than the sampling frequency, possibly only at start up.

The point of taking this roundabout way is the following: If the power frequency spectrum of the model, which is (h is the sampling interval)

$$\Phi(f) = h \, \lambda^2 / |1 + a_1 \, \exp(2\pi i f h) + \ldots + a_n \, \exp(2\pi n i f h)|^2,$$

approximates that of the speech, then the variance of $e(k)$ will be smaller than that of $y(k)$. Hence, one can use a shorter word length for $e(k)$, and thus transmit at a lower bit rate. If the speech spectrum has much of its power distributed at frequencies that are low compared to the sampling frequency, the saving of channel capacity is large.

In this case one wants only a 'data description', because one wants to reconstruct the data sequence. One does not primarily want to *forecast* the speech or to know about its *source*. However, if one would want to design the channel capacity, one will have to model sufficiently much of the source. That is another purpose, and it requires some kind of invariance assumption, *i.e.* the condition of Representability (so that one will know that the $a_1, \ldots, a_n$ parameters are valid as long as the channel is connected).

1.4   The experiment facilities

Needless to say, the result of identification depends critically on the experiment. This does not only mean the amount of data the experiment will produce, its accuracy, and how free it is of contamination. As important is how the experiment was carried out. It is possible to establish rules for how an experiment must be designed to be 'proper', *i.e.* to be able to yield data, that makes it possible to infer enough for the given purpose (see chapter 3 on "The experiment").

However, often one cannot design the experiment exclusively for the purpose of identification, but has to cope with a number of constraints on the experimentation. This may be the case, for instance, when the experiment is an expensive operation and therefore must be designed to satisfy other purposes besides that of identification. In the extreme, but not uncommon case, one is just given a set of measurements of object input and output, or, even worse, of variables related to those one wants to study, and asked to find a model from that. The situation is typical with large industrial processes, and in social, ecological, biological, and economic studies. In the latter cases proper experimentation might not only be expensive and difficult, but also unethical. Then it remains only to measure the variations of the variables as they occur, or use historical data.

Theoretically, one cannot make models from such data without a risk, since recording data without knowing what produced the input stimulus is not a 'proper experiment'. The risk lies in the fact that one must make assumptions (about the source of the stimulus), that one cannot check. However, in practice the argument that the experiment is not 'proper' may not be considered strong enough to stop an important project. And so the model designer may be forced to make the assumptions, proceed as if they were true, and take the risk that they are not. It is easy to envisage that those assumptions will be precisely the prerequisites for the identification software he/she has available. And thus the model designer may be led (by circumstances) into a number of 'pitfalls'.

The situation would still not be serious, if these risks were merely academic, or rare, or small. Unfortunately, it appears that a number of common technical and economic constraints are such that the risk of producing a wrong model is substantial. Some constraints may appear severe at first sight and still be manageable, others may seem quite harmless and still render identification impossible (Bohlin, 1987a).

―――――――――――――― **Example 1.3** ――――――――――――――

This in an example of historical data.

The following is data on the yearly inflation in Sweden from 1732 to 1977. The data shows an approximately logarithmic long–time

Price index (logarithmic scale)

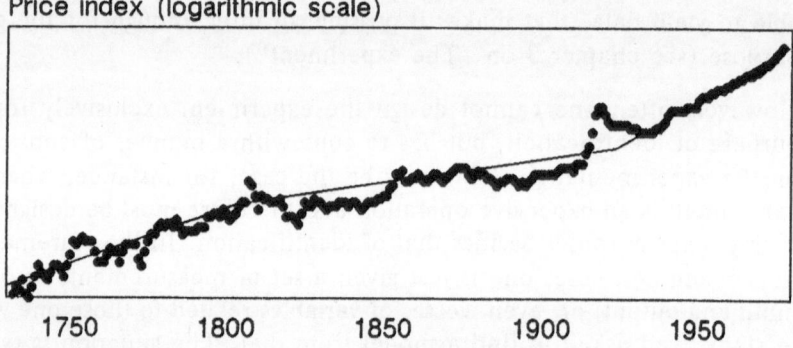

Fig. 1: Estimated yearly prices in Sweden.

growth with local disruptions during wartime. With the exception of those transients the growth is quite steady, even if one may possibly observe different slopes during shorter time intervals. One might further try and associate these different growth rates with the way the country was governed, whether by a sovereign king, or by an aristocratic, right–wing, center, or left–wing parliament. Or, one might simply observe that inflation always prevails in the long run, quite regardless of what government acted on what macro-economic theory, or without it. The latter might call for some interesting conclusions on the futility of economic government, but it would still not be proper identification to try and estimate the effect of government on inflation rate. Because data did not come from a proper experiment.

1.5   The model structure

The model structure (the form one writes the model in) is the result of 'modelling'. It is commonly constrained for two different reasons:

● To achieve *identifiability* from a given experiment — by employing enough apriori assumptions. A given experiment can only yield limited information. Hence, the set of 'all possible models' must be reduced to a set of 'admissible models' that are distinguishable by limited information. What more will be needed to describe the given object for the given purpose must come from apriori information.

● To limit programming and computing costs — by adopting the assumptions written into identification software.

These two reasons for constraining the model structure have different consequences in practice. Assumptions to ensure identifiability are unavoidable (with a given experiment), but also unfalsifiable from the experiment data only. This means that if one has made a wrong assumption, the model may pass all feasible tests and still be wrong (see section 4.6 on "The origin of 'pitfalls'"). In contrast, constraints for computational reasons are not unavoidable; one could get a more powerful computer or a less constrained program based on a more powerful method. And those constraints are falsifiable. For instance, if the object to be modelled is nonlinear, and the identification program assumes linearity, a proper falsification may still reveal that the linear model does not agree with data. Obviously, the first kind of error is the more serious one; the model is wrong, and one cannot know that, until the model has been applied and has failed its purpose.

---——————————————— **Example 1.4** ———————————————---

This very elementary example illustrates the relations between experiment, apriori information, identifiability, and falsifiability.

Assume from apriori that the pair of input and output variables c and d are connected by the affine relation $d = b\,c + v$.

If a given experiment yields one pair of data $(\mathbf{c},\mathbf{d})$, then one obviously cannot estimate both b and v, and the case is unidentifiable. If one assumes more apriori, for instance the value of v, then one can estimate b, and the case is identifiable.

If the experiment yields two pairs of data, then the case is identifiable regardless of whether one assumes v or not, and as long as one assumes there is an invariant affine relation.

If one assumes the value of v, then b can obviously be determined from either of the two pairs. However, if the value of v or the measurements or both are not exact, then the other data pair will not agree with the model, and the latter is falsified. Hence, the apriori assumption on v and errorless data can thus be put to test (because the double experiment is more informative than the single). Apriori information that can be put to test will be called 'hypotheses' in the sequel. If v is assumed unknown, the case is obviously unfalsifiable from two data pairs.

One can set up other hypotheses, for instance that the relation is quadratic, and falsify or verify those using more data.

In most practical cases at least the hypothesis of errorless data is wrong. This means that by collecting enough data one can falsify any deterministic model containing a given number of unknown parameters. In order to avoid this, one will have to provide for data error in the model structure. This will formally be detrimental to identifiability, but will obviously increase the chances of finding an unfalsified model. This becomes clear by trying to fit any polynomial of given order to more data in Example 1.4. That is possible only if a certain 'slack' is allowed between data and polynomial.

1.6   The philosophy

If one looks at the task of model design for an engineering purpose from a most fundamental point of view, it is not much different from the problem that concerns Natural Science. In both cases one wants to describe a piece of reality using mathematics. Hence, defining rules for how to do correct modelling and identification is akin to stating how to do Natural Science properly. The latter is a subject that has concerned philosophers for centuries. A short summary of modern ideas is given by Reckhow (1987). The principle of *Logical Empiricism* has most followers. It makes a good guideline to have its basic tenets of 'hypothesis', 'observation', and 'falsification' at the back of ones mind.

The principle of Logical Empiricism states that one should carry out the following formal procedure:

```
┌──────────────────────────────────────────────────┐
│            The principle of Logical Empiricism     │
└──────────────────────────────────────────────────┘
```

⌈ Repeat forever
│ ⌈ Refine the hypothesis
│ ⌈ Repeat until the hypothesis is falsified
│ │ ⌈ Refine the experiment and observe
│ │ ⌈ Test whether data falsify deducable consequences
⌊ ⌊ ⌊ of the hypothesis

——————— •••••• ———————

In addition, one should design the experiment and the test with the purpose of promoting falsification, but the hypothesis with the purpose of promoting verification.

> **Remark:** The scheme involves two kinds of logic (Reckhow, 1987): The derivation of observable consequences of the hypotheses is 'deductive' logic, *i.e.* reasoning from the general to the special. The refinement of hypotheses, however, requires 'inductive' logic, *i.e.* reasoning from the special (observations) to the general (a refined hypothesis). The latter logic is philosophically more debated.

> **Remark:** For a short survey of the deeper philosophical problems in model making and a list of references to the subject see Gaines (1977).

Needless to say, the principle of Logical Empiricism is very wide, and the actual way it will be used hinges on how one will interpret the verbal statements. Even so, it has the serious drawback in engineering applications that it does not explicitly consider the purpose of the model making. Chapter 4 will introduce an amendment for this, and specify the interpretations of the statements used in this book. In order to prepare for that, Chapter 2 will first introduce some means to cope with 'uncertainty'.

2 Randomness, probability, and likelihood

In most cases a physical object responds with different data on different occasions, even when subject to the same stimulus. The difference may be small enough to be negligible, but often it is not. Unless there is a change in response because the object depends on time t in a particular way that one would also like to model, one is faced with the following situation: One knows that response data is *random*, but one also expects that there is some *information* in the data that will be *invariant* and will hold also in the application phase. This information is obviously the most comprehensive 'model' one can ever get out of the particular data set, and any design for the application phase must be based on that invariant information. The problem of model making is then to *represent* the information in terms of what one knows, *i.e.* model structure (if the modelling part has been done), experiment specifications, and data, so that one can *compute* something. A way to find a computable representation is to use *probabilistic* concepts, *stochastic variables*, and the socalled *'Bayesian approach'*.

> **Remark:** The usefulness of Bayes' rule in formulating and solving all sorts of identification and control problems has been advocated by Peterka (1981). It is particularly useful in 'batch' identification — open and closed loop — (see section 1.3 on "The purpose"), where computations will be manageable in quite general cases. Bayes' rule also provides insight into the considerably more difficult case that identification is used recursively in conjunction with control in an 'adaptive' control system (Kramer and Sorensen, 1988a).

2.1 Bayes' idea

The idea of using probabilistic concepts for inference is based on that of conditional probability:

> **Recapitulation:** Let two stochastic variables x and y have a *joint probability density* $\{p[x,y]\}$, where (x,y) is in a suitable continuous

space. In order to interpret this, one may envisage a long series of *events* or *experiments* producing combinations (x,y) that occur randomly over the space of $\{x,y\}$. Then the frequency of occurrences of (x,y) within a small (dx,dy)–environment around (x,y) is $p[x,y]$ $dx\,dy$, so the quantity of $p[x,y]$ may be interpreted as the *density* of occurrences of (x,y) at (x,y). The distribution of x alone is the *marginal distribution* defined by the density $p[x] = \int p[x,y]\,dy$. If from a very long series of experiments one would pick out the subseries, where y occurred around the value y, one would get a series of events *conditional on y*. That series of events would have a *conditional distribution* defined by the density $p[x|y] = p[x,y]/p[y]$. It would be a distribution in the space of $\{x\}$ only, but with y as a fixed parameter.

A useful interpretation of the distribution $\{p[x]\}$ of a variable x is that it represents the 'information' one may have about x. The less dispersed that distribution is, the more accurate is ones information about x.

Now, if one wants to know more about x than given apriori by the distribution $\{p[x]\}$, one must make an *experiment* and observe a variable y that is related to x. When the experiment has been specified, the joint density $p[x,y]$ is defined and must be known (if one wants to apply Bayes' idea). The marginal distribution $\{p[x]\}$ is also called the *apriori* distribution, *i.e.* the distribution (information) of x *before* one has made any observations, or even decided what experiment to do and what variable to observe. When one has decided on the *experiment*, one gets the apriori distribution $\{p[x,y]\}$ of x and y together. When one has also *carried out* the experiment and *observed* y, then the information on x is the *aposteriori* distribution $\{p[x|y]\}$, *i.e.* the conditional distribution with the observed value y inserted. If x is actually related to y, the conditional distribution is less dispersed than the marginal distribution, so that one knows more about x after having observed y. The two distributions are shown in *Fig. 2*.

Generally, the aposteriori information is defined by the apriori information $\{p[x]\}$, the experiment specification $\{p[y|x]\}$ (which states how the result of the experiment will be related to x), and the observation data y. This follows from *Bayes' rule* (Sage, 1987a)
$$p[x|y] = p[x]\,p[y|x]/\int p[x]\,p[y|x]\,dx$$
Hence, this rule relates ones apriori information *and* the experiment data to what one wants to know, *viz.* the aposteriori distribution.

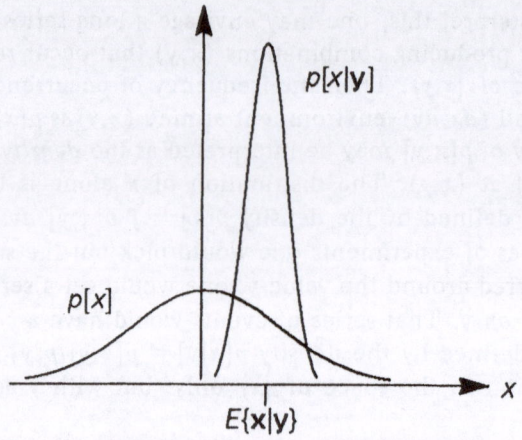

Fig. 2: Apriori and aposteriori distributions. The accuracy of **x** is higher after **y** has been observed.

An *estimate* of **x**, given **y**, may be computed as any characteristic of the distribution $\{p[x|y]\}$, for instance the *conditional expectation* (mean) $E\{x|y\} = \int x\, p[x|y]\, dx$, or the *mode*, *i.e.* the peak value of $p[x|y]$ with respect to x. The first estimate is the *Minimum–Variance* estimate and the second is the *Maximum Likelihood* estimate (Ljung, 1987b). The latter is much favoured in identification problems, simply because it is much easier to compute the peak value than the mean, when one has got complicated distributions in many dimensions. *Medians* and *quantiles* are also reasonable estimates; the latter one yields a domain such that there is a given probability that **x** will be in that domain. The latter is a way to express uncertainty that is concrete and easy to understand. Notice however that taking estimates always destroys information; the maximum information from an experiment with the outcome **y** is still the whole conditional distribution $\{p[x|y]\}$.

Remark: Fundamentally, a stochastic variable **x** is a mapping from a given space Ω into the space of possible values. This means that possible values are $x = x(\omega)$, where $\omega \in \Omega$. The space Ω has a given probability measure or distribution of ω attached to it, and this determines the distribution of the stochastic variable (Wilks, 1962).

One can interpret **x** as a function whose argument is some primitive random variable $\omega \in \Omega$. Each 'event' generates a particular value of ω (with the given probability), and the value $x = x(\omega)$ is the outcome of the event, the 'realization'. This definition of a stochastic variable fits nicely the way simulation programs operate: **x** is the simulation algorithm, its input ω is the output of a (pseudo–) random number generator, and its output x is the result of the simulation. Each 'event' is an execution of the program with a new random ω.

The same interpretation is the basis of the definition of stochastic algorithmic models (see section 5.3 on "Algorithmic models"), but the elements have other names: x is the 'model', ω is 'noise', and x is the 'model output'.

Notice in particular, that a stochastic variable is a *function* and does not have a value. It *generates* a value (from ω), and this value is a *realization*.

In abuse of that, this book uses the same notations (bold characters) for stochastic variables and their realizations, when there is no risk of misunderstanding. The motif is to simplify and save notations.

———————————— **Example 2.1** ————————————

This example illustrates the concepts of apriori and aposteriori probabilities and estimates.

Let x and y be stochastic variables satisfying the relation y = x + w, where x and w are independent gaussian variables with zero means and standard deviations σ_x and σ_w respectively. Assume that y is measured, and one wants to know x.

The *apriori distribution* of x is given by the density
$$p[x] = \exp\left(-\tfrac{1}{2}x^2/\sigma_x^2\right)/\sqrt{2\pi}\,\sigma_x^2$$
The *apriori estimate* is $\hat{x} = E\{x\} = 0$ and the *apriori variance* is $E\{x^2\} = \sigma_x^2$.

The joint distribution of x and y is gaussian
$$p[x,y] = [2\pi\,\sqrt{r_{11}\,r_{22} - r_{12}\,r_{12}}]^{-1}$$
$$\cdot\,\exp\left[-\tfrac{1}{2}(r_{22}\,x^2 - 2\,r_{12}\,x\,y + r_{11}\,y^2)/(r_{11}\,r_{22} - r_{12}\,r_{12})\right]$$

with zero means and covariances given by
$$r_{11} = E\{x^2\} = \sigma_x^2$$
$$r_{12} = E\{xy\} = E\{x^2 + xw\} = \sigma_x^2 + 0$$
$$r_{22} = E\{y^2\} = E\{x^2 + w^2 + 2xw\} = \sigma_x^2 + \sigma_w^2 + 0$$

The *aposteriori distribution* (after measuring y) is the conditional distribution with y inserted and defined by the density

$$p[x|y] = p[x,y]/p[y]$$

$$= [2\pi \sqrt{r_{11} r_{22} - r_{12} r_{12}}]^{-1}$$
$$\cdot \exp[-\tfrac{1}{2}(r_{22} x^2 - 2 r_{12} x y + r_{11} y^2)/(r_{11} r_{22} - r_{12} r_{12})]$$
$$\cdot \sqrt{2\pi r_{22}} \exp(\tfrac{1}{2}y^2/r_{22})$$

$$= [\sqrt{2\pi (r_{11} - r_{12}^2/r_{22})}]^{-1} \exp[-\tfrac{1}{2}(x - y \, r_{12}/r_{22})^2/(r_{11} - r_{12}^2/r_{22})]$$

$$= [\sqrt{2\pi/(1 + \kappa^2)}]^{-1} \sigma_w^{-1} \exp[-\tfrac{1}{2}(x - y/(1 + \kappa^2))^2 (1 + \kappa^2)/\sigma_w^2]$$

where $\kappa = \sigma_w/\sigma_x$.

The *aposteriori estimate* is $\hat{x}(y) = E\{x|y\} = y/(1 + \kappa^2)$. The *aposteriori variance* is $D^2\{x|y\} = E\{[x - \hat{x}(y)]^2|y\} = \sigma_w^2/(1 + \kappa^2) = \sigma_x^2 \kappa^2/(1 + \kappa^2)$. The latter is always between σ_w^2 and σ_x^2, which means that measuring y reduces the uncertainty of x, if $\sigma_w < \sigma_x$. If the apriori uncertainty $\sigma_x \to \infty$, then the estimate $\hat{x}(y) \to y$, and its standard deviation tends to that of the measurement error: $D\{\hat{x}(y)|y\} \to \sigma_w$.

Now, in order to apply Bayes' idea to the identification problem introduce the following general notations:

● The *control variables c* are those one intends to manipulate, and the *response variables d* are those one intends to observe for some purpose (for instance prediction or control) during the application.

● The *object* **S** is the piece of reality that produces the response variables *d*, when subject to the control variables *c*, and possibly other, uncontrolled variables.

● The *experiment system* or *data source* (**X,S**) is the system that produces *data* (**c,d**), *i.e.* values of c and d. The system consists of an *experimenter* **X** and the object **S**, producing *stimulus* c and *response* d respectively.

● The *experiment* (**X,S,T**) is defined by the experiment system (**X,S**) and the *experiment interval* **T**.

Remark: The definitions imply that what is 'input' and what is 'output' of the object is supposed to be given apriori by the identification problem. Whether or not one can actually decide this depends on what the purpose is, and also on how much freedom one has to decide on the experiment. If one cannot influence any of the variables, and has no other means to know apriori what part of the system that should be labelled 'experimenter', this means that one cannot distinguish the object from the experimenter. Then one has to model the combined system as a no–input system, and all variables are output.

Hence, the following cause–and–effect relationship holds: $(X,S) \mapsto (c,d)$, where the sign $\mapsto$ means that what is on the left side determines what is on the right. The dependence may be either deterministic, in which case X and S may be thought of as algorithms, and c and d as output data, or it may be stochastic, in which case the *probability density* of the right side depends deterministically on the left side. Thus, stochastic relations can be expressed as conditional probability densities $p[c,d|X,S]$, where c and d are the variables in the joint distribution corresponding to c and d. Since data (c,d) usually constitute a great number of real values, the dimension of the distribution is usually very large, and useful only for theory.

A number of special cases of identification are common, depending on what are the experiment conditions, what is known apriori, and what one wants to identify:

● *Predesigned stimulation*: c is set by the designer (X is known). Wanted is S.

● *Controlled random stimulation*: X is partly set by the designer, but may not be known completely. Wanted is S and possibly also X.

● *Spontaneous stimulation*: X is not set or controlled by the designer, and not known. Wanted is S and possibly also X.

● *No stimulation*: No X and c. Wanted is S.

A general assumption that is often made in identification theory, explicitly or implicitly, is that experimenter and object are separable. This means that the relation $(X,S) \mapsto (c,d)$ implies $X \mapsto c$ and $(c,S) \mapsto d$. It is a quite natural assumption, since it expresses the essence of identification, namely that one wants to model an object that responds with data d when subject to stimulus c, *independent* of what was the source of the stimulus. This is obviously essential, since when the model of S is put to use, the source of the stimulus will usually be other than X. One wants to separate X from S in the analysis, so obviously there must be some 'separability' condition. However, it is important to realize that the implication $(c,S) \mapsto d$ is indeed an assumption. If it does not hold in an actual case, then identification will give a false result. The logical and practical consequences of this will be discussed at length in the sequel.

Apply Bayes' idea

After these preliminaries, Bayes' idea of using conditional probabilities is applied to the process identification problem simply by expanding the unknown x to mean the experiment system (X,S), and the observed y to mean the whole data record (c,d). In doing this, one must regard (X,S) as a stochastic element (although possibly an extremely complex one). Hence, introduce (X,S) as denoting a general system and (c,d) as the data it would produce. The distinction between (X,S,c,d) and (X,S,c,d) is the same as that between the stochastic variables (x,y) and their possible values (x,y) (see the analysis above).

Disregarding for the moment the fact that the probability density of (X,S,c,d) is not well defined mathematically, it is possible to apply Bayes' rule to obtain the following relation

$$p[X,S|c,d] = p[c,d|X,S] \ p[X,S]/p[c,d] \qquad (2.1)$$

The general idea of '*the Bayesian approach*' to identification is that one may be able to use the right member of (2.1) to compute the aposteriori density of the system (X,S), and from that compute estimates of (X,S).

> **Remark:** This is not very practical, of course, since (X,S) is probably a too complicated element. But let's ignore also that for the moment, and regard the present analysis as a way to arrive at a general principle for identification.

> **Remark:** With the Bayesian approach one regards both the system (X,S) and data (c,d) as stochastic elements, whose relation is specified by their joint probability density. The rationale is that even if the system (X,S) need not actually be stochastic, it is *unknown* and can therefore be regarded as stochastic.

> **Remark:** If X is known completely, then one may use the alternative $p[S|X,c,d] = p[c,d|X,S] \ p[S|X]/p[c,d|X]$.

─────────────── **Example 2.2** ───────────────

This example is a Bayesian analysis of the identification of the generic linear gaussian system.

Assume that the object is
S: $d = c \, \Theta + \omega$,
$\dim(d) = N$, $\dim(\Theta) = \mathbf{n}$, and ω is gaussian with zero mean and unit covariance matrix. Assume also that the stimulus c is deterministic, $\dim(c) = (N, \mathbf{n})$.

Since ω is gaussian and orthonormal

$$\log p[d|\theta] = -\tfrac{1}{2} \|d - c\,\theta\|^2 - \tfrac{1}{2} N \log(2\pi)$$
$$= -\tfrac{1}{2} (d^T d - 2 \, d^T c \, \theta + \theta^T c^T c \, \theta) - \tfrac{1}{2} N \log(2\pi)$$

In order to express the fact that Θ is unknown, let its apriori density be

$$\log p[\theta] = -\tfrac{1}{2} \theta^T \Sigma_\theta^{-1} \theta - \tfrac{1}{2} \log \det(2\pi \, \Sigma_\theta)$$

and let $\Sigma_0 \to \infty$.

The conditional density is from Bayes' rule

$$\log p[\theta|d] = \log p[d|\theta] + \log p[\theta] - \log p[d]$$

It would be possible to evaluate the remaining density $p[d]$ on the right side by integrating $p[d|\theta]\,p[\theta]$, but the following is a shortcut for large R_θ: Notice that only the first term in $\log p[d|\theta] + \log p[\theta] - \log p[d]$ will depend on θ in the limit, and that $p[\theta|d]$ should integrate to unity. Then

$$\log p[\theta|d] = -\tfrac{1}{2} (d^T d - 2 \, d^T c \, \theta + \theta^T c^T c \, \theta) + constant(d)$$

By completing the squares and inserting $c = \mathbf{c}$, $d = \mathbf{d}$, one gets the aposteriori density

$$\log p[\theta|\mathbf{d}] = -\tfrac{1}{2} (\theta - \hat{\theta})^T R_\theta^{-1} (\theta - \hat{\theta}) - \tfrac{1}{2} \log \det(2\pi \, R_\theta)$$

where $\hat{\theta} = (\mathbf{c}^T \mathbf{c})^{-1} \mathbf{c}^T \mathbf{d}$, $R_\theta = (\mathbf{c}^T \mathbf{c})^{-1}$. The Bayesian estimate is $\hat{\theta}$, and its error covariance is R_θ.

───

2.2   The information contents of an experiment

The factors appearing in Bayes' rule (2.1) are

$$p[X,S|\mathbf{c},\mathbf{d}] = p[\mathbf{c},\mathbf{d}|X,S]\; p[X,S]/p[\mathbf{c},\mathbf{d}] \tag{2.2}$$

The left member is easy to interpret; it is the aposteriori density of $(\mathbf{X},\mathbf{S})$, when $(\mathbf{c},\mathbf{d})$ is known. The right member separates into three factors, two of which have intuitive meanings:

● $p[\mathbf{c},\mathbf{d}|X,S]$ is the density the experiment data (c,d) would have, if coming from the system (X,S), but evaluated for actual data $(\mathbf{c},\mathbf{d})$ coming from the 'true system' $(\mathbf{X},\mathbf{S})$. Intuitively, one can say that if the density is low, then (X,S) is not a good candidate for a model of $(\mathbf{X},\mathbf{S})$. This offers a way to reject models that do not accord with data (see section 8.1 on "Statistical tests").

● $p[X,S]$ is the probability density of the 'true system' $(\mathbf{X},\mathbf{S})$ at (X,S), after the experiment has been specified, but before anything has been observed. The values for all (X,S) represent all that is known apriori (if anything) about the true system.

● The meaning of $p[\mathbf{c},\mathbf{d}]$ is uninteresting for the determination of $(\mathbf{X},\mathbf{S})$ since it does not have (X,S) as argument. It is a normalization factor that makes the right member a density function and is computed from the marginal distribution of $p[X,S,c,d]$. It is a constant, when $(\mathbf{c},\mathbf{d})$ is known.

> **Remark:** So far the notation **S** is only a label for the physical object one wants to know about. However, since a theory on identification must necessarily involve the object, and must also be put into a mathematical framework, one has sometimes to regard **S** as a model, the 'true model'. That may be a controversial thing to do, since the physical object is obviously not a mathematical one, and probably not possible to describe exactly by any mathematics. Even if the existence of a 'true model' **S** may therefore be doubted from a philosophical point of view, the concept is used in text books on identification and will be useful as defining the source of data. Also, some conditions for proper identification will have to be expressed in terms of **S**, since they will unavoidably depend on what the real object is. A way out of the philosophical dilemma is to

think of **S** as a model, although possibly a prohibitively complex one. Regardless of this one may use Bayes' rule to evaluate the 'likelihood' of any tentative object defined by a model *M*. This means that there is no problem, until one has to define how *complex* the models *M* and **S** may be, *i.e.* the *spaces* in which *M* and **S** are elements. The trick is then to let the space of *M* be another than the (much larger) space of **S**. In this way one can never reach the 'true' **S**, and this satisfies one obvious condition on a realistic identification theory, namely that a physical process cannot be described exactly by any practical mathematics. The comment applies to the experimenter **X** also, when it is not defined by an exact algorithm.

Two things are worth emphasizing at this point:

The 'Likelihood function' is all the information

Since all one can ever know about a stochastic variable is its distribution, the conditional distribution (2.2) of (**X**,**S**) is all one can ever get out of data (**c**,**d**). Since that distribution is also computable from (**c**,**d**), it is well motivated to call the function $p[\bullet,\bullet|\mathbf{c},\mathbf{d}]$ the 'information of (**X**,**S**) contained in (**c**,**d**)'. That information concept is useful in modelling and identification. It is called the *'Likelihood function'*, and $p[X,S|\mathbf{c},\mathbf{d}]$ for a given (X,S) is called the *'likelihood'* of that particular (X,S).

Since the 'normalizing factor' $p[\mathbf{c},\mathbf{d}]$ is uninteresting for the estimation of (**X**,**S**), that factor is normally excluded from the definition of the Likelihood function, which becomes

$$L(X,S|\mathbf{c},\mathbf{d}) = p[\mathbf{c},\mathbf{d}|X,S]\, p[X,S] \qquad (2.3)$$

Since the first factor is then 'weighted' with the apriori density of (**X**,**S**) (which does not involve **c** and **d**), this is also called the *'weighted'* or *'biassed'* Likelihood function. The *'unbiassed'* Likelihood function is the first factor, and used in most practical cases. This means that what one usually means by the 'likelihood' of a system (X,S) is the probability density of data produced by (X,S), and with the actual data (**c**,**d**) inserted. For any (X,S) it is a non–negative real number.

> **Remark:** All inference may be based on the Likelihood function; it embodies all the information one has: The data (**c**,**d**) are the arguments, and the apriori information gives the form of the function. Notice that from (2.3) one can always (in principle) evaluate the likelihood for a given (X,S). But in order to define and compute

the Likelihood *function* $L[\bullet,\bullet|c,d]$ one must also specify the *space* of the argument (X,S), and that affects the form of the function and requires apriori information about (X,S). That information must be of such a form that one can assign a reasonable meaning to the 'density' $p[X,S]$ (see section 5.1 on "Parametrization"). The 'unbiassed' Likelihood function evades this problem, but also deprives the designer of a means for expressing apriori information. However, the most important part of the apriori information is still carried by the factor $p[c,d|X,S]$ expressing model structure.

Remark: So far has been assumed that there exists a density $p[X,S]$ for the apriori distribution. That can usually be assumed when the only unknown elements in (X,S) are a fixed number of real parameters. However, if also the number of parameters is unknown, then this number must be regarded as a stochastic variable, and since it is an integer, it does not have a probability density. In this case the unbiassed Likelihood function has no obvious meaning. It is otherwise motivated by complete apriori ignorance of (X,S), expressed by a constant value of $p[X,S]$, but only if the density exists. Otherwise, it is not a trivial problem to assign a function that corresponds to 'complete apriori ignorance'. The more general case is treated in section 2.3.2 on "Probabilistic models", and the problem of determining an unknown number of parameters is treated in chapter 9 on "Structure identification".

Remark: The proper interpretation of the Likelihood function and its role in statistical inference has created some controversy among statisticians. Kendall and Stuart (1969) review this briefly.

─────────────── **Example 2.3** ───────────────

This illustrates the distinction between the 'likelihood' and the 'Likelihood function'.

Let a system be defined by S: $d = c + \omega$, where ω is gaussian $(0,1)$. If S is given, its unbiassed likelihood is

$$L(S|c,d) = \exp[-\tfrac{1}{2}(d - c)^2]/\sqrt{2\pi}$$

which can be evaluated easily. However, defining the Likelihood *function* requires that one define a set in which S is a member. Such a set would be $\{S(a): d = a\,c + \omega | a \text{ real}\}$. The corresponding Likelihood function is

$$L[a|c,d,S(\bullet)] = \exp[-\tfrac{1}{2}(d - a\,c)^2]/\sqrt{2\pi}$$

where $|\mathbf{c},\mathbf{d},S(\bullet)$ indicates that the Likelihood function depends on the form of $S(\bullet)$, besides on $(\mathbf{c},\mathbf{d})$. The variable a is called 'parameter' and the fixed $S(\bullet)$ is called 'structure'. One can evaluate the function for all a, but not for all $S(\bullet)$, unless one first defines what is 'the set of all $S(\bullet)$'.

In an attempt to do that one might define

$\{S(a,n): d = P^n(a,c) + \omega | a \text{ reals, } n \text{ positive integers}\}$

where $P^n(a,c)$ is the polynomial with order n and coefficient vector a. The new Likelihood function is

$$L[a,n|\mathbf{c},\mathbf{d},S(\bullet,\bullet)] = \exp\{-\tfrac{1}{2}[\mathbf{d} - P^n(a,\mathbf{c})]^2\}/\sqrt{2\pi}$$

which can be evaluated for any given values of (a,n). There are now integer parameters n as well as real parameters a, but the likelihood still depends on the structure $S(\bullet,\bullet)$.

Even if polynomials can approximate all continuous functions, they still do not cover all possible models, *i.e.* all distributions $\{p[d|c]\}$, and one has to generalize further in order to achieve that. Approximating distributions by polynomials is not suitable, and there are better ways. In particular, expansion in 'Volterra series' (Monaco and Normand–Cyrot, 1987) is a way to approximate all nonlinear, deterministic, dynamic objects, and one may add all gaussian disturbances with rational spectra to that. However, even that does not approximate all objects one may encounter in practice (for instance, disturbances may not be additive), and in spite of that have already more parameters than one can usually handle. Hence, there will always remain something fixed/given/
assumed in the definition of the Likelihood function. In practice one may envisage that part, the 'structure' $S(\bullet,\bullet)$ as defining the *program* used for evaluating the likelihood, and the 'parameters' a and n as real and integer *variables*. By writing a more general program one can replace some 'structure' with 'parameters'. One can *change* the program, but one cannot *vary* it over 'the space of all possible programs'. That illustrates the difference between 'parameters' and 'structure'.

One gets only a description of the experiment

The aposteriori distribution gives only an 'experiment', or 'data description', *i.e.* any model derived from it will be valid during the particular experiment. One has also to know that the model will be valid for other experiments, at least for the stimulation prevailing during the application phase (see section 1.3 on "The purpose").

Remark: This may not be easy to ascertain, unless one has done a proper experiment (see chapter 3). Otherwise, it may not even be feasible, not even in principle. An unsurmountable obstacle in the general case is that one does not have a large number of experiments, which is what it takes to see if the model is independent of the experiment. One may not even be able to repeat the experiment once, even at any cost: If the time the experiment is carried out plays a role for the system's response, the calendar and clock readings t is a component in X. And time does not repeat; there is only one history.

─────────────── **Example 2.4** ───────────────

This example illustrates the difficulties with using historical data.

There is a well developed theory for the forecasting of stochastic processes with more or less apriori information. There has also been attempts of making macro–economic models based on 'black–box' identification techniques. The purpose has been to try and forecast important economic indicators and how they will respond to government decisions (Young *et al.*, 1973; Wall and Westcott, 1974).

The success of this has been uncertain. Since a national economy is an extremely slow process, it takes many years of data collection to obtain a basis for a sufficiently accurate model. Even if the estimated accuracy would be acceptable, there is still no guarantee that the model will yield a useful forecast. The model is based on historical data, while society changes, and the underlying processes creating the dynamics one wants to model may have changed entirely from the past time of data collection to the future time of forecasting.

Secondly, data contains the responses only to the particular decisions of the past government(s), which have (in the best case) been based on certain political ideas prescribing certain economic policies. If a 'black–box' model would be used for *changing* the policies from what has been used in the past (which would be the straightforward and naive purpose), it is no longer valid, since it has not been fitted to the responses of such tentative policies. However, it might possibly be used for evaluating the effects of a (small) number of past governments. The prospects of this would be most favourable in Great Britain, where the political power has been switching between the Conservative and the Labour parties for a long time, and in USA, where the administration has likewise switched between Republicans and Democrats. Attempts have also been made to model the British economy, and attempts have been made to evaluate the economic efficiencies of a number of US presidents. But, of course, no politician has to accept the results. There are plenty of arguments available to use against them.

In order to be able to ascertain that the experiment description one will get from Bayes' rule or the Likelihood function (describing the behaviour during the identification phase) will also be valid for the purpose (during the application phase), one must do a 'proper' experiment. To see what this means requires a closer investigation of what is 'statistical inference'.

2.3   Covariation and causality

There is a logical distinction between the concepts of *'covariation'* ('correlation', 'interdependence') and *'causality'* ('influence', 'dependence'). Two observed variables may *co–vary*, that is, vary together, without one being the cause of the other. It is conceivable that the covariation be an effect of a third, *'hidden variable'*, that influences both the observed variables.

What is observed as a result of an experiment is always a covariation (c and d vary together in a certain pattern). What is needed for an application is *causality*; it must be possible to state that c causes d to vary in a certain way, defined by a model. What is done by taking a model based on experiment data and accepting it for application, is *assuming* that the covariation observed in a particular experiment is indeed due to causality.

Therefore, the result may be wrong whenever the covariation observed between c and d is *not* a consequence of causality, *e.g.* when instead a third, hidden variable is the common cause of the variations in both c and d.

Failure to recognize this distinction and the role of hidden variables has caused much misunderstanding and much misuse of statistics in the past. Some of the consequences are quite amusing: Kendall and Stuart (1967) wrote in their great work on statistical theory, chapter 26:

> **Quotation:** "The relationship of crop–yields and rainfall is an example in which non–statistical considerations make it clear that there is an essential asymmetry in the situation: we say, loosely, that rainfall 'causes' crop–yield to vary, and we are quite certain that crops do not affect the rainfall, so we measure the dependence of yield upon rainfall.
>
> .....
>
> A statistical relationship, however strong and however suggestive, can never *establish* a causal connexion; our ideas on causation must come from outside statistics, ultimately from some theory or other. Even in the simple example of crop–yield and rainfall..., we had no *statistical* reason for dismissing the idea of dependence of rainfall upon crop–yield; the dismissal is based on quite different considerations. Even if rainfall and crop–yield were in perfect functional correspondence, we should not dream of reversing the 'obvious' causal connexion. We need not enter into the philosophical implications of this; for our purpose, we need only reiterate that statistical relationships, of whatever kind, cannot logically imply causation.

G. B. Shaw made this point brilliantly in his Preface to The Doctor's Dilemma (1906): 'Even trained statisticians often fail to appreciate the extent to which statistics are vitiated by the unrecorded assumptions of their interpreters. ... It is easy to prove that the wearing of tall hats and the carrying of umbrellas enlarges the chest, prolongs life, and confers comparatively immunity from disease. ... A university degree, a daily bath, the owning of thirty pairs of trousers, a knowledge of Wagner's music, a pew in church, anything, in short, that implies more means and better nurture... can be statistically palmed off as a magic spell conferring all sorts of privileges. ... The mathematician whose correlations would fill a Newton with admiration, may, in collecting and accepting data and drawing conclusions from them, fall into quite crude errors by just such oversights as I have been describing.'

Although Shaw was on this occasion supporting a characteristically doubtful case, his logic is valid. In the first flush of enthusiasm for correlation techniques, it was easy for early followers of Karl Pearson and Yule to be incautious. It was not until twenty years after Shaw wrote that Yule (1926) frightened statisticians by adducing cases of very high correlations that were obviously not causal: *e.g.* the annual suicide rate was highly correlated with the membership of the Church of England. Most of these 'nonsense' correlations operate through concomitant variation in time, and they had the salutary effect of bringing home to the statistician that causation cannot be deduced from any observed co-variation, however close."

――――――――――――― **Example 2.5** ―――――――――――――

Illustrating the effect of 'hidden variables'.

Let the output variable **d** be the number of drowning accidents per day during a summer season, and let the input variable **c** be the sales of ice cream during the same time. If one would fit a model to these data, one would find that drowning accidents will depend strongly on the consumption of ice cream, since data shows that both are high at the same time and low at the same time. The logical but stupid conclusion of the investigation would be that one should prohibit ice cream in order to bring down the number of drowning accidents.

In this case it is obvious that something is wrong, and it is also obvious that it is the weather that causes variations in both the sales of ice cream and the drowning accidents, that one cannot prohibit the weather from being hot, and thus promoting both swimming and ice cream consumption, and that it does not help to prohibit ice cream instead.

――――――――――――― **Example 2.6** ―――――――――――――

Illustrating the effect of 'hidden variables'.

When the art of computer control in the process industry was young, a big papermaking company launched a project with the aim of controlling the manufacturing by computer. One started by modelling the production process and used regression analysis to estimate possible dependences between important input variables and all other interesting variables that could be measured. The result was above expectations; one got excellent linear dependences between practically all variables. All was very encouraging, until someone discovered 60 Hz 'hum' on the ground connection of the data acquisition equipment. When the hum was eliminated, and the measurement and analysis repeated, the excellent correlations were gone. In this case the apparent dependences were caused almost entirely by the common hum component in the variables.

―――――――――――――― **Example 2.7** ――――――――――

The mechanism in *Fig. 3* has been designed exclusively to reveal the fundamental difference between the cases of using 'free' input and of forcing a predetermined input signal onto the object. Identify the mechanism by fitting a model of the form $d(k) = G\,c(k) + v(k)$, where k is discrete time, and G is a linear operator. The form of the mechanism inside the box is unknown, but it has a handle outside that can be manipulated, and the positions d and c can be viewed through the window.

Case 1: Spontaneous stimulation (no force is acting on c): Data is $\{\mathbf{c}(k) = \sin w(k),\ \mathbf{d}(k) = \sin w(k)\,|\,k=1,\dots,N\}$. The model will obviously be $d(k) = c(k)$.

Case 2: Forced stimulation (c is set to a given $C(k) \neq \sin w(k)$): Data is $\{\mathbf{c}(k) = C(k),\ \mathbf{d}(k) = \sin w(k)\,|\,k=1,\dots,N\}$. The model will be $d(k) = 0 \cdot c(k) + v(k)$.

The second model is the correct one for control purposes, since d cannot be controlled from c. The spring will yield, and only the 'hidden variable' w will affect d. The first model is the correct one for the purpose of describing data. None is any good for the purpose of forecasting. Notice that the difference is not a consequence of different input values, but on *how the experiment was carried out.* The given stimulation $C(k)$ in the second experiment may well

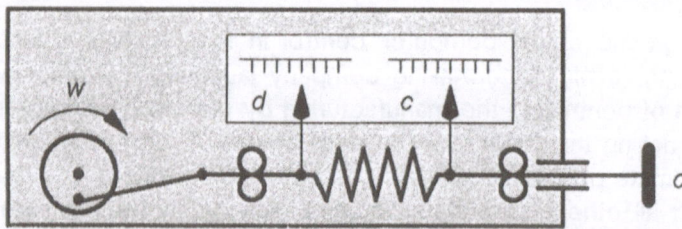

Fig. 3: Pedagogical mechanism, designed to reveal the difference between forced and spontaneous stimulation.

equal the sequence $d(k) = \sin w(k)$ recorded in the first experiment, as long as it is not the same as that in the second experiment.

In summary, *covariation* is what may be established from experiment data. *Causality* is what one must have in order to *use* the model. One must either *guarantee* the causality by making a proper experiment, or else *know* apriori that causality does indeed hold.

Hidden variables may play a role in system identification, when the experiment is not properly designed (see in particular section 3.5.2).

2.3.1 A theory for hidden variables

The probabilistic meaning of 'causality' and 'hidden variables' is easy to clarify by means of some exercise with conditional and marginal distributions: Denote by ω all 'hidden' variables that influence the experiment data (c,d). The probability density of (c,d) for the experiment interval T is

$$p[c,d|\mathbf{T}] = \int p[c,d|\omega,\mathbf{T}] \, dP[\omega|\mathbf{T}] = \int p[c|\omega,\mathbf{T}] \, p[d|c,\omega,\mathbf{T}] \, dP[\omega|\mathbf{T}]$$
(2.4)

The density of the same variables during the application interval T is

$$p[c,d|T] = \int p[c,d|\omega,T] \, dP[\omega|T] = \int p[c|\omega,T] \, p[d|c,\omega,T] \, dP[\omega|T]$$
(2.5)

Now, for the purpose of control one wants to be able to compute $p[c,d|T]$ for any controller $p[c|\omega,T]$, and, in particular, for one that is not influenced by hidden variables, even if the object is. Then from (2.5)

$$p[c,d|T] = p[c|T] \int p[d|c,\omega,T] \, dP[\omega|T]$$
(2.6)

which means that a useful identification should be able to compute the conditional density

$$p[d|c,T] = \int p[d|c,\omega,T] \, dP[\omega|T]$$
(2.7)

From the experiment one can possibly compute

$$p[d|c,\mathbf{T}] = \int p[d|c,\omega,\mathbf{T}] \, dP[\omega|c,\mathbf{T}] \tag{2.8}$$

To be able to equate those two conditional densities one must first make the obvious assumption that the object remains the same and is influenced by the same hidden variables (if any), *i.e.* $p[d|c,\omega,T] = p[d|c,\omega,\mathbf{T}]$, and $P[\omega|T] = P[\omega|\mathbf{T}]$. But this is not enough, since if $P[\omega|\mathbf{T}] \neq P[\omega|c,\mathbf{T}]$, then

$$p[d|c,T] = \int p[d|c,\omega,T] \, dP[\omega|T] = \int p[d|c,\omega,\mathbf{T}] \, dP[\omega|T]$$

$$\neq \int p[d|c,\omega,\mathbf{T}] \, dP[\omega|c,\mathbf{T}] = p[d|c,\mathbf{T}] \tag{2.9}$$

Hence, given invariance, it is not the different identification and application *times* that make the difference, but a possible statistician *dependence* between ω and c during the experiment. This means that the wanted conditional density $p[d|c,T]$ cannot be computed from the conditional data density $p[d|c,\mathbf{T}]$, *unless* the experiment is such that the inequality in (2.9) becomes equality. A sufficient condition for this is *separability* (see also section 3.2.3). To formalize this property, partition $\omega = (\omega',\omega'')$, where ω' and ω'' are independent. Then from (2.5)

$$p[c,d|T] = \int\int p[c|\omega',\omega'',T] \, p[d|c,\omega',\omega'',T] \, dP[\omega'|T] \, dP[\omega''|T]$$

$$= \int p[c|\omega',T] \, dP[\omega'|T] \int p[d|c,\omega'',T] \, dP[\omega''|T]$$

$$= p[c|T] \, p[d|c,T] \tag{2.10}$$

if

$$p[c|\omega',\omega'',T] = p[c|\omega',T]$$
$$p[d|c,\omega',\omega'',T] = p[d|c,\omega'',T] \tag{2.11}$$

that is, no hidden variable affects *both* stimulus and response.

In order to avoid the inconvenience of the explicit references to ω and T in the sequel, assume that invariance and separability do hold, and hence that all conditional densities are causal as well, unless ω or T are explicitly written out.

> **Remark:** Kashyap and Ramachandra Rao (1976) define 'causality' in a different way: A discrete–time variable $d(k)$ is said to depend causally on another variable $c(k)$, if d can be predicted better when one takes c into account than otherwise. This is obviously the case

whenever there is a statistical relation between $d(k+1)$ and $c(k), c(k-1)$,
..., so that $p[d(k+1)|d(k),d(k-1),\ldots,c(k),\ldots]$ defines a narrower distribution than does $p[d(k+1)|d(k),d(k-1),\ldots]$. Caines and Sethi (1979) define 'causality' in an essentially equivalent way. This definition includes also relations between d- and c-variables with the same time-value k. The latter may well mean that $d(k)$ depends 'causally' on $c(k)$ and at the same time $c(k)$ depends 'causally' on $d(k)$. The term 'instantaneous causality' is suggested for this case. Caines (1986) generally defines a relation between two variables c and d as 'causal', when the conditional distribution $p[d|c]$ differs from the marginal distribution $p[d]$. None of these definitions guarantees causality in the stronger sense used by Kendall and Stuart (1967) and in this book, and which is needed to guarantee that the model will be valid for the purpose of control. An explanation is that both Caines and Sethi (1979) and Kashyap and Ramachandra Rao (1976) are primarily concerned with forecasting stochastic vector–sequences without control variables, and for this purpose causality in the stronger sense is not required. This will show up later in this book; modelling for the less demanding purpose of forecasting does not have to satisfy the condition of an 'isolated' experimenter, which is in essence a condition of causality between control variables and response variables.

2.3.2 Probabilistic models

It is evident that one must enter *some* apriori assumptions about the object and experimenter into the identification procedure (unless one wants nothing more than a 'data description'). A general way to do this is by defining a *probabilistic model*, *i.e.* a function $p[y|x]$ specifying the probability density of a system's output variables y conditional on exogenous input variables x. What is known about the system is built into the function $p[\bullet|\bullet]$. What is not known is to be determined by identification. If this yields what one wants for a given purpose, then the system is 'identifiable' for that purpose (see chapter 6). If not, one must assume more or experiment more.

Two ways to exploit this are the following: Let (x,z) be all stochastic variables that might conceivable affect the output y. Then,

● Assumptions about *independence* of some of those variables are expressed by excluding them from the exogenous input, *i.e.* from those appearing to the right of the conditional sign ($|$). Thus, setting $p[y|x,z]$ = $p[y|x]$ means that one assumes apriori that z does not affect y.

● Assumptions about model *structure* are expressed by setting $p[y|x,z]$ = $\mu(y,x,z,a)$, where μ is a given function, and a unknown parameters. How to design suitable model structures is the topic of chapter 5 on "Modelling".

Notice that 'hidden variables' ω would also be in z, but usually cannot be dealt with by assuming independence, since they do in fact affect the output, $p[y|x,\omega] \neq p[y|x]$. However, the assumption of separability implies that $p[y|x] = \int p[y|x,\omega]\ dP[\omega|x] = \int p[y|x,\omega]\ dP[\omega]$, which means that $p[y|x]$ is a *causal* conditional density. Generally, whenever one uses Bayes' rule to write $p[x,y] = p[x]\ p[y|x]$ as a causal relation, one has implicitly assumed that there is no *common* hidden variable left out. A probabilistic model always defines a causal relation, in addition to a statistical dependence.

> **Remark:** The root of the problem analysed in this section is the following: When interpreting the results of an experiment, one must make sure that one has considered *all* that may have affected the result. One must do this to be able to ascertain separability (no common hidden variable). That is a possible problem only for improperly designed experiments (see chapter 3). But when separability has been established, one may safely forget about the hidden variables, and interpret density functions conditional only on the known variables as marginal densities, with hidden variables integrated out.

The probabilistic model carries apriori information.

A probabilistic model may contain little information or much: For instance, if one assumes only that the experiment system $(\mathbf{X},\mathbf{S})$ can be separated into experimenter and object, then $p[c,d|\mathbf{X},\mathbf{S}] = p[c|\mathbf{X}]$ $p[d|c,\mathbf{S}]$, where the left member is the probability density of the data coming from the experiment, and the right member consists of two probabilistic models. The equality contains two general assumptions, namely:

● The system $(\mathbf{X},\mathbf{S})$ is separable apriori in such a way that $\mathbf{X}$ will not affect the response $\mathbf{d}$ in other ways than through $\mathbf{c}$ (indicated by the fact that $\mathbf{X}$ is not among the arguments). If another experimenter X would produce the same stimulus $\mathbf{c}$, then the distribution of the response

would also be the same, since $p[c,d|X,S] = p[c|X] \, p[d|c,S] = p[c|X]$
$p[d|c,S] = p[c,d|\mathbf{X},\mathbf{S}]$.

● There is no common hidden variable. Hence, the stimulus **c** will
depend causally on the experimenter **X**, and the response **d** will depend
causally on the stimulus **c** and the object **S**.

... sometimes subtle

One may also express apriori assumptions (if any) on the true system
$(\mathbf{X},\mathbf{S})$ by specifying $p[X,S]$. To do this one has first to determine what
the 'density of a system' should mean mathematically. The latter is by
no means obvious. Probability densities are well–defined only if the
stochastic variable is continuous *and* there is no concentration of the
'probability mass' to subsets of zero volume. The latter difficulty can
possibly be evaded by allowing Dirac δ–functions to be parts of the den-
sity function. The data density $p[c,d|X,S]$ can probably be modelled in
this way. But in order to define the 'density' of the 'true system' $(\mathbf{X},\mathbf{S})$
one has first to define the space in which $(\mathbf{X},\mathbf{S})$ is a point. That space is
probably not a real space, as in Example 2.2. When the space has been
defined, one must specify a probability measure for practically all conceivable subsets $\mathfrak{B}$ of systems (actually a 'sigma algebra'). This will apparently create great mathematical and practical difficulties in a general
case. In the case of *parametric* systems (see section 5.1 on "Parametrization") the difficulties can be overcome. The system space will
be a mixed real and integer space, and the distribution will not be free
from mass concentration.

In any case one must be able to define the apriori probability measure
$P[\mathfrak{B}]$ for all $\mathfrak{B}$. But when that has been done, the aposteriori probability
that $(\mathbf{X},\mathbf{S})$ is in any given $\mathfrak{B}$ follows from Bayes' rule

$$P[\mathfrak{B}|\mathbf{c},\mathbf{d}] = p[\mathbf{c},\mathbf{d}]^{-1} \int \mathrm{Ind}\{(X,S) \in \mathfrak{B}\} \, p[\mathbf{c},\mathbf{d}|X,S] \, dP[X,S] \qquad (2.12)$$

where $p[c,d] = \int p[c,d|X,S] \, dP[X,S]$, and the integrals are defined in
the Lebesgue sense.

> **Remark:** If (X,S) are points in a real space, for instance only real
> parameters in the true system are unknown, then $dP[X,S] = p[X,S]$
> $d(X,S)$. Notice that P is always well defined; it is a nonnegative
> real number for all $\mathfrak{B}$. The set function $P[\bullet]$ is the *cumulative dis-*
> *tribution* over its argument set.

Remark: In most cases one may forget about the mathematical subtleties of defining $dP[X,S]$ and think *as if* $p[X,S]$ existed, and interpret it as a 'density'. Mass concentrations may be modelled by Dirac δ–functions. However, that will not necessarily solve all the mathematical problems with the definition of $dP[X,S]$, and as a consequence there will also be practical problems. For instance, the ML–estimate will be a dubious choice of estimation principle in such cases, since the Likelihood maximum (infinite) is obviously at a point of mass concentration. But that may still be far away from the bulk of the probability mass, and hence a poor estimate of the true system. However again, as soon as one (Lebesque) integrates over a set of (X,S), for instance to compute 'confidence intervals' or 'percentiles' for estimates, the use of δ–functions will make sense in practice.

Bayes' idea revisited

The general idea of Bayes' identification principle is to use the equation

$$dP[X,S|\mathbf{c},\mathbf{d}] = p[\mathbf{c},\mathbf{d}]^{-1}\, p[\mathbf{c},\mathbf{d}|X,S]\, dP[X,S] \tag{2.13}$$

to compute aposteriori probability densities (in the generalized sense) of the unknown $(\mathbf{X},\mathbf{S})$. In order to do that one needs *i*) probabilistic models to specify apriori information, *ii*) data, and *iii*) some knowledge of the experiment (*how* much will be answered in chapter 3). The equality sign ties what is assumed and known to what is wanted.

Bayes' rule is only a mathematical identity, but Bayes' approach to identification is to combine the rule by the introduction of apriori information through probabilistic models:

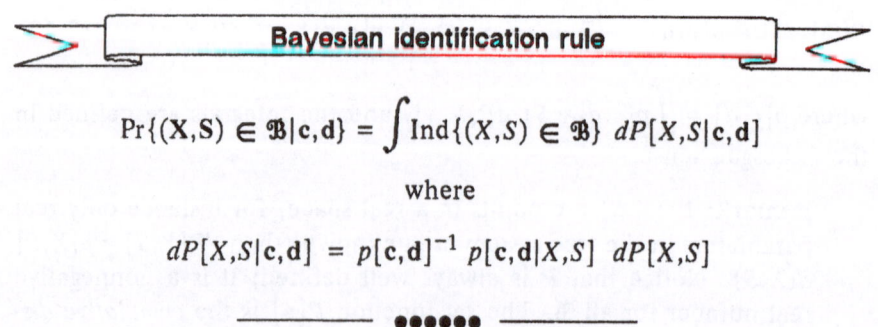

Bayesian identification rule

$$\Pr\{(\mathbf{X},\mathbf{S}) \in \mathfrak{B}|\mathbf{c},\mathbf{d}\} = \int \mathrm{Ind}\{(X,S) \in \mathfrak{B}\}\, dP[X,S|\mathbf{c},\mathbf{d}]$$

where

$$dP[X,S|\mathbf{c},\mathbf{d}] = p[\mathbf{c},\mathbf{d}]^{-1}\, p[\mathbf{c},\mathbf{d}|X,S]\, dP[X,S]$$

●●●●●●

The following chapters will be devoted to techniques for evaluating the right member in Bayes' rule *and* to means for setting up and checking experiments and model structures to render the left member useful for model design. Bayes' rule provides information on the *experiment system*, and one has still to ascertain that it also holds enough information for the particular purpose of the model design. The purpose is usually not that of describing the experiment.

3   The experiment

This chapter will investigate the concept of an 'experiment', what it means, and in particular what requirements it must satisfy to allow the designer to draw applicable conclusions from data obtained from the experiment. At the very best one may find a probabilistic model of the combined system of experimenter and object, but the problem is still to separate one from the other. This has to be done, whenever the purpose requires that one have a model of the object that is to be valid for other stimuli than generated by a particular experimenter.

The conditions involve X and S, of which at least the latter is unknown. Hence the conditions may or may not be possible to check objectively using some hard data. If not, the designer must guarantee the conditions from apriori information, realizing that if that information is wrong, so will the model be.

The experiment usually means applying some controlled stimulation to the object and observing the response d, all during a time $t \in T$. The stimulus c may be a vector time sequence (a set of 'test signals'). It may also be missing. In the latter case the experiment is not a controlled one, but a spontaneous, 'natural' process is observed. The experimentation interval T may be continuous, and it may be intermittent.

3.1   An introductory example

The following example illustrates the importance of defining the experiment and the purpose (Madanski, 1959).

———————————————— **Example 3.1** ————————————————

Assume one knows there is a linear relation between two time–dependent scalar variables z and u

$z(k) = a\,u(k)$

and that one can measure both z and u several times but with independent gaussian errors

$d_1(k) = u(k) + \sigma_1\,\omega_1(k)$

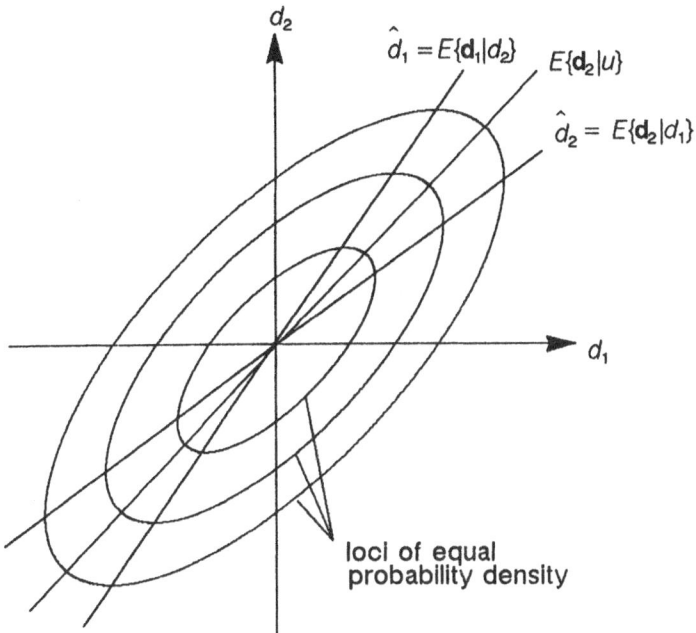

Fig. 4: Joint distribution p and conditional means E of gaussian variables d_1 and d_2.

$d_2(k) = z(k) + \sigma_2 \, \omega_2(k)$,
where ω_1 and ω_2 have zero means and unit variances.

One wants to estimate a. A simple way to do that is to fit the linear relation to data by the Least–Squares method, which means that one minimizes
$Q(a) = \Sigma_k \, [\mathbf{d}_2(k) - a \, \mathbf{d}_1(k)]^2$
with respect to a. The solution is well known to be
$\hat{a} = \Sigma_k \, \mathbf{d}_1(k) \, \mathbf{d}_2(k)/\Sigma_k \, \mathbf{d}_1(k)^2$
and $\hat{d}_2(k) \triangleq E\{\mathbf{d}_2(k)|d_1(k)\} = \hat{a} \, d_1(k)$ is the estimate of $\mathbf{d}_2(k)$ given $d_1(k)$, as illustrated by a straight line in *Fig. 4*.

Now, if one would write the known linear relation as $u(k) = z(k)/a$ (which is obviously the same assumption) and fit that to the same data by minimizing the loss function
$Q(a) = \Sigma_k \, [\mathbf{d}_1(k) - \mathbf{d}_2(k)/a]^2$
then one would get the estimate
$\hat{a} = \Sigma_k \, \mathbf{d}_2(k)^2/\Sigma_k \, \mathbf{d}_1(k) \, \mathbf{d}_2(k)$
and $\hat{d}_1(k) \triangleq E\{\mathbf{d}_1(k)|d_2(k)\} = d_2(k)/\hat{a}$. The corresponding straight line is also shown in *Fig. 4*. However, *the two lines are not the*

same, which means different estimates of a. And the difference does not depend on statistical error, since the lines will not be the same, even if the amount of data becomes infinite, and one should be able to estimate a exactly. So which one is the right one? Or are both wrong?

The answer depends on a condition that was not specified in the example, namely what caused the variables u and z to vary. If one assumes that u was first generated by an independent source of gaussian variables with zero means and variance σ_3^2 (for instance a noise generator), then one can estimate a as follows:

Apparently $\mathbf{d}_1(k)$ and $\mathbf{d}_2(k)$ are jointly gaussian distributed with zero means and covariances

$r_{11} \triangleq E\{\mathbf{d}_1(k)^2\} = \sigma_3^2 + \sigma_1^2$
$r_{22} \triangleq E\{\mathbf{d}_2(k)^2\} = a^2 \sigma_3^2 + \sigma_2^2$
$r_{12} \triangleq E\{\mathbf{d}_1(k)\ \mathbf{d}_2(k)\} = a\ \sigma_3^2$

Given sufficient data one can evaluate the left hand sides with arbitrary accuracy to get r_{11}, r_{22}, r_{12}, and then solve the last and first equations to get the unbiassed estimate

$\hat{a} = r_{12}/\sigma_3^2 = r_{12} /[r_{11} - \sigma_1^2] \rightarrow r_{12}/\sigma_3^2 = a$

if the variance σ_1^2 of the measurement error is known.

The two least–squares estimates are (if $\dot{a} > 0$)

$\hat{a} = r_{12}/r_{11} \rightarrow a\ \sigma_3^2/(\sigma_3^2 + \sigma_1^2) < a$
$\hat{a} = r_{22}/r_{12} \rightarrow (a^2\ \sigma_3^2 + \sigma_2^2)/(a\ \sigma_3^2) > a.$

Hence, none of the two least–squares estimates is unbiassed, and the line in between in *Fig. 4* is the correct one, *if the purpose of the identification is to find out the value of a.*

However, if one wants to use this information for making some *decision* on how to control the system, then a piece of information is still missing: One must know which of the variables u and z that caused the other to vary. In other words, one must know which way causality goes, or indeed, whether there is any causality at all. One must know that, because one must be able to ascertain that if one were to manipulate one variable, the other would follow. And that is by no means guaranteed by having established the relation $z = a\ u$. If z would actually be the cause of u, and one would manipulate u, then z would not follow u.

The meaning of 'causality' was discussed at length in section 2.3 on "Covariation and causality". However, its proper interpretation is so important, that a further elaboration is motivated.

─────────────────────── **Example 3.2** ───────────────────────

Ohm's law states that there is a relation $v = R\,i$ between current i and voltage v through a resistor. One should be able to estimate the resistance R with arbitrary accuracy, for instance as above. But this says nothing about whether current caused voltage, or vice versa. That depends on whether the circuit is voltage or current driven. It is also possible that it is driven by some other voltage, or current, or changing magnetic field, or temperature, or pressure, or any other physical phenomenon that is able to induce voltage and current. In those cases it is possible that neither current nor voltage causes the other to vary. In any case one cannot establish this, by measuring current and voltage, and without knowing more about the circuit than Ohm's law. A student in electrical engineering might say that, primarily, voltage causes current to flow, but that is more information than provided by Ohm's law. In less well known cases one may not have the corresponding information. Also, the circuitry may be such that one can manipulate the current, but not the voltage, and then one would have little use of the added information that "voltage causes current".

───

Now, return to Example 3.1 and assume that the purpose is to control d_2, and that one has established that u causes z and hence d_2 (probably one would rather control z, but it is not accessible for feedback). Then there is still another element left undefined in this example, namely *how the control will be carried out*. If one plans to control d_2 by manipulating u, then the model one needs is the conditional distribution $p[d_2|u]$. Hence, the model $z = a\,u$ derived from the experiment depicted in the upper system in *Fig. 5* (and yielding an unbiassed estimate of a) is the correct one for the purpose.

However, if one would not be able to access u for direct control by the computer used for the experiment, one may have to access u indirectly by controlling d_1. One would achieve that by the cascaded control system depicted in the lower system in *Fig. 5*. Then the model one needs is $p[d_2|d_1]$, so that the first of the two biassed least–squares estimates yields the correct model for that purpose!

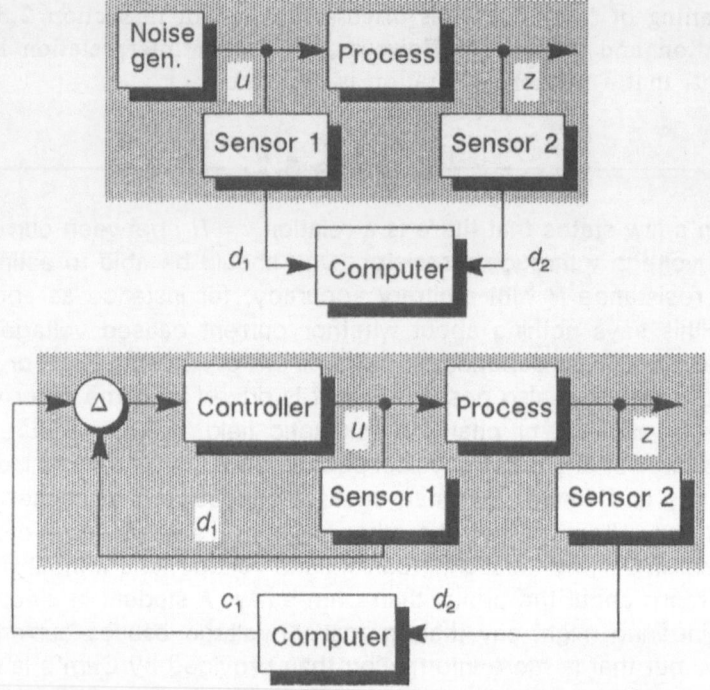

Fig. 5: Different systems for identification and application. The model will not be applicable, not even if the controller is perfect, so that $c_1 = d_1$.

--------------------------------- **Example 3.1** ---------------------------------

(continued)

It is not difficult to compute the two conditional distributions, which are the correct models for the two purposes: From $d_2 = a\,u + \sigma_2\,\omega_2$ follows $p[d_2|u] = \phi(d_2 - a\,u, \sigma_2^2)$, where $\phi(d_2 - m, \sigma^2)$ is the gaussian distribution with mean m and variance σ^2. A little more analysis of the joint distribution

$$p[d_1, d_2]$$
$$= [2\pi \sqrt{r_{11}\,r_{22} - r_{12}\,r_{12}}]^{-1}$$
$$\cdot \exp[-\tfrac{1}{2}(r_{22}\,d_1{}^2 - 2\,r_{12}\,d_1\,d_2 + r_{11}\,d_2{}^2)/(r_{11}\,r_{22} - r_{12}\,r_{12})]$$

yields the conditional distribution

$$p[d_2|d_1] = \phi\{d_2 - a\,d_1\,\sigma_3^2/(\sigma_3^2 + \sigma_1^2),$$
$$[\sigma_3^2\,(a^2\,\sigma_1^2 + \sigma_2^2) + \sigma_1^2\,\sigma_2^2]/[\sigma_3^2 + \sigma_1^2]\}.$$

which has the gain parameter $a \, \sigma_3^2/(\sigma_3^2 + \sigma_1^2)$ of the first least-squares estimate.

Remark: The fact that one gets different gain parameters for the lower and upper systems in *Fig. 5* reflects the fact that the systems are not the same, not even if the 'controller' is perfect. Sensor 1 is involved in different ways.

One concludes from the examples that what is correct inference from experiment data is not only dependent on the *experiment*, but also on the *purpose* of the modelling. And this suggests a basic rule for experiment design: Stimulation of the object during the identification phase should be carried out *on the same system* as when the model is put to use in the application phase.

3.2 Requirements for proper experimentation

At a closer look on the experimentation process there are three 'actors' involved:

● The *experimenter* X. It is the process that generates the stimulus **c**, according to some deterministic or random rule, and records the response **d**. It is usually a computer with output channels to *actuators* (final controlling elements) and input channels from *sensors* (measuring transducers). It can also be any other device that carries out equivalent tasks, or a human operator, that uses the brain to decide on the stimulus, muscles to activate controlling elements, and some of 'the five senses' to register responses.

● The *process* (*proper*) **P**. This is the part of the real world that is subject to the stimulus **c**, and whose response **d** one wants to describe for some purpose. It includes the physical actuators and the sensors, since **c** and **d** are basically data and not physical variables (that is, they carry information and not energy). See Bohlin (1987a) on how to define the variables in practical cases.

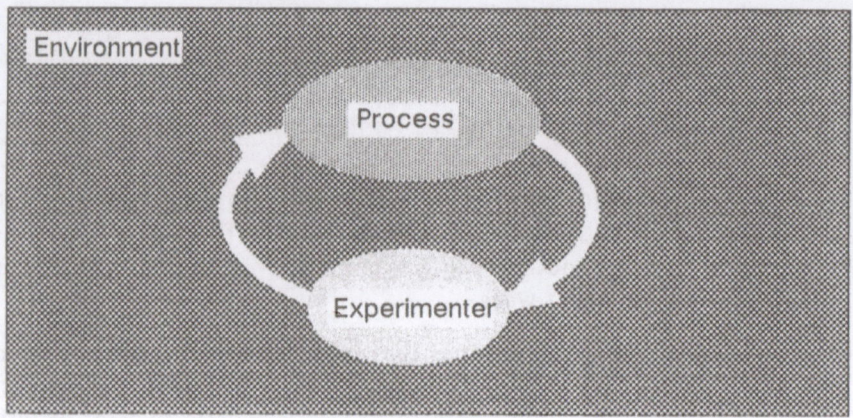

Fig. 6: Illustrating an experiment. The experimenter is in contact with the process through channels for stimulus and response. The environment is possibly in contact with both process and experimenter, affecting their output.

● The *environment* E. This is anything else that may affect the process **P** and/or the experimenter **X**, and that causes that the response **d** is not a consequence of the stimulus **c** alone. It may also affect the stimulus **c**.

The general experiment condition is illustrated in *Fig. 6*. The white parts, stimulus and response, represent what is completely known. The grey 'process' is usually known only in part, and the even greyer 'environment' is even more unknown. Also the 'experimenter' may be in a shade of gray, in particular if it is not a computer.

The experiment data (c,d) has the conditional distribution $\{p[c,d|X,E,P,T]\}$, and that distribution is the most one can get out of the particular experiment. One wants the distribution of the variables (c,d) during an *application*. Under the assumptions that the process and the environment will be the same, this distribution is $p[c,d|X,E,P,T]$, where X is the stimulus (the control) prevailing in the application phase T. It follows that one can treat the process and the environment together, and hence $S = (E,P)$ is the object to identify. The element **S** is usually interpreted as the 'system' in the literature (Ljung, 1987a; Söderström and Stoica, 1989), but it would be more appropriate to interpret it as 'the surrounding world'. The delimiting of **S** is *defined* by the choice of variables c and d. This means that instead of cutting out a piece or reality and labelling it **S**, one should cut out the piece of 'the experimenter' (since it is easier to delimit) and regard the rest of the world as the object **S**. Because it is there, when the model is later used for decisions affecting the same surrounding world. The fact that 'the

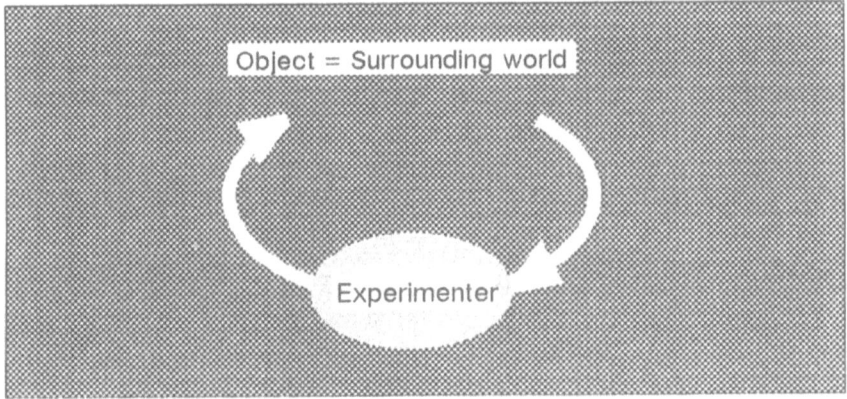

Fig. 7: Illustrating the proper interpretation of an experiment. The experimenter cannot distinguish between the process proper and the environment.

rest of the world' is of course impossibly complex is not actually a problem (not even in principle), since one needs to model only the aspects that the experimenter can sense and manipulate and hence learn something about. The experiment is then illustrated by *Fig. 7*.

Generally, there is no guarantee that the distribution function $p[c,d|\mathbf{X},\mathbf{S},\mathbf{T}]$ one may be able to compute from the results of the experiment says anything about that one wants, which is $p[c,d|X,S,T]$ (a conditional distribution $p[\bullet|y]$ for one particular value y says nothing about the distribution $p[\bullet|y']$ for another value y'). The distribution during the application phase is given by $p[c,d|X,S,T] = p[c|X,S,T] \, p[d|c,X,S,T]$. Even if one can make the controlling process $p[c|X,S,T]$ independent of $\mathbf{S}$, the responses would be determined by $p[d|c,X,S,T]$, and not by the computable response $p[d|c,\mathbf{X},\mathbf{S},\mathbf{T}]$ during the experiment.

Since there are three quantities to the right of the 'condition' sign (|) that are different during the application, *viz.* (c,X,T), the designer has basically to ascertain that the model holds also *outside* the range of those variables $(\mathbf{c},\mathbf{X},\mathbf{T})$ during the experiment. Obviously it is not true generally.

3.2.1 Sufficient stimulation

To ascertain that the stimulus during the experiment covers the range of control variables (if any) during the application the designer must re-

quire that the experimenter be sufficiently stimulating. To formalize
that idea, define this first:

Definition of the Range of Model Validity: Let $\mathcal{X}$ be the set of admissible controllers X acting during the application time interval T. Then the *Range of Model Validity* $\mathcal{R}(\mathcal{X})$ is the set of possible variable values (c,d) during the application. That is, $\mathcal{R}(\mathcal{X}) = \{c,d|p[c,d|X,S,T] > 0,$ some $X \in \mathcal{X}\}$

> **Remark:** If the controller during the application is the same as the experimenter (*e.g.* the purpose is prediction, or monitoring a given system), then $\mathcal{X} = \mathbf{X}$.

The following condition must hold:

Condition: Sufficient Stimulation

The experiment $(\mathbf{X},\mathbf{S},\mathbf{T})$ and the set $\mathcal{X}$ of controllers X during the application must be such that the support of the distribution of data covers the Range of Model Validity. That is, if $\mathbf{T}$ is the experiment interval, then

$$p[c,d|\mathbf{X},\mathbf{S},\mathbf{T}] > 0, \text{ all } (c,d) \in \mathcal{R}(\mathcal{X}).$$

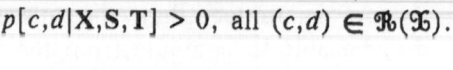

> **Remark:** The condition says that there must be a positive probability that any stimulus c will appear during the identification phase, at least for such data sequences that may possibly be generated by the controller X acting during the application. Thus the experiment should excite at least all modes of the object that may be excited when the model is put to use. Occasionally, the controller X is known during the application, and is the same as $\mathbf{X}$ during the experiment, for instance in cases of adaptive control, and then the condition of Sufficient Stimulation would always hold. However, the proper interpretation of X in that case is not obvious (see Example 6.8). Generally, the controller X is not known in the application phase, and one has to ensure the condition for a set $\mathcal{X}$ of possible controllers.

> **Remark:** If the user has a choice of experimenter, there are methods to design optimal experimenters for given purposes (Goodwin, 1982; Goodwin and Payne, 1977).

3.2.2 Reproducibility

In essence, the condition of Sufficient Stimulation ensures that the c–variable will vary enough to ensure the validity of the model within a sufficiently wide domain for the application. It is not possible to require something similar for the T variable. If there are disjoint experiment and application phases, the intervals **T** and T are definitely different — there is only one time. In principle, one cannot say anything for sure about how the object will behave after it has been observed. It is conceivable that it may change properties altogether.

Hence, it should be clear that one must have some outside knowledge, on which to base a reasonable assumption that the object will remain the same, at least when the experiment is repeated (see section 1.3 on "The purpose"). This property is called *reproducibility,* and it is so fundamental in Natural Science, that experiments that are not reproducible are generally considered by the scientific community to be of no value. For instance, experiments on 'supernatural' phenomena may usually be rejected by this motivation.

The following is a way to formalize the condition of reproducibility that will suit the purpose of identification:

Condition: Reproducibility

The system $(\mathbf{X},\mathbf{S})$ must be such that for all repetitions of the experiment $(\mathbf{X},\mathbf{S},T)$ with equal lengths of T, the distribution of data $(\mathbf{c},\mathbf{d})$ will be the same. That is, if **T** and T are disjoint intervals, then.

$$p[c,d|\mathbf{X},\mathbf{S},\mathbf{T}] = p[c,d|\mathbf{X},\mathbf{S},T], \text{ all } (c,d) \in \mathfrak{R}(\mathbf{X}),$$

——————— •••••• ———————

Notice that reproducibility can be tested, by repeating the experiment. However, for identification based on only one experiment it is an *assumption*, and hence the designer must ascertain that the above *condition* is satisfied.

The following intuitively rather obvious lemma will be needed in the sequel:

Lemma of the Range of Model Validity: If the Experimenter X is *Sufficiently Stimulating* and the condition of *Reproducibility* is satisfied, then $\Re(X) \supseteq \Re(\mathfrak{M})$.

Proof:
From the definition of the Range of Model Validity and the condition of Reproducibility follows
$$\Re(X) \equiv \{c,d | p[c,d | X,S,T] > 0\} = \{c,d | p[c,d | X,S,T] > 0\}.$$
But from the condition of Sufficient Stimulation follows that $p[c,d | X,S,T] > 0$ for all $(c,d) \in \Re(\mathfrak{M})$. Hence $\Re(\mathfrak{M}) \subseteq \Re(X)$. ∎

3.2.3 Separability

The following example will illustrate a basic problem in the separation of the experiment system into experimenter and object:

─────────────── **Example 3.3** ───────────────

The example demonstrates the effect of mutual surroundings (again it is not immaterial how the stimulus was generated).

Define an object and experimenter by the relations
S: $d = s\,c + v_1$
X: $c = x\,v_2 + \alpha\,v_1$
where s is an unknown gain, x is a known amplitude parameter of the stimulus c, and v_1, v_2 are independent gaussian variables with zero means and standard deviations σ_1 and σ_2. Interpret the random variables as follows: v_2 is generated by a noise generator inside the experimenter, and v_1 is a disturbance of the object from the environment. That disturbance may affect also the experimenter by the 'leaking factor' α, since the experimenter is not isolated from the environment.

The logarithm of the distribution of experiment data (d,c) is

$$\log p[d,c] = -\tfrac{1}{2} \log \det(2\pi R) - \tfrac{1}{2}(d\ c)\,R^{-1}\,(d\ c)^T$$

$$= -\log(2\pi) - \tfrac{1}{2}\log(r_{11}\,r_{22} - r_{12}\,r_{12})$$
$$-\tfrac{1}{2}[r_{22}\,d^2 - 2\,r_{12}\,d\,c + r_{11}\,c^2]/(r_{11}\,r_{22} - r_{12}\,r_{12})$$

$$= -\log(2\pi) - \tfrac{1}{2}\log(r_{11}\,r_{22} - r_{12}\,r_{12})$$
$$-\tfrac{1}{2}[(d - c\,r_{12}/r_{22})^2\,r_{22}/(r_{11}\,r_{22} - r_{12}\,r_{12})] - \tfrac{1}{2}c^2/r_{22}$$

where
$$R = \{r_{ij} | i,j = 1,2\}$$
$$r_{11} \triangleq E\{\mathbf{d}^2\} = E\{[(1+s\alpha) \, v_1 + s \times v_2]^2\} = (1+s\alpha)^2 \, \sigma_1{}^2 + s^2 \, x^2 \, \sigma_2{}^2$$
$$r_{12} \triangleq E\{\mathbf{dc}\} = E\{[(1+s\alpha) \, v_1 + s \times v_2][\alpha \, v_1 + x \, v_2]\}$$
$$= \alpha(1+s\alpha) \, \sigma_1{}^2 + s \, x^2 \, \sigma_2{}^2$$
$$r_{22} \triangleq E\{\mathbf{c}^2\} = E\{[\alpha \, v_1 + x \, v_2]^2\} = \alpha^2 \, \sigma_1{}^2 + x^2 \, \sigma_2{}^2.$$

Now, provided the experiment is repeated sufficiently many times, one can compute the distribution from data, and hence also its conditional distribution

$$\log p[d|c]$$
$$= -\tfrac{1}{2} \log(2\pi) - \tfrac{1}{2} \log(r_{11} - r_{12}{}^2/r_{22})$$
$$- \tfrac{1}{2}[(d - c \, r_{12}/r_{22})^2/(r_{11} - r_{12}{}^2/r_{22})]$$

which is the best object model one would get out of the experiment. It depends on σ_2, if $\alpha \neq 0$. For instance, the estimated gain would be

$$\hat{s} = E\{d|c\} = r_{12}/r_{22} = [\alpha(1+s\alpha) \, \sigma_1{}^2 + s \, x^2 \, \sigma_2{}^2]/[\alpha^2 \, \sigma_1{}^2 + x^2 \, \sigma_2{}^2] \neq s.$$

In an application the object would react with gain s, and not with the estimated gain $\hat{s}$.

The result in the example should not be surprising, considering the fact that the surroundings affect both experimenter and object through *other* and *unknown* channels than those for data. This makes it a source of a 'hidden variable', as discussed in section 2.3 on "Covariation and causality", and the influence of that variable is as depicted in *Fig. 8*. Since v_1 means a second and unknown channel between c and d in addition to that of the object, no inference can distinguish between those channels, and the model will describe the combined influences of both, and not only that of the object. The upper graph shows also the direction of the influence (causation), but remember that statistical inference can detect only dependence, *i.e.* it can never see more than the lower graph. And since v_1 is unknown, the channels $c - d$ and $d - v_1 - c$ cannot be distinguished. This shows even more clearly why the 'side channel' via the hidden variable will ruin the model making, when v_1 is not observed.

The lesson of the example is that one must see to it that such unrecorded influence cannot occur. In other words, one must *isolate* the experimenter from influence of the surroundings. The straightforward way to do this is to experiment in open loop, *i.e.* to generate the stimulus *c* independent of everything else.

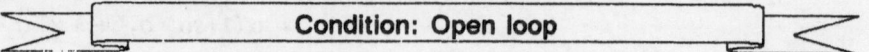

Condition: Open loop

The system (**X**,**S**) must be separable apriori into an experimenter **X** and an object **S**, in such a way that **S** will not affect the control variable *c*, and **X** will not affect the response variable *d* in other ways than through *c*. That is,

$$p[c,d|\mathbf{X},\mathbf{S},\mathbf{T}] = p[c|\mathbf{X},\mathbf{T}]\ p[d|c,\mathbf{S},\mathbf{T}], \text{ all } (c,d) \in \mathcal{R}(\mathbf{X})$$

———————— •••••• ————————

Remark: Notice that since **X**, and **S** have been defined to contain all phenomena that may affect *c* and *d*, they also contain the sources of possible hidden variables. Hence, an open loop means that there are no *common* hidden variables, so that $p[c|\mathbf{X},\mathbf{T}]$ and $p[d|c,\mathbf{S},\mathbf{T}]$ are probabilistic models, *i.e.* describe causal dependences.

Remark: The open loop is not the only interesting case of an isolated experimenter. In fact, **X** must be isolated from **S**, but not from *d*. This means that closed–loop experiments may still be sufficiently isolated to be feasible for identification (see section 3.5).

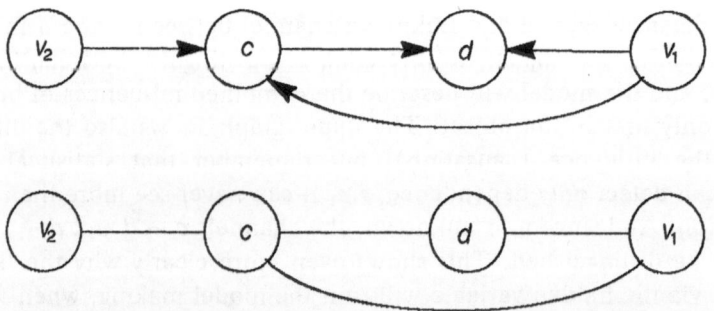

Fig. 8: Signal flow graphs of the example. The upper graph depicts causality, and the lower graph dependence. The latter is what statistical inference can possibly detect.

Proper experiments make applicable models

The concepts of Separability and an Isolated experimenter will be developed further in connection with dynamic systems (section 3.4). Here will be shown that under the three conditions stated so far, including that of an open loop, a model based on experiment data will actually be valid also for applications. A similar result holds for the case of a closed loop, but the open loop case is simpler and will therefore be investigated first. This requires two more definitions:

Definition of Representability and Applicability: An object S at application time T' is *Representable*, if it can be substituted by a model of the object at identification time **T** for all admissible controllers $X \in \mathcal{X}$. That is, $p[c,d|X,S,T'] = p[c|X,T] \, p[d|c,S,T]$, all $X \in \mathcal{X}$, all $(c,d) \in \mathcal{R}(\mathcal{X})$. If a model is sufficiently accurate, it will also be *Applicable* (see section 1.3 on "The purpose").

> **Remark:** Notice that a Representable object does not have to be time invariant. It may depend on time t as described by $p[d|c,t,S,T]$, if t is known (from a clock reading). Alternatively, time may be regarded as a component of the control variable c. The point is that the dependence of time may be possible to determine during the identification phase, and this dependence may remain valid during the application phase.

Lemma of the Open Loop: If the experiment is *Reproducible* and *Sufficiently Stimulating*, and the experiment and the application (control) are carried out in *Open Loop*, then the object is *Representable*, and a model that is sufficiently accurate will also be *Applicable*.

> **Proof:**
> From the conditions of *i)* an Open–Loop experiment, *ii)* Reproducibility, and *iii)* Sufficient Stimulation follows that:
> $p[c|X,T] \, p[d|c,S,T] = p[c,d|X,S,T]$
> $= p[c,d|X,S,T] = p[c|X,T] \, p[d|c,S,T] > 0$,
> all $(c,d) \in \mathcal{R}(X)$. Hence, all factors are positive, and
> $p[d|c,S,T] = p[c|X,T] \, p[d|c,S,T]/p[c|X,T]$.
> Since the left member is a probability density, it integrates to one, and
> $p[d|c,S,T] = p[d|c,S,T]$, all $(c,d) \in \mathcal{R}(X)$.
> From the lemma of the Range of Model Validity follows
> $\mathcal{R}(\mathcal{X}) \subseteq \mathcal{R}(X)$.
> Hence, $p[d|c,S,T] = p[d|c,S,T]$, all $(c,d) \in \mathcal{R}(\mathcal{X})$.
> Finally, since the control is carried out in open loop
> $p[c,d|X,S,T] = p[c|X,T] \, p[d|c,S,T]$

$$= p[c|X,T]\, p[d|c,\mathbf{S},\mathbf{T}] \text{ all } X \in \mathfrak{X}, \text{ all } (c,d) \in \mathfrak{R}(\mathfrak{X}),$$

which is the criterion of a Representable object. Hence, a sufficiently accurate model of $p[d|c,\mathbf{S},\mathbf{T}]$ is Applicable.   ∎

Remark: Since the explicit references to the experiment and application intervals $\mathbf{T}$ and T have now served their purposes, these notations will generally be omitted in the sequel.

In order to investigate the possibility of obtaining separability in a closed loop, it is necessary first to clarify the meaning of a 'dynamic system'.

3.4   Dynamic systems

Causality plays a fundamental part in 'dynamic systems'; in fact, it is the property of a causal dependence in time (cause comes before effect) that *defines* such systems. Causality also determines the proper modelling of experiments in *closed loop*, *i.e.* cases where the experimenter output c depends on the object's response d, which depends on c, as usual.

In experiments on dynamic objects the stimulus and response data are *sequences*
$$c_N = \{c(k)|k=1,\ldots,N\}, \ d_N = \{d(k)|k=1,\ldots,N\}.$$
Also the experimenter and the object may depend on k, so that there are sequences
$$X_N = \{X(k)|k=1,\ldots,N\}, \ S_N = \{S(k)|k=1,\ldots,N\}.$$
If k is associated with time, one can define an experiment with *repeated* stimulation as follows:

Procedure: Let $c(k-1)$ be the stimulus and $d(k)$ the response in step #k. Carry out a sequence of stimulation and observation, resulting in data with the following probability densities:
$k = 1$:
Response: $p[d(1)|X(1),S(1)]$
Stimulus: $p[c(1)|d(1),X(1),S(1)]$
$k = 2$:
Response: $p[d(2)|c(1),d(1),X(2),X(1),S(2),S(1)]$
Stimulus: $p[c(2)|c(1),d(2),d(1),X(2),X(1),S(2),S(1)]$
.....
Generally, the k:th stimulus has the density
$$p[c(k)|c(k-1),\ldots,c(1),d(k),\ldots,d(1),X(k),\ldots,X(1),S(k),\ldots,S(1)]$$
$$\triangleq p[c(k)|c_{k-1},d_k,X_k,S_k]$$
and that of the response is

$$p[d(k)|c(k-1),\dots,c(1),d(k-1),\dots,d(1),X(k),$$
$$\dots,X(1),S(k),\dots,S(1)] \triangleq p[d(k)|c_{k-1},d_{k-1},X_k,S_k].$$

The subscripts of X_k and S_k will play no role in the sequel, and will therefore be dropped for convenience. This does not mean that $X(k)$ and $S(k)$ are necessarily independent of k, but that this dependence is supposed to be included in the notations X and S.

Notice that if the density $p[c(k)|c_{k-1},d_k,X,S]$ of the stimulus actually depends on d_k, the experiment is carried out *in closed loop*.

Bayes' chain rule yields a dynamic model

The probability density of the whole data sequence (c_N,d_N) is $p[c_N,d_N|X,S]$. By applying repeatedly Bayes' rule for conditional probabilities one gets Bayes' *chain rule*:

$$p[c_N,d_N|X,S] = p[c(N),d(N)|c_{N-1},d_{N-1},N,X,S] \ p[c_{N-1},d_{N-1}|X,S]$$

$$= \prod_{k=1}^{N} p[c(k),d(k)|c_{k-1},d_{k-1},k,X,S]$$

$$= \prod_{k=1}^{N} p[c(k)|c_{k-1},d_k,k,X,S] \ p[d(k)|c_{k-1},d_{k-1},k,X,S] \tag{3.1}$$

Now, if one has sufficiently much data, one might possibly be able to estimate the distribution $\{p[c_N,d_N|X,S]\}$. This requires that a number of conditions be satisfied (see chapter 6 on "Large–sample theory"), but it is clear that one can never get more. By formally evaluating marginal distributions one can obtain conditional distributions, in particular $p[c(k),d(k)|c_{k-1},d_{k-1},k,X,S]$ $=$ $p[c_k,d_k|X,S]/p[c_{k-1},d_{k-1}|X,S]$. However, one cannot interpret the result as a probabilistic model of the dynamic system of experimenter and object based only on Bayes' rule. The latter yields a great number of factorizations of which (3.1) is one. For instance, it would be possible to let Bayes' chain rule run forwards in the k–sequence and obtain a model with only anticipatory dependence.

One must take into account the purpose of the modelling. If that purpose includes forecasting or control, then one must be able to compute $p[c(k),d(k)|c_{k-1},\mathbf{d}_{k-1},k,X,S]$ from given $(\mathbf{c}_{k-1},\mathbf{d}_{k-1})$, so that the factorization (3.1) is the interesting one. However, if one would want a model for the purpose of interpolating in a data sequence, or to estimate an unrecorded prehistory, other factorizations are needed.

Remark: Notice again that the probabilistic models in (3.1), for instance $p[d(k)|c_{k-1},d_{k-1},k,X,S]$, hold the apriori information of causality. A corresponding conditional density can be computed from any sufficiently long data sequence. But so can any other conditional density, for instance $p[d(k)|c_K,d_{k-1},k,X,S]$ for $K > k$. The difference is that there is no reason for wanting to compute the latter density, since one does not believe apriori that $d(k)$ will depend on future stimulus $c(k+1),...,c(K)$. Still the *data* $\mathbf{d}(k)$ may be statistically dependent on future data $\mathbf{c}(k+1),...,\mathbf{c}(K)$ — there might be feedback during the experiment.

What is cause and what is effect in a closed loop?

In the definition of a dynamic system the causation implied is
$$... \; d_{k-1} \mapsto c_{k-1} \mapsto d_k \mapsto c_k \mapsto d_{k+1} \mapsto c_{k+1} \mapsto ...$$
which says that among variables with the same index k the response $d(k)$ causes the stimulus $c(k)$, and not the opposite. This might seem against logic, and would also have been easy to change by factorizing the joint conditional density as

$$p[c(k),d(k)|c_{k-1},d_{k-1},k,X,S]$$
$$= p[c(k)|c_{k-1},d_{k-1},k,X,S] \; p[d(k)|c_k,d_{k-1},k,X,S]$$

That would have changed the cause–and–effect relationship to
$$... \; c_{k-1} \mapsto d_{k-1} \mapsto c_k \mapsto d_k \mapsto c_{k+1} \mapsto d_{k+1} \mapsto ...$$
However, it would only mean a change of indexing convention — replacing c_k with c_{k-1} brings the original sequence back. In a closed loop there is obviously causal dependences both ways, and one can index in any way, without changing the actual causality relations. The convention is adapted to practice: After a stimulus $c(k)$ there is usually a delay before the response $d(k+1)$ is registered. Then there is another delay before the experimenter can decide on the next stimulus $c(k+1)$. However, if the experimenter is a computer, as is the rule in sequential experimentation, the second delay is the far shorter. Hence, the necessary delay in the loop is allocated to the object (by the indexing) and not to the experimenter. *Fig. 9* illustrates the relation in the particular case that the response $d(k)$ is a sampled continuous variable y and $c(k)$ feeds a zero–order hold circuit with output u.

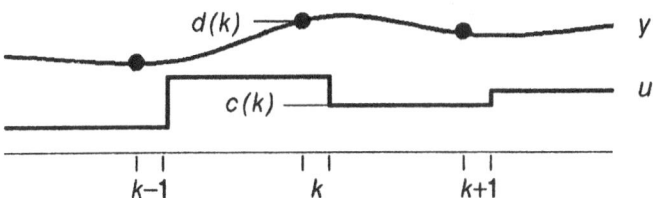

Fig. 9: Illustrating the definition of variables when the experiment is carried out by a computer.

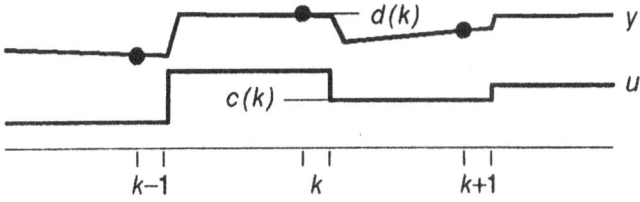

Fig. 10: Illustrating the case of a fast response.

Fast responses are sometimes a mathematical trouble

Fig. 10 illustrates the case when the object's response is almost immediate. The relations between continuous and discrete variables are the same, but notice that the discrete stimulus $c(k)$ is the value of the continuous variable immediately *after* the change, and the response $d(k)$ the value immediately *before* the change. This means that a continuous 'unit' model $y = u$ will be sampled as $d(k) = c(k-1)$. The discrete–time model will have an apparent delay, even if the continuous–time model has none.

That would suggest that one index the discrete–time variables, so that the model would be $d(k) = c(k)$ instead. However, this would lead to difficulties in cases of closed loop, as the following example will illustrate:

> **Example:** Describe the fast object by $d(k) = g\,c(k) + v(k)$, where $v(k)$ is a disturbance, and apply the proportional controller $c(k) = r(k) - d(k)$. This makes a system of algebraic (non–recursive) equations, a so–called 'algebraic loop'. In order to compute the response one has to solve those equations. That may be quite difficult in more complicated cases (see also section 5.5 on "Implicit and explicit models"). In the present case the solution is
> $d(k) = r(k)\,g/(1 + g) + v(k)/(1 + g)$
> Hence the output variable $d(k)$ follows the reference $r(k)$ and the disturbance $v(k)$ immediately, which is to be expected from a fast

loop. And if the gain g is high, it follows the reference closely, while suppressing the disturbance effectively, which is also to be expected. Hence, there is nothing obviously 'forbidden' (so far) about having algebraic loops.

With the convention of a delay in the forward path, the closed system will be $d(k) = g\ c(k-1) + v(k)$, $c(k) = r(k) - d(k)$. It has the explicit solution

$$d(k) = \sum_{l=1}^{k} (-g)^{k-l} \left[r(l-1) + v(l) \right]$$

To get a fair comparison between the cases, assume the sampling rate is high, so that $r(l)$ and $v(l)$ change but slowly for arguments near k. Then

$d(k) \approx r(k)\ g/(1 + g) + v(k)/(1 + g)$,

which agrees with the previous solution.

However, the solution holds only for $|g| < 1$. For larger g–values the system is unstable. That important property of the actual closed system was lost in the first analysis with the algebraic loop!

In conclusion, the convention of modelling the closed loop with a delay in the forward path is motivated partly by convenience, both computational and theoretical; one can compute the variables recursively, one at a time, also in complicated cases, and one can use Bayes' chain rule to simplify the theory. But it may also describe the dynamic properties better.

> **Remark:** Regardless of this, one may still write a dynamic model as $p[d(k)|c_k, d_{k-1}, k, X, S]$, and thus allow a direct influence of $c(k)$ on $d(k)$, *as long as one takes care to avoid algebraic loops.* Sometimes one may also do it *at the cost* of algebraic loops. A direct term may have other numerical advantages (see section 4.4.3 on "Internal models"). That depends on the application. In the sequel the immediate–response form $p[d(k)|c_k, d_{k-1}, k, X, S]$ will be used, whenever it does not complicate the theory, and since it is a simple and sometimes useful generalization of the lagged–response form $p[d(k)|c_{k-1}, d_{k-1}, k, X, S]$.

The (unbiassed) Likelihood function of a source model (X, S) describing experiments involving repeated stimulation and observation of a dynamic object (a 'closed loop') is

$$L[X, S | c_N, d_N] = \prod_{k=1}^{N} p[c(k), d(k) | c_{k-1}, d_{k-1}, X, S]$$

$$= \prod_{k=1}^{N} p[\mathbf{c}(k)|\mathbf{c}_{k-1},\mathbf{d}_k,k,X,S] \; p[\mathbf{d}(k)|\mathbf{c}_{k-1},\mathbf{d}_{k-1},k,X,S] \qquad (3.2)$$

3.4.1 Large samples

The asymptotic properties of the Likelihood function $L(X,S|\mathbf{c}_N,\mathbf{d}_N)$ for large values of the sample length N are of particular interest. More specificly, one has good use for the following:

● The limiting function when $N \rightarrow \infty$. The properties of this function determine whether it will be worth while to try and increase the length of the experiment in case a present one will not yield the desired accuracy, or whether no length will do.

● Asymptotic approximations for large but finite N. They may be used for computing the model accuracy for finite N, and for all kinds of investigations of model properties and tests of validity.

> **Remark:** Mainly, one has to use large–sample approximations, since they are the only forms one can cope with in practice. However, this is not a too serious limitation of the theory, since in cases of considerable random disturbance one usually needs long samples anyhow, to be able to infer anything with sufficient accuracy to satisfy the given purpose. A possible difficulty is that one usually cannot compute how large N has to be for the asymptotic approximation to be a valid one (Ljung, 1987a). In practice one tends to ignore this, and assume (without further evidence) that if the model is accurate enough for the purpose — and one can estimate that — then the accuracy of the *estimate* of the model accuracy will also be good enough.

For infinite N the Likelihood function (2.3) becomes singular, and has to be rewritten into a form that may tend to a continuous limit. But that is easy to do. Introduce

$$Q(X,S|\mathbf{c}_N,\mathbf{d}_N) \triangleq - \log L(X,S|\mathbf{c}_N,\mathbf{d}_N)/N$$

$$= - \frac{1}{N} \sum_{k=1}^{N} \log p[\mathbf{c}(k),\mathbf{d}(k)|\mathbf{c}_{k-1},\mathbf{d}_{k-1},k,X,S] \qquad (3.3)$$

The last member has the form of a sample average of the logarithm of the probabilistic model. This still does not mean that the average will always converge, but at least it will not generally converge to infinity, as will $\log p[\mathbf{c}_N, \mathbf{d}_N | X, S]$.

Practically all identification methods for complicated cases are based on the Maximum Likelihood estimate, *i.e.* of finding the peak value of the aposteriori density of $(\mathbf{X}, \mathbf{S})$ for large N, or they can be interpreted as if they were. This means maximizing the Likelihood function $L(X, S | \mathbf{c}_N, \mathbf{d}_N)$, and hence minimizing the function $Q(X, S | \mathbf{c}_N, \mathbf{d}_N)$ with respect to (X, S). This makes it possible to regard $Q(X, S | \mathbf{c}_N, \mathbf{d}_N)$ as a general loss function for the identification problem, the '*Likelihood loss function*'. However, in order to minimize the function, one has first to determine the *space* in which one may vary (X, S). This is the same as defining the *model structure* and is the whole purpose of the 'modelling' part of the model design.

─────────────── **Example 3.4** ───────────────

Find the loss function and the ML–estimate for the following case with a simple parametric structure:
X: $c(k) = \omega'(k)$
S: $d(k) = \Theta\, c(k) + \omega''(k)$
where $\omega'(k)$ and $\omega''(k)$ are Independent gaussian random variables with zero means and unit variances. The only unknown Is Θ. Hence, define
S: $d(k) = \theta\, c(k) + \omega'(k)$

Obviously the joint distribution of $(\mathbf{c}(k), \mathbf{d}(k))$ Is gaussian with zero mean and covariance matrix given by $r_{11} = 1$, $r_{12} = \Theta$, $r_{22} = \Theta^2 + 1$. The distribution for a two–dimensional gaussian variable Is given by

$$p[c, d] = [2\pi\, \sqrt{r_{11}\, r_{22} - r_{12}\, r_{12}}\,]^{-1}$$
$$\cdot\ \exp[-\tfrac{1}{2}(r_{22}\, c^2 - 2\, r_{12}\, c\, d + r_{11}\, d^2)/(r_{11}\, r_{22} - r_{12}\, r_{12})].$$

Since variables for different k are Independent, this yields the loss function

$$Q(\theta | \mathbf{c}_N, \mathbf{d}_N) = -\log L(\theta | \mathbf{c}_N, \mathbf{d}_N)/N = -\frac{1}{N} \sum_{k=1}^{N} \log p[\mathbf{c}(k), \mathbf{d}(k)]$$

$$= \frac{1}{N} \sum_{k=1}^{N} [(\theta^2 + 1)\, \mathbf{c}(k)^2 - 2\, \theta\, \mathbf{c}(k)\, \mathbf{d}(k) + \mathbf{d}(k)^2]/2 + \log(2\pi)$$

This is a quadratic function in θ, which is minimum for the argument

$$ML(\theta|\mathbf{c}_N,\mathbf{d}_N) = \sum_{k=1}^{N} \mathbf{c}(k)\,\mathbf{d}(k)/\sum_{k=1}^{N} \mathbf{c}(k)^2$$

Investigate also the limiting function. Since $\mathbf{c}(k)$ and $\mathbf{d}(k)$ are independent for different k, the 'law of large numbers' (Wilks, 1962) yields immediately that the sample average tends to its mean, and

$$\begin{aligned}
Q(\theta|&\mathbf{c}_\infty,\mathbf{d}_\infty) \\
&= E\{[(\theta^2 + 1)\,\mathbf{c}(k)^2 - 2\,\theta\,\mathbf{c}(k)\,\mathbf{d}(k) + \mathbf{d}(k)^2]/2\} + \log(2\pi) \\
&= [(\theta^2 + 1)\,r_{11} - 2\,\theta\,r_{12} + r_{22}]/2 + \log(2\pi) \\
&= [(\theta^2 + 1) - 2\,\theta\,\Theta + \Theta^2 + 1]/2 + \log(2\pi) \\
&= (\theta - \Theta)^2/2 + \log(2\pi)
\end{aligned}$$

Hence, the loss function tends to a limiting function, which has a unique minimum for the true parameter value.

The satisfactory properties in the example are by no means general. There are a number of basic conditions that have to be satisfied in order to ascertain a successful outcome of identification. They will be treated in the following chapters.

3.5 Experiments on dynamic objects

The first two conditions for a proper experiment, *viz.* Sufficient Stimulation and Reproducibility, are the same for a dynamic system, but now the data is $(\mathbf{c}_N,\mathbf{d}_N)$. The condition of an Open Loop will also apply in the same way. However, if one wants to ensure separability by having an isolated experimenter also in a closed loop, this can be achieved with a weaker condition replacing that of the Open Loop:

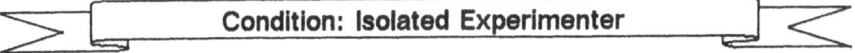

Condition: Isolated Experimenter

The experimenter **X** cannot influence the object's response $d(k)$ in other ways than by the past stimulus c_{k-1}, and the object **S** cannot influence the stimulus $c(k)$ in other ways than by the past and present response d_k. That is,

$$p[c(k)|c_{k-1},d_k,k,\mathbf{X},\mathbf{S}] = p[c(k)|c_{k-1},d_k,k,\mathbf{X}]$$
$$p[d(k)|c_{k-1},d_{k-1},k,\mathbf{X},\mathbf{S}] = p[d(k)|c_{k-1},d_{k-1},k,\mathbf{S}]$$

for all k and all $(c_k,d_k) \in \mathfrak{R}_k(\mathbf{X})$.

━━━━━━━ •••••• ━━━━━━━

Notice that since $\mathbf{X}$ and $\mathbf{S}$ have been defined to contain all phenomena that may affect c_N and d_N, they also contain the sources of possible hidden variables. Hence, the condition of an Isolated Experimenter means that there are no *common* hidden variables, so that $p[c(k)|c_{k-1},d_k,k,\mathbf{X}]$, $p[d(k)|c_{k-1},d_{k-1},k,\mathbf{S}]$ are probabilistic models, *i.e.* describe causal dependencies.

Remark: The condition may be illustrated by the signal flow graph in *Fig. 11*. There must be no interdependence between experimenter and object, either than through the channels for data.

Remark: A sufficient condition is the 'Natural condition of control' (Peterka, 1981), *viz.* $p[c(t)|c_{k-1},d_{k-1},\theta] = p[c(t)|c_{k-1},d_{k-1}]$, which says that the controller does not depend on the unknown object, except via past data.

Remark: Having only the second of the two requirements is not enough, as demonstrated by Example 3.3. The response d in the example does not depend on x; it depends on a variable v_1, that also influences the stimulus c. Still the case is unidentifiable (if one does not know α). The condition eliminates hidden *covariation* in the data by prohibiting hidden *causality* both ways.

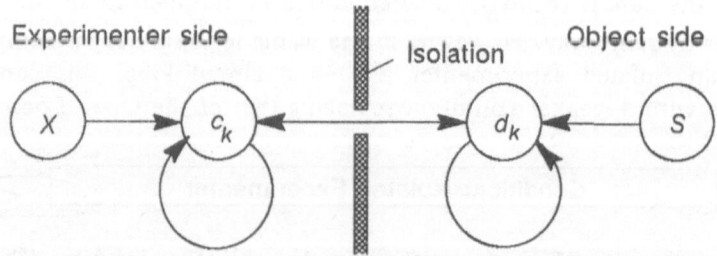

Fig. 11: Signal flow graph illustrating the condition of an isolated dynamic experiment. All causal or other dependence must pass through the data channel.

Remark: Notice that the condition of an Open Loop is a special case of an Isolated Experimenter. In the open loop case $p[c(k)|c_{k-1},d_k,k,\mathbf{X},\mathbf{S}] = p[c(k)|c_{k-1},k,\mathbf{X}]$. This implies $p[c_K,d_K|\mathbf{X},\mathbf{S}] = p[c_K|\mathbf{X}]\,p[d_K|c_K,\mathbf{S}]$, all $(c_K,d_K) \in \mathfrak{R}_K(\mathbf{X})$, which is the condition of an Open Loop. The difference is that an Isolated Experimenter allows experiments in 'closed loop'.

Remark: Notice also that if identification is carried out as part of an 'adaptive control', then it would seem that the experimenter/controller is adapted to the object, and $p[c(k)|c_{k-1},d_k,k,\mathbf{X},\mathbf{S}]$ would not be independent of $\mathbf{S}$. Hence this case would not be 'isolated'. However, to analyse adaptive control correctly, one must take into account not only the structure of the controller being adapted, but also the *tuning routine* used to adapt the controller. The latter would introduce dependence only of (k,c_k,d_k), and hence $p[c(k)|c_{k-1},d_k,k,\mathbf{X},\mathbf{S}]$ would not depend on $\mathbf{S}$, when the other arguments are given. In fact, any adaptive controller is still a controller, and isolated from the object it controls. Another thing is that analysis of identification under adaptive control (for instance of the identifiability properties) will be difficult, even in quite simple cases of identification and tuning algorithms.

Remark: The condition of an Isolated Experimenter is obviously satisfied, if the experiment is carried out by a computer or other inanimate device physically separated from the object, with the exception of channels for transmitting stimulus and response data $(\mathbf{c},\mathbf{d})$. It may or may not be the case, if the experiment is carried out by a human operator. He/she might be influenced by the way the object responds, and let that influence the stimulus, voluntarily or involuntarily. That would violate the requirement of an Isolated Experimenter. Notice that the important difference is not that the computer would follow a fixed rule, while the stimulus designed by a human operator would be more random and possibly without a rule well defined in advance. The computer may also generate the stimulus randomly, and by a rule that may change with the response, or that may be too complex to seem 'well–defined', or may be given by an 'Artificial Intelligence' program that may behave very 'human–like'. The important difference is *whether or not there are 'side channels' that carry information that causes statistical dependence.*

Isolated experiments are illustrated by *Fig. 12*. To emphasize that the experimenter is Isolated, it has been labelled 'computer', and also surrounded with a white zone. It has still been painted grey to illustrate

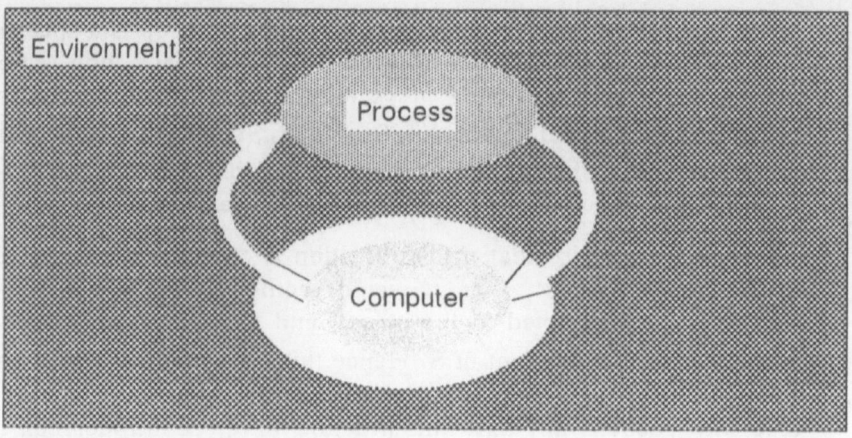

Fig. 12: Illustrating an Isolated experiment. The experimenting computer is in contact with the process only through channels transmitting stimulus and response data. It is in contact with the environment only through the process.

that one does not have to know the stimulation program to be able to identify correctly. The case when the process is not clearly distinguished from the environment apriori (like in 'black–box' identification) may be illustrated conveniently by painting the 'process' box in the same shade of grey as the 'environment'.

Proper experiments should be isolated in this way, and the sequel holds for Isolated experiments.

Applicability revisited

The condition of an Isolated Experimenter implies that

$$p[c(k),d(k)|c_{k-1},d_{k-1},k,\mathbf{X},\mathbf{S}]$$
$$= p[c(k)|c_{k-1},d_k,k,\mathbf{X}]\, p[d(k)|c_{k-1},d_{k-1},k,\mathbf{S}] \tag{3.4}$$

which means that if one can determine $p[d(k)|c_{k-1},d_{k-1},k,\mathbf{S}]$ at all from the experiment, the result will be independent of the experiment.

> **Remark:** This also suggests a way to test the condition of Isolation: If it will be possible to make two sufficiently different experiments on the same object, then one can compute two models and test whether they are significantly different. If they are, at least one of the experiments was not Isolated (or Reproducibility does not hold).

Furthermore, the distribution of data during an *application* (the basis for making decisions) will be

$$p[c(k),d(k)|c_{k-1},d_{k-1},k,X,\mathbf{S}]$$
$$= p[c(k)|c_{k-1},d_k,k,X]\ p[d(k)|c_{k-1},d_{k-1},k,\mathbf{S}] \tag{3.5}$$

and this means that one can apply another stimulating process $p[c(k)|c_{k-1},d_k,k,X]$ to the model and the combined system will be modelled correctly. This means that the lemma of the Open Loop may be replaced by a more general one for the closed loop. To formulate this, first rewrite the definition of Representability and Applicability:

Definition of Dynamic Representability and Applicability: A dynamic object $(\mathbf{S},T)$ at application time T is *Representable*, if it can be substituted by a model of the object $(\mathbf{S},\mathbf{T})$ at identification time for all admissible controllers $X \in \mathfrak{X}$. That is,

$$p[c_K,d_K|X,\mathbf{S},T] = \prod_{k=1}^{K} p[c(k)|c_{k-1},d_k,k,X,T]\ p[d(k)|c_{k-1},d_{k-1},k,\mathbf{S},T],$$

all $X \in \mathfrak{X}$, all k, all $(c_K,d_K) \in \mathfrak{R}_K(\mathfrak{X})$. If a model is sufficiently accurate, it will also be *Applicable* (see section 1.3 on "The purpose").

Lemma of the Closed Loop: If the experiment is *Reproducible* and *Sufficiently Stimulating* and the experimenter and the controller are *Isolated*, then the object is *Representable*, and a model that is sufficiently accurate will also be *Applicable*.

Proof:
From the conditions of i) an Isolated Experimenter, ii) Reproducibility, and iii) Sufficient Stimulation follows that for application intervals T of the same length N as the identification interval $\mathbf{T}$:

$$\prod_{k=1}^{N} p[c(k)|c_{k-1},d_k,k,X,\mathbf{T}]\ p[d(k)|c_{k-1},d_{k-1},k,\mathbf{S},\mathbf{T}]$$
$$= p[c_N,d_N|X,\mathbf{S},\mathbf{T}]$$
$$= p[c_N,d_N|X,\mathbf{S},T]$$
$$= \prod_{k=1}^{N} p[c(k)|c_{k-1},d_k,k,X,T]\ p[d(k)|c_{k-1},d_{k-1},k,\mathbf{S},T] > 0$$

all $(c_N,d_N) \in \mathfrak{R}_N(X)$.
Hence, all factors are positive, and

$$p[d(k)|c_{k-1},d_{k-1},k,\mathbf{S},\mathbf{T}]$$
$$= p[c(k)|c_{k-1},d_k,k,X,T]$$
$$\qquad\qquad \cdot\ p[d(k)|c_{k-1},d_{k-1},k,\mathbf{S},T]/p[c(k)|c_{k-1},d_k,k,X,\mathbf{T}].$$

Since the left member is a probability density, it follows that it integrates to one, and

$$p[d(k)|c_{k-1},d_{k-1},k,\mathbf{S},\mathbf{T}] = p[d(k)|c_{k-1},d_{k-1},k,\mathbf{S},T]$$

all k, all $(c_k,d_k) \in \Re_k(\mathbf{X})$.

From the definition of the Range of Model Validity and the condition of Sufficient Stimulation follows $\Re_k(\mathfrak{M}) \subseteq \Re_k(\mathbf{X})$. Hence,

$$p[d(k)|c_{k-1},d_{k-1},k,\mathbf{S},\mathbf{T}] = p[d(k)|c_{k-1},d_{k-1},k,\mathbf{S},T]$$

all k, all $(c_k,d_k) \in \Re_k(\mathfrak{M})$.

Since this holds also for as many consecutive intervals of length N as are needed to cover an application interval of length K, it follows that for Isolated controllers,

$$p[c_K,d_K|X,\mathbf{S},T]$$

$$= \prod_{k=1}^{K} p[c(k)|c_{k-1},d_k,k,X,T] \; p[d(k)|c_{k-1},d_{k-1},k,\mathbf{S},T]$$

$$= \prod_{k=1}^{K} p[c(k)|c_{k-1},d_k,k,X,T] \; p[d(k)|c_{k-1},d_{k-1},k,\mathbf{S},\mathbf{T}]$$

all $X \in \mathfrak{M}$, all k, all $(c_k,d_k) \in \Re_k(\mathfrak{M})$,

which is the criterion of a Representable object. Hence, a sufficiently accurate model of $p[d(k)|c_{k-1},d_{k-1},k,\mathbf{S},\mathbf{T}]$ is Applicable. ∎

3.5.1 Closed *vs* open loop

The distributions one wants to describe by a model, those of the object output, given input, and of the input, given the feedback signal, if any, are

$$p[d_K|c_K,\mathbf{S}] \triangleq \prod_{k=1}^{K} p[d(k)|c_{k-1},d_{k-1},k,\mathbf{S}]$$

$$p[c_K|d_K,\mathbf{X}] \triangleq \prod_{k=1}^{K} p[c(k)|c_{k-1},d_k,k,\mathbf{X}] \tag{3.6}$$

The distribution one can possible compute from an experiment in closed loop is an estimate of $p[c_K,d_K|\mathbf{X},\mathbf{S}]$. And, subject to a number of further conditions, it might even be possible to get an unbiassed and sufficiently accurate estimate from a sufficiently long sample, but never more (see chapter 6 on "Large–sample theory").

Now, by the condition of an Isolated Experimenter

$$p[c_K, d_K | X, S]$$

$$= \prod_{k=1}^{K} p[c(k) | c_{k-1}, d_k, k, X] \, p[d(k) | c_{k-1}, d_{k-1}, k, S]$$

$$= p[c_K | d_K, X] \, p[d_K | c_K, S] \tag{3.7}$$

If the experiment is further Sufficiently Stimulating then, in the limit, one can determine $p[d_K | c_K, S]$ from the equation (3.7) together with $p[c_K | d_K, X]$. One can also estimate $p[d_K | c_K, X, S]$ as the marginal distribution of $p[c_K, d_K | X, S]$. Hence, it is tempting — but a *'pitfall'* — to try and use the equation

$$p[d_K | c_K, X, S] = p[d_K | c_K, S] \tag{3.8}$$

to determine $p[d_K | c_K, S]$. This would avoid having to compute the feedback model $p[c_K | d_K, X]$ in case one does not want it. But the result would be *wrong*, even in the limit, since

$$p[d_K | c_K, X, S] = \prod_{k=1}^{K} p[d(k) | c_K, d_{k-1}, k, X, S]$$

$$\neq \prod_{k=1}^{K} p[d(k) | c_{k-1}, d_{k-1}, k, X, S]$$

$$= \prod_{k=1}^{K} p[d(k) | c_{k-1}, d_{k-1}, k, S] = p[d_K | c_K, S] \tag{3.9}$$

In spite of the causality between the stimulus and response sequences, the data $d(k)$ may still depend statistically on (covary with) current and future stimulus $c(k+l)$, $l \geq 0$, via the feedback. The result will be that what one thinks is a model of the object, the 'forward path' $c_K \mapsto d_K$, may instead be one of the inverse of the 'feedback path' $d_K \mapsto c_K$. That depends on which of the paths has the strongest causal dependence. And in industrial control this is usually the feedback path, since that is determined by a deterministic controller.

———————————— **Example 3.5** ————————————

Illustrating incorrect and correct identification in closed loop.

Let the experiment system be
S: $d(k) = d(k-1) + a\, c(k-1) + \omega_1(k)$

X: $c(k) = -d(k) + \mathbf{b}\,\omega_2(k)$

where $\{\omega_1(k)\}$ and $\{\omega_2(k)\}$ are independent and gaussian with zero means and unit variances. The parameters **a** and **b** are unknown.

The density function of the data is from (3.7):

$$\log p[c_K, d_K | X, S] = -\tfrac{1}{2} \sum_{k=1}^{K} [d(k) - d(k-1) - \mathbf{a}\,c(k-1)]^2$$

$$-\tfrac{1}{2} \sum_{k=1}^{K} [c(k) + d(k)]^2\,\mathbf{b}^{-2} - K \log(2\pi\,\mathbf{b})$$

Assume that one has sufficiently much data (c_N, d_N) to compute $\log p[c_K, d_K | X, S]$ with arbitrary accuracy. Then, if one would (tempting but incorrectly) use the equation (3.8) for identifying $p[d_K | c_K, S]$, one would get the solution $p[d_K | c_K, S] = p[d_K | c_K, X, S]$.

The right member can be evaluated from $\log p[c_K, d_K | X, S]$ by 'completing the squares'. The result will be

$$\log p[d_K | c_K, X, S] = \sum_{k=1}^{K} \log p[d(k) | c_K, d_{k-1}, X, S]$$

$$= -\tfrac{1}{2} N \log[2\pi/(1 + \mathbf{b}^{-2})] - \tfrac{1}{2} \sum_{k=1}^{K} (1 + \mathbf{b}^{-2})\{d(k)$$

$$- [d(k-1) - \mathbf{b}^{-2}\,c(k) + \mathbf{a}\,c(k-1)]/(1 + \mathbf{b}^{-2})\}^2$$

Hence, the dynamic model would be

$$\log p[d(k) | c_k, d_{k-1}, S]$$

$$= -\tfrac{1}{2} \log[2\pi/(1 + \mathbf{b}^{-2})] - \tfrac{1}{2}(1 + \mathbf{b}^{-2})\{d(k)$$

$$- [d(k-1) - \mathbf{b}^{-2}\,c(k) + \mathbf{a}\,c(k-1)]/(1 + \mathbf{b}^{-2})\}^2$$

This corresponds to the model
S: $d(k)$

$$= (1 + \mathbf{b}^{-2})^{-1}\,d(k-1) - (1 + \mathbf{b}^2)^{-1}\,c(k) + (1 + \mathbf{b}^{-2})^{-1}\,\mathbf{a}\,c(k-1)$$
$$+ \omega(k)/\sqrt{1 + \mathbf{b}^{-2}}$$

The model is obviously different from S, unless $\mathbf{b}^{-2} = 0$. For low values of **b** it tends to $d(k) = -c(k) + \mathbf{b}\,\omega(k)$, which is the inverse of the feedback path. For all **b**–values $d(k)$ has a direct dependence on $c(k)$. If the feedback has a more reproducible response than the control object, then $\mathbf{b}^2 \ll 1$, and the dependence on $c(k)$ dominates over those on $c(k-1)$ and $d(k-1)$. Admittedly, a direct dependence is not in accordance with the correct model structure.

However, one might not know the correct structure, and think that the object actually has a very fast response.

Now, suppose one does know the correct structure. Then one could try the ML–estimate (Ljung, 1987a), evaluated *as if* the loop were open:

$$\hat{a} = \arg \max \, p[\mathbf{d}_N|\mathbf{c}_N, a] = \arg \min \, Q(a|\mathbf{c}_N, \mathbf{d}_N)$$

$$= \arg \min \, \frac{1}{N} \sum_{k=1}^{N} [\mathbf{d}(k) - \mathbf{d}(k-1) - a \, \mathbf{c}(k-1)]^2$$

For long samples the loss function tends to
$$Q(a|\mathbf{c}_\infty, \mathbf{d}_\infty)$$
$$= E\{[\mathbf{d}(k) - \mathbf{d}(k-1)]^2\} - 2 \, a \, E\{[\mathbf{d}(k) - \mathbf{d}(k-1)] \, \mathbf{c}(k-1)\}$$
$$+ a^2 \, E\{\mathbf{c}(k-1)^2\}$$

where the covariances are obtained from the data distribution. The optimum tends to

$$\hat{a} \rightarrow E\{[\mathbf{d}(k) - \mathbf{d}(k-1)]\}/E\{\mathbf{c}(k-1)^2\}$$
$$= E\{\mathbf{a} \, \mathbf{c}(k-1)^2 + \mathbf{c}(k-1) \, \omega_1(k)\}/E\{\mathbf{c}(k-1)^2\} = \mathbf{a}$$

The difference from the first approach is that by fixing the structure to S: $d(k) = d(k-1) + a \, c(k-1) + \omega_1(k)$ one has excluded all solutions having a direct term, as being contrary to apriori knowledge.

The correct way of identifying the object when one does *not* know the structure (but has an arbitrary accurate data distribution) is using equ. (3.7):

$$\sum_{k=1}^{K} \log \, p[c(k)|c_{k-1}, d_k, k, X] + \sum_{k=1}^{K} \log \, p[d(k)|c_{k-1}, d_{k-1}, k, S]$$

$$= \log \, p[c_K, d_K|X, S]$$

$$= -\frac{1}{2} \sum_{k=1}^{K} [c(k) + d(k)]^2 \, \mathbf{b}^{-2}$$

$$-\frac{1}{2} \sum_{k=1}^{K} [d(k) - d(k-1) - a \, c(k-1)]^2 - K \log(2\pi \, \mathbf{b})$$

Since this holds for all (c_K, d_K), it follows that

$$\log \, p[c(k)|c_{k-1}, d_k, k, X] = -\frac{1}{2}[c(k) + d(k)]^2 \, \mathbf{b}^{-2} - \frac{1}{2} \log(2\pi \, \mathbf{b}^2)$$
$$\log \, p[d(k)|c_{k-1}, d_{k-1}, k, S]$$
$$= -\frac{1}{2}[d(k) - d(k-1) - a \, c(k-1)]^2 - \frac{1}{2} \log(2\pi)$$

Hence, it is theoretically feasible to compute $p[d(k)|c_{k-1},d_{k-1},k,\mathbf{S}]$ in closed loop also without knowing the structure of $\mathbf{S}$, provided the data distribution is known.

The simpler fitting equation (3.8) holds only in open loop: From (3.7)

$$p[d_K|c_K,\mathbf{S}] = p[d_K|c_K,\mathbf{X}.\mathbf{S}] \, p[c_K|\mathbf{X},\mathbf{S}]/p[c_K|d_K,\mathbf{X}] \qquad (3.10)$$

Now, if the whole sequence c_K is set independently of d_K, *i.e. the loop is open*, then

$$p[c_K|d_K,\mathbf{X}] = \prod_{k=1}^{K} p[c(k)|c_{k-1},d_k,k,\mathbf{X}]$$

$$= \prod_{k=1}^{K} p[c(k)|c_{k-1},k,\mathbf{X}] = p[c_K|\mathbf{X}] \qquad (3.11)$$

and from (3.10) and (3.11)

$$p[c_K|d_K,\mathbf{X}] = p[d_K|c_K,\mathbf{X},\mathbf{S}] \, p[c_K|\mathbf{X},\mathbf{S}]/p[c_K|\mathbf{X}] \qquad (3.12)$$

Since $p[d_K|c_K,\mathbf{S}]$ and $p[d_K|c_K,\mathbf{X},\mathbf{S}]$ are probability densities in the space of d_K, and the last two factors do not depend on d_K, they must be equal, and the equation (3.8) holds in open loop. Hence by using equation (3.8) one has assumed implicitly that the loop is open, and gets an error when it is not.

Notice that the error is a consequence of ignoring the fundamental difference between 'covariation' and 'causality', which was discussed at length in section 2.3. The error due to 'hidden variables' has the same fundamental origin and led to the condition of an Isolated Experimenter (section 6.3). Both introduce false dependences through other channels than one has provided for in the model structure (via $\mathbf{X}$ for non–isolated experiments, and via future c-values for closed loops). Hence the errors.

However, unlike the case of a non–isolated experiment, the closed loop can be handled by proper identification. Instead of using the equation (3.8) for the open loop one should use the corresponding equation for the closed system of controller and object:

$$p[c_K,d_K|\mathbf{X},\mathbf{S}] = \prod_{k=1}^{K} p[c(k)|c_{k-1},d_k,k,\mathbf{X}]\ p[d(k)|c_{k-1},d_{k-1},k,\mathbf{S}] \quad (3.13)$$

Then there is a channel in the model to take up the statistical dependence in the data between response and future stimulus, namely $p[c(k)|c_{k-1},d_k,k,X]$, and the fitting algorithm does not have to force that dependence into the much more unwielding channel $p[d(k)|c_{k-1},d_{k-1},k,S]$ (as in open loop identification).

3.5.2 Cases of proper experiments

It was stated in section 3.2 on "Requirements for proper experimentation" that $\mathbf{X}$ and $\mathbf{S}$ should be defined in such a way that they include the sources of possible hidden variables. This means in practice that one should provide for sufficiently many random variables to model unknown phenomena. In order to be able to assume that the random variables affecting the object and experimenter models are also *independent*, one has to satisfy the condition of an Isolated Experimenter. However, that condition is formulated as a relation between marginal distributions (with hidden variables integrated out), and this tends to hide the fact that the hidden variables are still there.

In order to illuminate the *interpretation* of the condition of Isolation in practice, introduce again the hidden variables ω explicitly. Then the analysis of the effect of hidden variables (section 2.4 on "Covariation and causality") yields a number of experiments that are isolated:

● **Automatic/Manual:** $p[c(k)|c_{k-1},d_k,\omega_k,k,\mathbf{X}]$ does not depend on (d_k,ω_k): This is the case of experiment in open loop. The condition holds in either of the cases
1) the c_N–sequence is predetermined (hence the stimulus is clearly independent of everything that may happen during the experiment),
2) the stimulus is set during the experiment (for instance by an operator), but without regard to the response of the object,
3) the stimulus is generated by a device (such as a computer or a signal generator) that is clearly separated from the object.

● **Feedback:** $p[c(k)|c_{k-1},d_k,\omega_k,k,\mathbf{X}]$ does not depend on ω_k: This is the case of feedback during the experiment. The condition says that feedback is allowed from the response only, and not from any other source. The condition may be violated, if a human operator intervenes with the experiment by adjusting c, for instance in order to amend an unfavourable trend in an object variable, other than d or c. However, unplanned human intervention is harmless if it is done for purposes not connected (statistically) with the object behaviour.

● **Spontaneous:**

$$p[d(k)|c_{k-1},d_{k-1},\omega_k,k,\mathbf{S}] = p[d(k)|c_{k-1},d_{k-1},\omega'_k,k,\mathbf{S}]$$

$$p[c(k)|c_{k-1},d_k,\omega_k,k,\mathbf{X}] = p[c(k)|c_{k-1},d_k,\omega''_k,k,\mathbf{X}]$$

where $\omega_k = (\omega'_k,\omega''_k)$, and ω'_k and ω''_k are independent: This is the case of *spontaneous stimulation, i.e.* when the object is stimulated by 'natural input' or 'operating signals' the source of which is ω'. It is then required that the same source do not influence the object's response d in other ways than via the input c. That is, the experiment should be *Isolated*. With spontaneous stimulation this may be difficult to guarantee in practice, since all the sources of input variation are usually not known.

4 The identification problem

The first two chapters have introduced two general principles on which to base the inference from data, which is the essence of identification. One has the Principle of Logical Empiricism, which yields a general procedure for refinement and falsification of hypotheses, and one has Bayes' Rule, which provides a way to formalize the information in experiment data, and in this way ties what one wants to know to what one can observe from experiments.

Both principles are based on the fiction that there is some 'truth' to be searched for, even if it may be impossible to find it. Bayes' rule tries to calculate the *'likelihood'* of that 'truth', namely of the data source (X,S), and recognizes the basic 'impossibility' in the practical impossibility of getting enough data to make 'likelihood' become 'certainty'. The Logical Empiricism implicitly defines 'truth' as a hypothesis that can never be falsified by any observation, but since the latter is impossible to prove in practice, the procedure, in principle, never *stops*.

In practice, this means that it will not stop until the researcher can no longer get the funds for prolonged research. When one is making models for an engineering purpose, the case is different on the important point that the procedure should preferably stop for other reasons. One can generally distinguish between two approaches to model design:

● **The scientist's rule:** Proceed until the model explains all data.

● **The engineer's rule:** Proceed until the model satisfies a given purpose.

This chapter formulates two general procedures for identification with and without purpose, including rules for the stopping of the procedures. It also points out the fundamental causes of 'pitfalls' in identification, *i.e.* reasons why the stopping rules may yield wrong results, and the procedure therefore stop prematurely.

4.1   Validation and falsification

The stopping will depend on whether or not the model is to be designed for a well–defined purpose *and* one can *verify* that it does satisfy the purpose. To emphasize that in model design for engineering purposes one is usually content with a *purposive* model, the term 'validation' will be used instead of 'verification', since 'valid for a given purpose' is a weaker quality than 'true'. Notice that the term 'falsification' is still useful — it is easier to falsify than verify. If one will be able to validate a model for a given purpose, then one can still use the basic refinement and falsification procedure of Logical Empiricism. The concept of 'truth' may be kept as a fiction, which may help intuition, but with no effect on the identification practice.

In order to assign meanings to the concepts of 'validation' and 'falsification', rewrite the basic refinement and falsification procedure slightly to suit the task of modelling for an engineering purpose:

Model making by 'the engineer's rule':
  Repeat until there is a valid model
    Refine the model structure (hypothesis)
    Repeat until the structure is falsified or there is a valid model
      Refine the experiment and observe
      Test whether there is a valid model
      Test whether data falsifies the structure

It has an outer and an inner loop, and one must be able to stop both. This calls for the following definitions:

Definitions of Validation and Falsification: *Validation* is a test procedure, the outcome of which determines whether a model satisfies its purpose. *Falsification* is a test procedure, the outcome of which determines whether a model or structure is contradicted by data. *Validatability* and *falsifiability* are the qualities of an identification problem that determine whether or not there exist such tests, so that one can stop the recursive procedure.

Notice that a 'validated' model may be simpler than one that agrees with all data, and may therefore be 'falsified'. There is no paradox here; one may need a complex model indeed to describe a large industrial plant, as revealed by data from extensive experimentation, but a much simpler model may still be both necessary and sufficient to control it. From a decision–theoretical viewpoint Schneeweiss (1987) stresses the difference between 'abstraction' (finding a realistic model structure, validated empirically), and 'relaxation' (finding a good–enough model for decision). The latter is usually simpler than one that describes pertinents aspects of real objects, even if one takes into account only those aspects that can be supported by data.

> **Remark:** If the problem is not validatable or not falsifiable, then the procedure will stop when one runs out of experiments, or structures, or both. This is not an unusual case, even in engineering; often one has got only one set of data and one structure to fit it to (and often no well–defined purpose), and hence there will be no sequential procedure. So, in practice there is always a final model. If the problem is validatable, one will *know* whether or not the final model is good enough for the purpose. If it is not validatable, the model may still be good, but one cannot know that.

> **Remark:** If the purpose of making the model is not defined, or one cannot validate the model, but one has a way to falsify a model (that is generally easier), then one can do no better than finding the simplest model that is not contradicted by data from a given experiment. This yields 'the scientist's rule', which is obtained by leaving out all reference to the attribute 'valid'.

> **Remark:** The basic difference between falsification and validation may be illustrated by that between systematic and random errors. A model with large random errors in the parameters is usually not satisfactory, even if the limited data cannot reject it. It can possibly be improved by more accurate parameter estimates obtained from fitting to more data. A model with large systematic errors cannot be improved by more data. To improve it one has to change the inadequate model structure responsible for the systematic error. The *precision* of a model usually increases indefinitely with the amount of data, but the *accuracy* only to a limit set by the model structure.

> **Recapitulation:** The distinction between 'precision' and 'accuracy' is the following: A high–precision estimate is one that is reproducible and determined with many digits. It may still have a poor accuracy, *i.e.* be far from the true parameter value. In fact, the trivial 'estimate' zero (based on the 'assumption' that there is no influence) has infinite precision, but in general very poor accuracy.

Remark: For a short tutorial on validation and falsification problems see Ljung (1982a).

────────────────── **Example 4.1** ──────────────────

This example illustrates the need for the concepts of validation and falsification and the distinction between them.

Assume that one wants to identify an object that is apriori known to be linear, and of the form $d(k) = \theta\, c(k) + \lambda\, \omega(k)$, where d and c are scalars, and that one also has a well-defined fitting criterion, namely the minimization of prediction error variance. The optimal model will be

$$\hat{\theta} = (\mathbf{c}_N{}^T \mathbf{c}_N)^{-1} \mathbf{c}_N{}^T \mathbf{d}_N$$
$$\hat{\lambda}^2 = \mathbf{d}_N{}^T [I - \mathbf{c}_N\, (\mathbf{c}_N{}^T \mathbf{c}_N)^{-1}\, \mathbf{c}_N{}^T]\, \mathbf{d}_N / (N-1)$$
$$E(\theta - \hat{\theta})^2 = \lambda^2\, (\mathbf{c}_N{}^T \mathbf{c}_N)^{-1},$$

where $\mathbf{c}_N$ and $\mathbf{d}_N$ are the data vectors, and N is the length of the sample. Assume for simplicity that the input $\mathbf{c}_N$ is deterministic and $\mathbf{c}_N{}^T \mathbf{c}_N = N$ (for instance a PRBS). Then the prediction error variance will have the expected value

$$E[\mathbf{d}(N+1) - \hat{\theta}\, \mathbf{c}(N+1)]^2 = E[(\theta - \hat{\theta})\, \mathbf{c}(N+1) + \lambda\, \omega(N+1)]^2$$
$$= \lambda^2\, [\mathbf{c}(N+1)^2/N + 1].$$

It decreases with increasing sample length N. Hence, in order to know when to stop experimenting one must obviously specify with what accuracy the model has to predict in order to be 'valid'. That depends on the purpose. Hence, one needs another specification, one of *sufficient accuracy*, that depends on the purpose and does not follow from the fitting criterion (prediction error). The latter is also given by the purpose, but is obviously not enough. In this simple example the added specification is slight and obvious, but the point is that model structure, experiment data, and fitting criterion do not define an identification problem. There must also be a *validation* criterion.

Secondly, the structure may be wrong. In that case the prediction error may never become sufficiently small to satisfy the purpose (if in the example one would be unaware of an input c, the prediction error variance would be $\lambda^2\, [\theta^2\, \mathbf{c}(N+1)^2 + 1]$, and would not decrease at all for large N). In order to avoid having to experiment forever, trying in vain to improve the prediction error, one must have a means to *falsify* the wrong model structure. This requires another criterion, which does not follow from any of the fitting or validation criteria.

In conclusion, one has to have *two* criteria for stopping an identification procedure, neither of which follows from a fitting criterion. 'Validation' and 'falsification' are needed to stop identification, when it is either successful or hopeless. The values of the logical variable *good_model* are three: 'true', 'false', and 'undecided'.

Remark: It will be argued below (section 4.3 on "Fitting") that if one has defined criteria for validation and falsification, one does not need also to specify a fitting criterion.

The task is now to formulate the stopping rules in mathematical terms.

4.2   Model structure, data descriptions, and purposive models

The object **S**, the experimenter **X**, the stimulus **c**, and the response **d**, were introduced in chapter 2 to formalize the idea of Bayes and of the 'likelihood' of a given system (X,S). By this idea, any system (X,S) was regarded as a tentative 'model' of $(\mathbf{X},\mathbf{S})$, although with different likelihood. However, in order to use the idea for making inference about the system $(\mathbf{X},\mathbf{S})$, and also to be able to state under what conditions the resulting models will be valid, one must formalize more.

Define a *model M* as a known system producing the same kind of variables (c,d) as the data source $(\mathbf{X},\mathbf{S})$, *i.e.* taking values in the same space as data $(\mathbf{c},\mathbf{d})$. Introduce the following sets of models M:

● *The model structure*: $\mathcal{M} = \{$all 'tentative' models$\}$. It is the result of modelling, and specifies both what are the facts and what is assumed and hypothesized about the data source $(\mathbf{X},\mathbf{S})$. Models in $\mathcal{M}$ must not be too complex to handle. However, the structure can possibly be falsified, if data says otherwise than the result of modelling.

Clarification: The distinction between 'facts', 'assumptions', and 'hypotheses' is the following: It is understood that 'hypotheses' are to be *tested* against data, while 'facts' and 'assumptions' are not. This means that if an assumption is wrong, that will not be detected in the identification process, and the result may be wrong, without any warning that this be the case. It is therefore important to test all the results of the modelling that *can* be tested, and thus making as many as possible of the 'assumptions' become 'hypotheses'. However, some assumptions cannot fundamentally be tested against data. When such assumptions are also wrong they will be 'pitfalls' (see section 4.6). Notice that facts cannot be wrong by definition — 'facts' that are wrong would be 'assumptions'.

Special cases: If the modelling assumes separate processes for experimenter and object, then $\mathcal{M} = \{M^c, M^d\}$. If the experimenter is further assumed to be known, then $\mathcal{M} = \{M^c, M^d | M^c = \mathbf{X}\}$. If the modelling assumes a dynamic system, then $\mathcal{M} = \{M(k) | k=1,\ldots,N\}$ and $c_{k-1}, d_{k-1}, M(k) \mapsto c(k), d(k)$. If the system is further assumed separated, then $\mathcal{M} = \{M^c(k), M^d(k) | k=1,\ldots,N\}$ and $c_{k-1}, d_k, M^c(k) \mapsto c(k)$; $c_{k-1}, d_{k-1}, M^d(k) \mapsto d(k)$.

● *The data descriptions*: $\mathcal{B}(\mathbf{c,d}) = \{$all 'unfalsified' models$\}$, *i.e.* those who are not contradicted by data. The set depends on data, and may include other models than those in $\mathcal{M}$. Since data depends on the experiment, $\mathcal{B}$ depends indirectly on $\mathbf{X}$.

Remark: The notation $\mathcal{B}$ comes from the fact that all models in the set describe the 'behaviour' of the system. The name comes from the fact that models in $\mathcal{B}$ describe the particular data from the particular experiment, but do not necessarily describe the system that produced the data, or the object of identification. For instance, a guarantee that the behaviour will be the same when the experiment is repeated will need additional evidence than data, and a guarantee that the object will be described by the same model when subject to other stimulus, will require still more evidence (see chapter 3).

● *The purpose*: $\mathcal{G}(\mathbf{X,S}) = \{$all 'purposive' models$\}$, *i.e.* those who satisfy a given purpose. The set generally depends on the unknown $(\mathbf{X,S})$. It may include other models than those in $\mathcal{M}$.

Special case: If one is only interested in identifying the object proper, then $\mathcal{G}(\mathbf{X,S}) = \mathcal{G}(\mathbf{S})$, regardless of whether $\mathbf{X}$ is known or not.

It is assumed that $\mathcal{M}$, $\mathcal{G}$, and $\mathcal{B}$ are subsets of the set of 'all models' M. It will also be convenient (and no loss of generality) to regard the system $(\mathbf{X,S})$ as a 'model' in the same set. However, it is not assumed that $(\mathbf{X,S})$ is in $\mathcal{M}$, in order to take into account the reasonable preconcep-

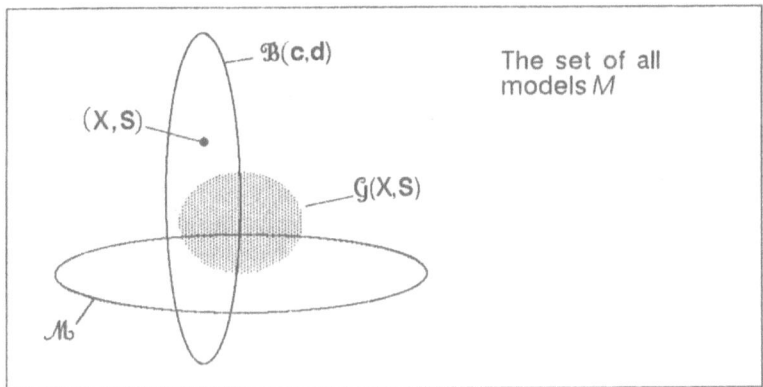

Fig. 13: Venn diagram illustrating the relations between 'tentative models' $\mathcal{M}$, 'unfalsified models' $\mathcal{B}$, and 'purposive models' $\mathcal{G}$. The purposive set (painted grey) is unknown.

tion that the 'true data source' usually cannot be modelled exactly (for one thing, it would be prohibitively complex). For the same reason, (X,S) is normally not in $\mathcal{G}$; a 'purposive' model must be possible to handle mathematically. However, (X,S) is likely to be in $\mathcal{B}$, since the 'model' that actually produced the data would be likely to agree with data. The relations are illustrated by the Venn diagram in *Fig. 13*.

> **Remark:** The set $\mathcal{M}$ of tentative models may or may not have elements in common with $\mathcal{B}$ or $\mathcal{G}$. Notice that a model in $\mathcal{B}$ may not be 'purposive', and one in $\mathcal{G}$ need not necessarily be 'unfalsified'. In other words, a model does not have to agree with all available data to be good enough for an independent purpose, and, conversely, it may agree with data and still be useless.

> **Remark:** The achievable precision of the result of identification is given by the width of the set $\mathcal{M} \cap \mathcal{B}$, which depends on the apriori information and the quality of experiment data. A better experimenter X, *e.g.* a longer data record or more stimulation of the object reduces $\mathcal{B}$. Better apriori information reduces $\mathcal{M}$.

> **Remark:** The purpose, and hence the set $\mathcal{G}$ may or may not depend on the object S. It does, obviously, if the purpose is one of controlling the object. It does not, for instance, if one is making a model of the source of a data sequence for the purpose of encoding the data to economize transmission to a receiver (see Example 1.2). This case leads to interesting questions as to whether there are universally best choices of $\mathcal{M}$, but they are outside the scope of this book (Rissanen and Langdon, 1981; Rissanen, 1983).

Remark: Generally, $\mathcal{G}$ depends on how one intends to use the model, and that ties it to the two basic requirements of Representability and Accuracy (Section 1.3 on "The purpose"). Loosely, the model should respond in an adequate way to stimulus during the application phase (and this may or may not mean other stimulus than during the experiment), and it should respond with sufficient accuracy, also that depending on the application. The aspect of $\mathcal{G}$ that can usually be validated is the *sufficient accuracy*; the identification procedure is stopped when the accuracy is enough for the purpose (see chapter 7 on "Validation techniques"). The *representability* is more involved, and is ensured mainly by making a correct *experiment* (see chapter 3).

When does data contradict a model?

When data $(\mathbf{c},\mathbf{d})$ comes from a deterministic system (or in practice nearly deterministic), then one can always decide whether or not a given model M is in $\mathfrak{B}$, for instance by simulating the model, and see whether the output (c,d) agrees with $(\mathbf{c},\mathbf{d})$. This means in practice that one has first to specify a level γ of approximation, and the width of the set $\mathfrak{B}(\gamma|\mathbf{c},\mathbf{d})$ will depend on that level. The higher level, the more models will be tolerated as 'unfalsified by data'.

When the output is not reproducible enough to be described by a deterministic model, a comparison between model and data will be difficult. In that case the data depends on some random process ω, besides on $(\mathbf{X},\mathbf{S})$, and a simulated output from a stochastic model M will likewise depend on another random process w (from a random number generator). Hence, $(\omega,\mathbf{X},\mathbf{S}) \mapsto (\mathbf{c},\mathbf{d})$, while $(w,M) \mapsto (c,d)$. Since w has no relation to ω, the outputs (c,d) and $(\mathbf{c},\mathbf{d})$ may deviate much, even if $M = (\mathbf{X},\mathbf{S})$. If disturbances do affect the data much, then comparing the outputs, by whatever criterion, will be a measure of how much (c,d) deviates from $(\mathbf{c},\mathbf{d})$, and will not necessarily say anything about how much M deviates from $(\mathbf{X},\mathbf{S})$.

In this case the designer has to provide another criterion (than closeness of output) for determining what should be regarded as 'unfalsified models'. Suitable statistical criteria are defined in chapter 8 on "Falsification techniques", and one may conceive others. But all, including output approximation error, can be expressed by means of a measure of 'distance' $B(M|\mathbf{c},\mathbf{d})$ between model and data and a 'tolerance level' γ. This generates a parametric family of sets of unfalsified models by

$$\mathfrak{B}(\gamma|c,d) = \{M|B(M|c,d) \leq \gamma\} \tag{4.1}$$

The falsification rule is $M \notin \mathfrak{B}(\gamma|c,d)$ or, equivalently, $B(M|c,d) > \gamma$, where γ is given.

The purpose must be expressed in terms of data

A validation rule cannot be defined immediately using the set $\mathfrak{G}(X,S)$ of purposive models, since it generally depends on the unknown object. However, Bayes' rule provides a statistical relation between unknown source (X,S) and known data (c,d), and that can be used to define a parametric validation rule based on probability as parameter: Define the Bayesian validation criterion

$$V(M|c,d) \triangleq \Pr\{M \in \mathfrak{G}(X,S)|c,d\}$$
$$= \int \mathrm{Ind}\{M \in \mathfrak{G}(X,S)\} \ dP[X,S|c,d] \tag{4.2}$$

where $dP[X,S|c,d]$ is obtained from Bayes' rule (2.5). This defines the parametric validation rule $V(M|c,d) \geq \gamma$. In analogy with the set of unfalsified models one can also define a parametric family of sets of 'validated models':

$$\mathfrak{V}(\gamma|c,d) = \{M|V(M|c,d) \geq \gamma\} \tag{4.3}$$

Both $\mathfrak{B}$ and $\mathfrak{V}$ are obviously instrumental in stopping the sequential identification procedure. They also define fitting criteria! (See section 4.3.)

In general, it is difficult to evaluate the integral in (4.2) for a given purpose $\mathfrak{G}(X,S)$. In such cases the designer will have to specify $V(M|c,d)$ heuristically, using his/her insight into the particular identification problem. Generally, this too is a difficult problem (see section 9.4 on "Designing the criterion").

A popular approach that takes a particular purpose into account uses the sample variance of prediction errors (Ljung, 1987a). This is reasonable, when one wants to have a model whose response simulates the data, or a model that predicts well (for forecasting or control). It is often used also when one has no other well-defined purpose and no validation rule.

4.3   Fitting

Obviously, identification should result in a model in the set
$\mathcal{M} \cap \mathcal{V}(\gamma|c,d)$, or at least in the set $\mathcal{M} \cap \mathcal{B}(\gamma|c,d)$, in case there is no
validation rule. However, picking one in the set requires a principle for
determining an 'identifier' $\mathcal{I}$: $(c,d) \mapsto \hat{M}$. The latter is usually formulated as the solution of a *fitting* problem. 'Fitting' means selecting a
model $\hat{M} \in \mathcal{M}$ by minimizing a function that evaluates the 'loss' of any
model M when its output is confronted with the experiment data.

> **Remark:** A number of common parameter estimation techniques
> are not based on an explicit criterion, such as the 'Instrumental
> Variable' (Söderström and Stoica, 1989) and the 'Extended Kal-
> man Filter' methods (Ljung, 1979). Their motivations are instead
> the ease of computing and the favourable statistical convergence
> properties of the resulting estimates for long samples.

One may choose to regard the fitting criterion (or the identifier $\mathcal{I}$) as a
fourth basic element of the identification problem (in addition to
$\mathcal{M}, \mathcal{B}, \mathcal{G}$), which has to be specified apriori. Many text books do that
(Eykhoff, 1974; Ljung, 1987a; Söderström and Stoica, 1989).
However, there are two approaches to identification that avoid having
to specify the fitting criterion quite out of the air:

● **The Most Purposive (MP) model:**
If the purpose of the identification is well defined, one may use Bayes'
idea and the 'engineer's rule'. It follows from the latter that one should
select a model $\hat{M}$ in $\mathcal{M}$ that makes $\Pr\{\hat{M} \in \mathcal{G}(X,S)|c,d\}$ maximum, so
that it has the highest probability of being purposive. In this way a
natural fitting criterion is defined uniquely by the purpose $\mathcal{G}$ and one
needs no fourth problem–defining element. The fitting gain will be
$V(M|c,d)$, defined by (4.2) or in any other way. The model is

$$\hat{M} = \arg\max\{V(M|c,d)|M \in \mathcal{M}\} \tag{4.4}$$

> **Remark:** The Minimum Prediction Error model is a special case,
> for the purpose of forecasting or control (Ljung, 1987a). If one ac-
> tually wants a model that predicts with minimum error — which is
> reasonable for design of feedback control — then one might argue
> that this *defines* the purpose of the identification, and that, there-
> fore, the fitting criterion may serve as a substitute for $\mathcal{G}$, and not as

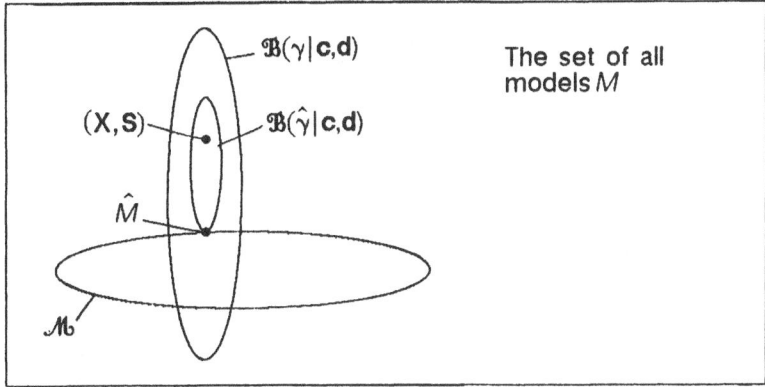

Fig. 14: Venn diagram illustrating the principle of selecting the model that passed as 'unfalsified' with the lowest tolerance.

a fourth independent element. This is a reasonable view to take, in case one has no other well-defined purpose or no validation rule. However, a fitting criterion can never provide a basis for deciding on a sufficient accuracy; for that one has still to specify a whole set $\mathcal{G}$ of purposive models.

● **The Most Falsifiable (MF) model:**

If no purpose is defined, then one must take the 'scientist's approach' and base the definition of a 'good' model on hypothesis and observation only, *i.e.* on $\mathcal{M}$ and $(\mathbf{c},\mathbf{d})$. Again the sequential refinement and falsification procedure yields a fitting criterion: If $\mathcal{B}(\gamma|\mathbf{c},\mathbf{d})$ is defined for different tolerance levels γ, one should select the model that stays unfalsified for the lowest tolerance level $\hat{\gamma}$. That model is

$$\hat{M} = \arg \min\{B(M|\mathbf{c},\mathbf{d})|M \in \mathcal{M}\} \tag{4.5}$$

> **Remark:** The principle is an adaptation to a more general one by Popper (see for instance Reckhow, 1987): "Among hypotheses that pass severe tests choose the one that is most falsifiable". Applied to the problem of selecting a model $\hat{M} \in \mathcal{B}(\gamma|\mathbf{c},\mathbf{d})$ the principle selects the one that passes the most severe test, which means the one with the smallest set $\mathcal{B}(\gamma|\mathbf{c},\mathbf{d})$, which means the lowest tolerance level γ.

The MF-criterion is illustrated in *Fig. 14*. In this case, the fitting criterion is defined by $\mathcal{B}$. One may use a fitting criterion to define $\mathcal{B}$ instead, as in (4.1), but the point is that the criteria for fitting and for falsification need not *both* be specified apriori to define an identification problem. In addition, both may be defined without reference to a prespecified fitting criterion (or distance measure B), for instance using the theory of statistical hypothesis testing (see chapter 8 on "Falsification techniques").

Similar conclusions apply to the MP–criterion: If a purpose is defined (and the problem is validatable), then the fitting criterion is also defined. It will further be independent of the definition of the set $\mathcal{B}$.

Hence, fitting model to data is a problem that is not necessarily fundamental in identification. This may seem surprising, since, historically, identification theory started with analysing the properties of models fitted to data in various ways (Åström and Eykhoff, 1971). Also the theories behind most of the available identification methods are based on explicit fitting criteria, including the very powerful MPE method (Ljung, 1987a). The equally powerful ML method (although technically equivalent in many practical cases) is more suited to the approaches taken in this book.

> **Remark:** Using the analysis in section 8.3 on "Conditional falsification of models" to compute an optimal set $\mathcal{B}(\gamma|c,d)$ yields that $B(M|c,d)$ is a monotone function of $-\log L(M|c,d)$, equ. (8.16). The MF model becomes the Maximum–Likelihood (ML) model. This lends some 'legitimacy' to the ML principle of fitting model to data, in addition to the usual motivation that it provides maximally efficient estimation for long data samples. Another name for the (biassed) Maximum–Likelihood estimate is the *Maximum a Posteriori* (MAP) estimate (Ljung, 1987a).

4.4   Basic identification procedures

In order to design an algorithm for the basic sequential refinement and falsification procedure in section 4.1, it remains to replace the natural-language statements by something mathematically more well defined. Thus, the statement

[Test whether data falsifies the structure

may be replaced by

[If $\mathcal{M} \cap \mathcal{B}(\gamma|c,d) = \emptyset$, then indicate *falsified* else *unfalsified*

However, since the test criterion $\mathcal{M} \cap \mathcal{B}(\gamma|c,d) = \emptyset$ is equivalent to $B(M|c,d) > \gamma$ for all $M \in \mathcal{M}$, it is not in a convenient form for evaluation by a computer as it stands, and will have to be rewritten further. Obviously, it is equivalent to testing only the MF–model, since that model has the lowest value of $B(M|\mathbf{c},\mathbf{d})$. This yields

> Falsification procedure:
>> Compute the MF model: $\hat{M} \leftarrow \arg\min\{B(M|\mathbf{c},\mathbf{d})\,|\,M \in \mathcal{M}\}$
>> If $B(\hat{M}|\mathbf{c},\mathbf{d}) > \gamma$, then indicate *falsified* else *unfalsified*

This form is suitable for evaluation by a computer. It brings fitting into the identification procedure as a byproduct of the falsification of the model structure $\mathcal{M}$.

The statement

> Test whether there is a valid model

will be similar to that of the falsification procedure:

> Validation procedure:
>> Compute the MP model: $\hat{M} \leftarrow \arg\max\{V(M|\mathbf{c},\mathbf{d})\,|\,M \in \mathcal{M}\}$
>> If $V(\hat{M}|\mathbf{c},\mathbf{d}) \geq \gamma$, then indicate *validated* else *unvalidated*

The MP principle may of course be replaced by another principle, or by any optimal or non–optimal way of obtaining a model to validate. This may save computing when such models turn out to be good, but may also make a successful validation somewhat less probable.

There are design alternatives for the procedures

The falsification and validation algorithms may also be written

> Falsification procedure:
>> Initialize M and search until stop
>>> If $B(M|\mathbf{c},\mathbf{d}) \leq \gamma$, then indicate *unfalsified* and stop
>>> Compute a new M to reduce $B(M|\mathbf{c},\mathbf{d})$
>>> If no reduction, then indicate *falsified* and stop

> Validation procedure:
>> Initialize M and search until stop
>>> If $V(M|\mathbf{c},\mathbf{d}) \geq \gamma$, then indicate *validated* and stop

> [Compute a new M to improve $V(M|\mathbf{c},\mathbf{d})$
> [If no improvment, then indicate *unvalidated* and stop

The alternatives would obviously produce the same test results, but with different amounts of computation. The latter would depend on the outcome of the test. For instance, if the structure $\mathcal{M}$ will be *unfalsified*, then the alternative falsification procedure will exploit the fact that it is sufficient to test whether $B(M|\mathbf{c},\mathbf{d}) \leq \gamma$ for *some* M. It will not be necessary first to carry a search procedure for a model $\hat{M}$ to convergence, unless the outcome will be *falsified*. The loop will stop before $\hat{M}$ is computed.

The alternatives will obviously be more efficient, when the structure (and the model) is right the first time. If, on the other hand, falsification and validation of each particular model would be a more time consuming task than selecting the next model in the search (which is likely, see chapters 7 and 8 on "Validation" and "Falsification techniques"), and the first trial structures are inadequate, then it would be a waste of time to falsify each model in the loop. A rule of thumb for the designer would be to favour the alternative only in cases of much reliable apriori information. Since the emphasize in this book is on uncertain apriori information, the original alternative will be used in the sequel.

> **Remark:** It would obviously be possible to conceive a number of other falsification and validation schemes, the results of which would be logically equivalent, but with different strategies to reach those results, and therefore not equally time–efficient. The best choice between them would still depend on circumstances that are unknown to the designer, but schemes are conceivable where the designer (or the computer) would use information from the outcomes of earlier tests to change strategy.

One gets a natural stopping rule also for the fitting

As indicated, the weakness of the falsification scheme is that it requires first a fitting procedure, that will run to convergence before the optimal model is tested, which logically would not be necessary in all cases. However, the requirement on 'convergence' (the stopping rule) need not be severe. It is determined by the tolerance level γ, which is a rather imprecise parameter to start with, and it would obviously be no point in requiring 'convergence' with a much higher precision. It follows also that it is the value of the loss function alone (and not for instance the magnitude of changes in M or closeness to the optimum) that should determine the stopping of the loop.

Secondly, the validation loop does not require convergence, when its outcome is 'validated'. Irrespective of this, it may or may not be desirable to have a final model that is 'valid' with a higher γ–value. However, as in the case of falsification, the 'convergence' does not have to be pushed further than required by the purpose. If the latter would require that M be actually needed with a high accuracy, that requirement will reflect on the function value $V(M|\mathbf{c},\mathbf{d})$, and still the closeness of M to the optimum will not have to be considered separately.

The meaning of 'refine' must be defined by the modelling

With those qualifications, it would be possible to insert the validation and falsification procedures, and thus arrive at two mathematically defined algorithms for identification by the 'engineer's' and 'scientist's rules'. It remains however to define what should be understood by 'refining' the model structure and experiment respectively. In principle, the elements $\mathcal{M}$ and $\mathbf{X}$ should have been specified apriori. This is known as *model structure determination* and *experiment design*, and is not considered here as being a part of the 'identification' task. It belongs to the 'modelling' task.

The point of the sequential procedure is now that one does not have to determine some 'certified' $\mathcal{M}$ and $\mathbf{X}$ apriori, based on facts and assumptions, but only a sequence of *tentative* structures and experiments, based on *hypotheses*. Which the identification will then be able to test against data.

Hence the procedure requires that one specify apriori two *expanding* sequences of model structures $\mathcal{M}(1),\mathcal{M}(2),\dots,\mathcal{M}(\bar{n})$ and experimenters $\mathbf{X}(1),\mathbf{X}(2),\dots,\mathbf{X}(\bar{m})$, *i.e.* such that model structures $\mathcal{M}(n)$ with higher values of n are always more general than those with lower values, and such that experimenters $\mathbf{X}(m)$ with larger values of m always produce additional data. Hence, $\mathcal{M}(n+1) \supset \mathcal{M}(n)$, and since adding data from more experiments can never increase the set of unfalsified models, $\mathcal{B}(m+1) \subseteq \mathcal{B}(m)$. Preferably, the first model structure should be crude and the first experiment should be inexpensive.

In principle the sequence is arbitrary otherwise, but must of course contain some feasible structure $\mathcal{M}(n)$. The widest model structure is $\bar{\mathcal{M}} =$

$\mathcal{M}(\overline{n})$, and the most extensive experimenter is $\overline{\mathbf{X}} = \mathbf{X}(\overline{m})$. They will play important roles in validation and falsification.

Remark: The sequence $\{\mathcal{M}(n)\}$ also provides a distinction between 'assumptions' and 'hypotheses'. What can be falsified are the 'complexity' numbers n, which therefore represent the 'hypotheses', while $\mathcal{M}$ represents the form of the model, based on 'assumptions' and 'facts'. Thus, theory requires that one order the hypotheses in a sequence, according to how reliable they appear apriori. For instance, the higher–complexity model structures contain increasingly unreliable hypotheses.

─────────────── **Example 4.2** ───────────────

The example illustrates the concepts of 'purposive' models, and of 'expanding' structures and experiments.

Let the object be

S: $d(k) = 0.8\, d(k-1) + c(k-1) + \omega(k)$

where $\{\omega(k)\}$ is a white gaussian sequence with zero mean and unit variance. Let the given purpose of identification be to design a stable minimum–variance controller for **S**.

A reasonable sequence of expanding structures is defined by

$\mathcal{M}(1):\ d(k) = d(k-1) + c(k-1) + \omega(k)$, (one member only)
$\mathcal{M}(2):\ d(k) = d(k-1) + \theta_1\, c(k-1) + \omega(k)$
$\mathcal{M}(3):\ d(k) = \theta_2\, d(k-1) + \theta_1\, c(k-1) + \omega(k)$

where each structure involves one more real parameter.

A reasonable sequence of experimenters is defined by $\mathbf{X}(m)$: $\{c(k)\,|\,k=1,\dots,100\,m\}$ which is gaussian and white with zero mean and unit variance, and $m = 1,2,\dots,$ *i.e.* each new experiment means taking 100 more measurements.

The minimum variance controller for the model structure $\mathcal{M}(3)$ is $c(k) = -d(k)\,\theta_2/\theta_1$. Inserting that into the object yields

$d(k) = (0.8 - \theta_2/\theta_1)\, d(k-1) + \omega(k)$

which is stable if $|0.8 - \theta_2/\theta_1| < 1$. It is reasonable also to require that θ_1 and θ_2 have the correct signs. Hence the set of purposive models is

$\mathcal{G} = \{\theta_1, \theta_2\,|\,-0.2 < \theta_2/\theta_1 < 1.8, \theta_1 > 0\}$

The set is depicted in *Fig. 15.* It depends on the object **S.**

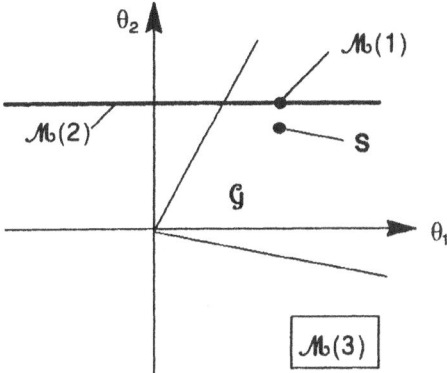

Fig. 15: The domain $\mathcal{G}$ of purposive models and a sequence $\mathcal{M}(1)$, $\mathcal{M}(2)$, $\mathcal{M}(3)$ of expanding struc-tures in parameter space.

4.4.1 Two general algorithms

The clarifications of the meanings of the natural–language statements in the procedure for **Identification by 'the engineer's rule'** make it possible to express the procedure as follows. The result is expressed in a natural algorithmic pseudo–language (generally, the symbols $b \leftarrow a$ mean that "*a* is assigned to *b*", while $a \mapsto b$ means that "*b* is computed from *a*"):

Identification according to 'the engineer's rule':

Initialize: $n,m \leftarrow 0$; indicate *unvalidated* and *unfalsified*

Repeat until outer stopping rule

Refine model structure: $n \leftarrow n+1$

Repeat until inner stopping rule

If *unfalsified*, then refine the experiment and observe:

$m \leftarrow m+1$; $X(m) \mapsto (\mathbf{c},\mathbf{d})$

Test whether there is a valid model:

Compute the MP model:

$\hat{M} \leftarrow \arg\max\{V(M|\mathbf{c},\mathbf{d})|M \in \mathcal{M}(n)\}$

If $V(\hat{M}|\mathbf{c},\mathbf{d}) \geq \gamma$, then indicate *validated* else *unvalidated*

$\quad$ [If *validated*, then stop inner and outer loops

$\quad$ [Test whether data falsifies the structure:

$\quad\quad$ [Compute the MF model:

$\quad\quad$ [$\hat{M} \leftarrow$ arg min$\{B(M|\mathbf{c},\mathbf{d})|M \in \mathcal{M}(n)\}$

$\quad\quad$ [If $B(\hat{M}|\mathbf{c},\mathbf{d}) > \gamma$, then indicate *falsified* else *unfalsified*

$\quad$ [If *falsified*, then stop inner loop

A narrative of the identification procedure would be the following: Start with $n = m = 1$, fit a model $M \in \mathcal{M}(1)$, and test if it is acceptable, *i.e.* check if $M \in \mathcal{V}(1)$. Probably it is not. If the set $\mathcal{M}(1) \cap \mathcal{V}(1)$ is not empty, because the first and simplest model structure $\mathcal{M}(1)$ is so restricted that it holds no purposive model, probably the model cannot be validated because the computable set $\mathcal{V}(1)$ of validated models is an uncertain (and therefore conservative) estimate of the set $\mathcal{G}$ of purposive models. This is a likely consequence of the insufficient data from the simplest experimenter $\mathbf{X}(1)$, since this will mean a small $\mathcal{V}(1)$.

Next, check if $\mathcal{M}(1) \cap \mathcal{B}(1)$ is empty, to see if the structure $\mathcal{M}(1)$ can be falsified immediately. Probably it cannot. Even if the set $\mathcal{M}(1)$ is actually too narrow, the set $\mathcal{B}(1)$ of unfalsified models is probably so large that it will not be able to falsify $\mathcal{M}(1)$. This, again, is a consequence of the insufficient data, which will widen $\mathcal{B}(1)$.

Try then to reduce $\mathcal{B}(m)$ and/or increase $\mathcal{V}(m)$, by applying better experimenters $\mathbf{X}(2),\mathbf{X}(3),\ldots$ until one of two things happens: 1) either $\mathcal{V}(m)$ becomes a certain enough estimate of $\mathcal{G}(\mathbf{X},\mathbf{S})$, while there is also a purposive model in $\mathcal{M}(1)$, or 2) the reduction of $\mathcal{B}(m)$ falsifies $\mathcal{M}(1)$. In the first case the procedure stops with a valid model. In the second case the procedure repeats from $\mathcal{M}(2)$.

The procedure is illustrated in *Fig. 16*.

> **Remark:** The logic of bringing the sequence of experiments and tests (the inner loop) to the point of falsification, before one expands the structure, is to minimize the set of models that are *not* falsified.

The 'scientist's rule' will be similar, except that it does not include a validation procedure. This means that the outer stopping rule will be *unfalsified* instead of *validated*. The procedure would stop either at the largest available m (one runs out of experiment facilities) or at the largest available n (one runs out of explanations). A similar ending occurs

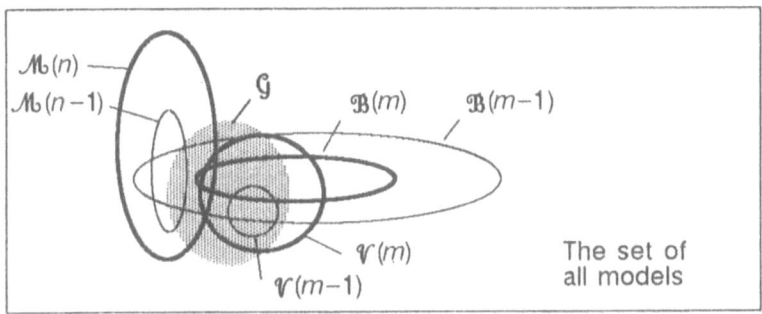

Fig. 16: Illustrating the rule for selecting n and m to determine the proper complexity of the model structure and the proper extent of the experiment.
$\mathcal{M}(n{-}1) \cap \mathcal{V}(m{-}1)$ is empty, $\mathcal{M}(n{-}1) \cap \mathcal{B}(m{-}1)$ is not empty.
⇒ No validated model, but the structure is not falsified. Get more data!
$\mathcal{M}(n{-}1) \cap \mathcal{V}(m)$ and $\mathcal{M}(n{-}1) \cap \mathcal{B}(m)$ are empty.
⇒ Structure is falsified. Find a better one!
$\mathcal{M}(n) \cap \mathcal{V}(m)$ is not empty.
⇒ There is a valid model in $\mathcal{M}(n)$. Choose one in $\mathcal{V}(m)$ and end identification!

when identification is not validatable, for instance, the purpose has not been defined in a way that one can check. The procedure becomes:

Identification according to 'the scientist's rule':

  Initialize: $n,m \leftarrow 0$; indicate *unfalsified*

  Repeat until outer stopping rule

    Refine model structure: $n \leftarrow n{+}1$

    Repeat until inner stopping rule

      If *unfalsified*, then refine the experiment and observe:
      $m \leftarrow m{+}1$; $X(m) \mapsto (\mathbf{c},\mathbf{d})$

      Test whether data falsifies the structure:

        Compute the MF model:
        $\hat{M} \leftarrow \arg\min\{B(M|\mathbf{c},\mathbf{d})\,|\,M \in \mathcal{M}(n)\}$

        If $B(\hat{M}|\mathbf{c},\mathbf{d}) > \gamma$, then indicate *falsified* else *unfalsified*

      If *falsified*, then stop inner loop

    If *unfalsified*, then stop outer loop

Generalization: In case one has several alternative ideas of what might be a suitable model structure, and does not know which one

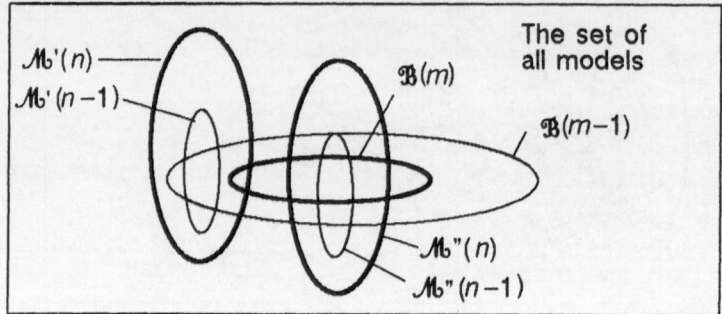

Fig. 17: Illustrating the case of two alternative sequences of structures ("Scientist's rule").

to choose, it is possible to make several sequences $\mathcal{M}(1), \mathcal{M}(2), \ldots$ and arrive at several sets of models $\mathcal{M}(n) \cap \mathcal{B}(m)$. Then one can choose between them, using any criterion of preference. The principle of *parsimony* is suitable here, *i.e.* preferring the simplest alternative that has not been falsified. *Fig. 17* illustrates the case of two alternatives.

4.5 Conditions for Bayesian validation

Obviously, all conceivable purposes are not validatable under all circumstances; there must be conditions that involve the true object, and that therefore are the responsibility of the designer. Since it must first of all be possible to express the set $\mathcal{G}(X,S)$ of purposive models mathematically, it is possible to define the set only for arguments (X,S) in the widest model structure $\bar{\mathcal{M}}$. This means that it must be reasonable to conceive a one-to-one mapping of the 'true model' $(X,S) \mapsto (M,E)$, where $M \in \bar{\mathcal{M}}$ may be interpreted as the 'best' model, in some not yet specified sense, and $E \notin \bar{\mathcal{M}}$ as the 'modelling error'. The E–component is introduced to formalize the unavoidable difference in practice between reality and the best model in $\bar{\mathcal{M}}$. Then, the purpose must depend only on M, $\mathcal{G}(X,S) = \mathcal{G}(M)$. For instance,

$$\mathcal{G}(M) = \{M | G(M,M) \geq 0, M \in \bar{\mathcal{M}}\}, \; M \in \bar{\mathcal{M}} \tag{4.6}$$

Now, the Bayesian validation rule (4.2) requires that one compute

$$V(M|\mathbf{c},\mathbf{d}) = \Pr\{M \in \mathcal{G}(\mathbf{X},\mathbf{S})|\mathbf{c},\mathbf{d}\} = \Pr\{M \in \mathcal{G}(\mathbf{M})|\mathbf{c},\mathbf{d}\}$$

$$= \int \mathrm{Ind}\{M \in \mathcal{G}(\mathbf{M})\}\ dP[\mathbf{M},\mathbf{E}|\mathbf{c},\mathbf{d}]$$

$$= \int\int \mathrm{Ind}\{M \in \mathcal{G}(\mathbf{M})\}\ dP[\mathbf{M}|\mathbf{c},\mathbf{d}]\ dP[\mathbf{E}|\mathbf{M},\mathbf{c},\mathbf{d}]$$

$$= \int \mathrm{Ind}\{M \in \mathcal{G}(\mathbf{M})\}\ dP[\mathbf{M}|\mathbf{c},\mathbf{d}] \qquad (4.7)$$

This is valid irrespective of how $(\mathbf{X},\mathbf{S})$ is mapped into model and error. However, the problem is to compute the conditional probability $P[\mathbf{M}|\mathbf{c},\mathbf{d}]$. The difficulty is that Bayes formula yields $P[\mathbf{M},\mathbf{E}|\mathbf{c},\mathbf{d}]$. If evaluated for model structures $\bar{\mathcal{M}}$, the result is the distribution $P[\mathbf{M}|\bar{\mathcal{M}},\mathbf{c},\mathbf{d}] = P[\mathbf{M}|\mathbf{E}{=}\emptyset,\mathbf{c},\mathbf{d}]$, which is generally not the wanted marginal distribution $P[\mathbf{M}|\mathbf{c},\mathbf{d}]$, but the distribution *conditional* on 'no modelling error'. It will be necessary to *assume* that the two distributions are equal. Hence the following condition ensures validatability in Bayes' sense:

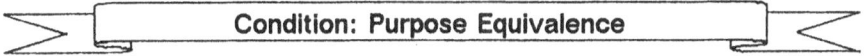

Condition: Purpose Equivalence

The system $(\mathbf{X},\mathbf{S})$, the purpose $\mathcal{G}$, and the widest model structure $\bar{\mathcal{M}}$ are such that there exists an equivalent model $\mathbf{M} \in \bar{\mathcal{M}}$, that is

$$\mathcal{G}(\mathbf{X},\mathbf{S}) = \mathcal{G}(\mathbf{M})$$
$$P[\mathbf{M}|\mathbf{c},\mathbf{d}] = P[\mathbf{M}|\bar{\mathcal{M}},\mathbf{c},\mathbf{d}]$$

————— •••••• —————

The condition $P[\mathbf{M}|\mathbf{c},\mathbf{d}] = P[\mathbf{M}|\bar{\mathcal{M}},\mathbf{c},\mathbf{d}]$ is satisfied if the statistical 'orthogonality' condition $P[\mathbf{M},\mathbf{E}|\mathbf{c},\mathbf{d}] = P[\mathbf{M}|\mathbf{c},\mathbf{d}]\ P[\mathbf{E}|\mathbf{c},\mathbf{d}]$ holds. This defines the 'best' model $\mathbf{M}$ and relates it to the 'true model' $(\mathbf{X},\mathbf{S})$.

The two conditions mean that 1) the purpose must depend only on such properties $\mathbf{M}$ of the true source that can be modelled, and 2) these properties must be selected in such a way that there would be no more (purposive) information to obtain from trying to estimate $\mathbf{E}$.

This should formalize what an engineer would actually do in practice, facing the fact that one can never model reality exactly. That is, to base the analysis on the best model structure he/she can conceive, trusting

that the remaining differences will not affect any decisions of importance for the purpose.

─────────────────── **Example 4.3** ───────────────────

This is a case where the condition of Purpose Equivalence is not
satisfied, and an attempted validation would produce an incorrect
result.

Consider the system in Example 2.2:

S: $d = c \Theta + \omega$,

$\dim(d) = N$, $\dim(\Theta) = n$, and ω is gaussian with zero mean and unit
covariance matrix. The stimulus c is deterministic, $\dim(c) = (N,n)$.

Assume (incorrectly) that the widest structure is

$\mathcal{M}_b(\bar{n})$: $d = c_1 \theta_1 + \omega$

where $\dim(\theta_1) = \bar{n} < n$, and $c = (c_1 \ c_2)$. $\Theta = \operatorname{col}(\Theta_1, \Theta_2)$.

Specify that the purpose is to estimate the $\bar{n}$ first components Θ_1
of Θ with a given minimum error norm, so that

$\mathcal{G}(S) = \{\theta_1 | \|\theta_1 - \Theta_1\|^2 < \gamma^2\}$

From Example 2.2 the aposteriori density is

$\log p[\theta|d] = -\frac{1}{2} (\theta^T c^T c \theta - 2 d^T c \theta + d^T d) + constant(d)$
$= -\frac{1}{2} (\theta - \hat{\theta})^T R^{-1} (\theta - \hat{\theta}) + constant(d)$

where $\hat{\theta} = R \begin{bmatrix} c_1^T \\ c_2^T \end{bmatrix} d$, $\quad R = \begin{bmatrix} c_1^T c_1 & c_1^T c_2 \\ c_2^T c_1 & c_2^T c_2 \end{bmatrix}^{-1}$

The marginal distribution is given by

$\log p[\theta_1|d]$
$= -\frac{1}{2} (\theta_1 - \hat{\theta}_1)^T [c_1^T c_1 - c_1^T c_2 (c_2^T c_2)^{-1} c_2^T c_1] (\theta_1 - \hat{\theta}_1)$
$\hspace{8cm} + constant(d)$

The conditional distribution for $\theta_2 = 0$ is given by

$\log p[\theta_1|\theta_2 = 0, d] = -\frac{1}{2} (\theta_1 - \hat{\theta}_1)^T (c_1^T c_1)^{-1} (\theta_1 - \hat{\theta}_1) + constant(d)$

Since the conditional density is used in the validation of $\hat{\theta}_1(\bar{n})$, while
the marginal distribution affects the *validity* of $\hat{\theta}_1(\bar{n})$, the fact that

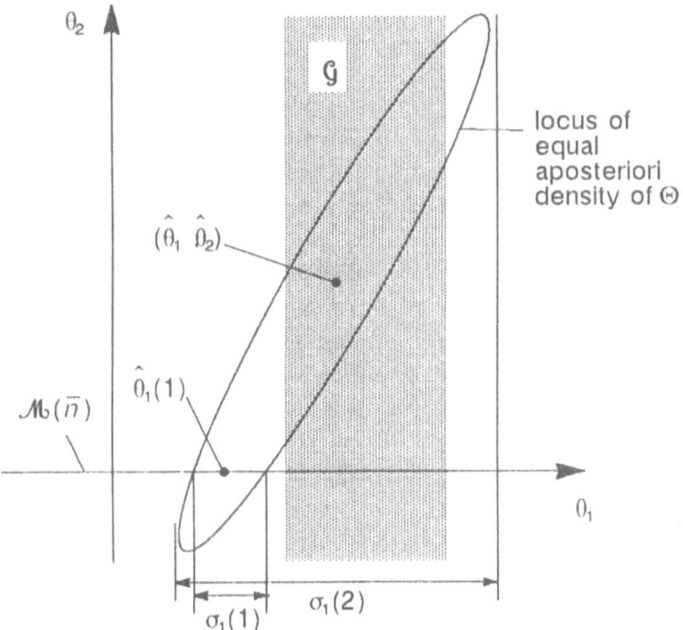

Fig. 18: Illustrating the case when the condition of Purpose Equivalence is not satisfied. The model $\hat{\theta}_1(1)$ is validated erroneously.

the two distributions differ constitutes a 'pitfall'. Only if $c_1^T c_2 = 0$ do the distributions agree. Hence, one can avoid this particular 'pitfall' in this particular case by choosing the stimulus c orthogonal and thus satisfying the condition of Purpose Equivalence.

The case is illustrated in *Fig. 18* for $\bar{n} = 1$, $n = 2$. The variances of $\hat{\theta}_1(\bar{n})$ and $\hat{\theta}_1(n)$ are

$$\sigma_1^2(\bar{n}) = [c_1^T c_1]^{-1}$$

and

$$\sigma_1^2(n) = [c_1^T c_1 - c_1^T c_2 (c_2^T c_2)^{-1} c_2^T c_1]^{-1}$$

respectively.

The validatability condition may be difficult to interpret

It may not be so difficult to understand the nature of the problem caused by correlated input signals in Example 4.3. If several input signals affect the same measured output, it is more difficult to distinguish between their individual effects if they are correlated, than if they are not. However, there is a more devious 'pitfall' hidden in the example: Even if input are uncorrelated, the validatability condition of Purpose Equivalence may not be satisfied in practice. In fact, the example hinges on the assumption that the noise term ω has a known variance. In practice, the variance would be unknown, and this changes things. The following example will illustrate this.

─────────────── **Example 4.4** ───────────────

Illustrating the difficulty of interpreting the condition of Purpose Equivalence

Consider the same system as in Example 4.3:

S: $d = c_1 \Theta_1 + c_2 \Theta_2 + \sigma \omega$, $\sigma = 1$.

but the model structure

$\mathcal{M}(\bar{n})$: $d = c_1 \theta_1 + \lambda \omega$

Assume also the most favourable case that c_1 and c_2 are orthogonal

$c_1{}^T c_1 = N I$, $c_2{}^T c_2 = N I$, $c_1{}^T c_2 = 0$

Hence, in obvious generalization of Example 4.3

$\log p[d|\theta_1,\theta_2,\lambda] = -\frac{1}{2} N \log(2\pi \lambda^2) - \frac{1}{2} \|d - c_1 \theta_1 - c_2 \theta_2\|^2 \lambda^{-2}$

and

$\log p[\theta_1,\theta_2,\lambda|d]$
$= -\frac{1}{2} N \log(2\pi \lambda^2) - \frac{1}{2} \|\theta - \hat{\theta}\|^2 \lambda^{-2} + \log p[\theta_1,\theta_2,\lambda] - \log p[d]$

where $\hat{\theta} = c^T d/N$.

However, the conditional distribution calculated as in Example 4.3

$p[\theta_1|\theta_2=0,\lambda=\sigma,d] = \exp[-\frac{1}{2} N \log(2\pi \sigma^2) - \frac{1}{2} \|\theta_1 - \hat{\theta}_1\|^2 \sigma^{-2}]$

differs from the marginal distribution

$$p[\theta_1|\mathbf{d}] = \int\int p[\theta_1,\theta_2,\lambda|\mathbf{d}] \, d\theta_2 \, d\lambda$$

which is not readily computable, not even if the apriori parameter distribution $p[\theta_1,\theta_2,\lambda]$ has a simple form.

In order to explain this evaluate the mean and variance of the estimation error:

$$E\{\hat{\theta}_1\} = N^{-1} E\{\mathbf{c}_1^T \mathbf{d}\} = N^{-1} E\{\mathbf{c}_1^T (\mathbf{c}_1 \Theta_1 + \mathbf{c}_2 \Theta_2 + \sigma \omega)\} = \Theta_1$$
$$E\{[\hat{\theta}_1 - \Theta_1][\hat{\theta}_1 - \Theta_1]^T\} = N^{-2} \sigma^2 E\{\mathbf{c}_1^T \omega \omega^T \mathbf{c}_1\} = N^{-1} \sigma^2 I$$

This is independent of whether or not there is a $\mathbf{c}_2 \Theta_2$–term, and whether or not the noise variance is known.

The probability that the model $d = c_1 \hat{\theta}_1 + \lambda \omega$ is purposive is therefore

$$\mathrm{Pr}\{\|\Theta_1 - \hat{\theta}_1\|^2 < \gamma^2|\mathbf{d}\} = \mathrm{Chi\_square}\,[N \, \sigma^{-2} \, \gamma^2, \bar{n}]$$

since $N \, \sigma^{-2} \, \|\Theta_1 - \hat{\theta}_1\|^2$ is Chi–square distributed with $\bar{n}$ degrees of freedom.

However, it will not be possible to evaluate $\mathrm{Pr}\{\|\Theta_1 - \hat{\theta}_1\|^2 < \gamma^2\}$ in this way, since σ is unknown. A calculation must be based on the model structure $\mathcal{M}_b(\bar{n})$. Maximum–Likelihood estimation of Θ_1 and λ yield $\hat{\theta}_1 = \mathbf{c}_1^T \mathbf{d}/N$, as before, and $\hat{\lambda}^2 = N^{-1} \|\mathbf{d} - \mathbf{c}_1 \hat{\theta}_1\|^2$. But

$$\hat{\lambda}^2 = N^{-1} \|\mathbf{c}_1 (\Theta_1 - \hat{\theta}_1) + \mathbf{c}_2 \Theta_2 + \sigma \omega)\|^2$$
$$\rightarrow N^{-1} E\|\mathbf{c}_1 (\Theta_1 - \hat{\theta}_1) + \mathbf{c}_2 \Theta_2 + \sigma \omega)\|^2$$
$$= N^{-1} E\{\mathrm{Tr}[(\Theta_1 - \hat{\theta}_1)^T \mathbf{c}_1^T \mathbf{c}_1 (\Theta_1 - \hat{\theta}_1) + \Theta_2^T \mathbf{c}_2^T \mathbf{c}_2 \Theta_2$$
$$+ \sigma^2 \omega^T \omega]\}^2$$
$$= N^{-1} \{N \, \mathrm{Tr}[N^{-1} \sigma^2 I] + N \|\Theta_2\|^2 + N \sigma^2\}$$
$$\rightarrow \|\Theta_2\|^2 + \sigma^2$$

The aposteriori distribution of Θ_1 conditional on $\mathcal{M}_b(\bar{n})$ is

$$\log p[\theta_1|\lambda,\mathbf{d}] = -\tfrac{1}{2} N \log (2\pi \, \lambda^2) - \tfrac{1}{2} \|\theta_1 - \hat{\theta}_1\|^2 \, \lambda^{-2}$$

and hence the calculated probability that $\hat{\theta}_1$ is valid becomes

$$\Pr\{\|\theta_1 - \hat{\theta}_1\|^2 < \gamma^2|\lambda, \mathbf{d}\}$$
$$= \Pr\{N^{-2}\,\hat{\lambda}^2\,\|\theta_1 - \hat{\theta}_1\|^2 < N^{-2}\,\hat{\lambda}^2\,\gamma^2|\lambda, \mathbf{d}\}$$
$$\to \text{Chi\_square}\,[N\,\gamma^2/(\|\Theta_2\|^2 + \sigma^2), \bar{n}]$$

But this is a lower value than the correct probability
Chi_square $[N\,\gamma^2/\sigma^2, \bar{n}]$.

The explanation is that since the estimated noise variance $\hat{\lambda}^2$ includes also that of the modelling error $c_2\,\Theta_2$, the estimated uncertainty of $\hat{\theta}_1$ will be larger than the actual value, while the value of $\hat{\theta}_1$ is in fact independent of whether there is a error or not, as long as it is orthogonal. A sufficiently accurate $\hat{\theta}_1$ may therefore be rejected, if the modelling error is not negligible compared with the noise. Fortunately, however, there is no 'pitfall' in this case, since an inaccurate model will not be accepted.

The example shows that there may still be reason to try and bring down the modelling error, in order to be able to trust the result of validation (see also section 7.3 on "Two pitfalls".

4.6 The origin of 'pitfalls'

'Pitfalls' are cases where identification would produce an incorrect model and at the same time be unable to diagnose that this is the case. The error will not reveal itself, until the model has been put to use and failed obviously, with more or less severe consequences. The 'pitfall' is the worst possible outcome of identification; the model is wrong, and one thinks it is right.

Since it is implicitly assumed that theory and programming will be correct, and also that the unfortunate outcome is not just due to extremely bad luck (since there is chance involved), the origin of 'pitfalls' are the *assumptions* and the *insufficiently tested hypotheses*. The meanings of the three elements in $M \in \mathcal{Mb}(n)$ play an important role:

● *Assumptions* are represented by $\mathcal{Mb}$. Satisfying the assumptions requires that the designer will be able to acknowledge that certain *condi-*

tions do hold for the actual system of experimenter and object. When they do not, this creates a number of 'pitfalls'.

> **Examples:** The general conditions of Reproducibility, Sufficient Stimulation, Open Loop, Isolated Experimenter, and Purpose Equivalence have been introduced earlier, and more will follow. Since they are extremely important for the success of identification, each one is marked in the same way by a symbolic stripe.

● *Hypotheses* are represented by *n*. Satisfying the hypotheses requires that one can derive and execute *tests* for the hypotheses. When one is unable to do so, this creates more 'pitfalls'.

● *Models* are represented by *M*. Finding a model requires that one carry out a fitting task, usually involving a search. This does not create any new 'pitfalls'. For instance, the practical problems of premature stopping of the search, or of reaching the wrong minimum, do not constitute 'pitfalls'. A wrong model will either be falsified or uncertain, and thus automatically sorted out.

A validation rule hinges on the model structure

When there is a validation rule, there is basically only one cause for 'pitfalls', namely that the criterion does not specify the true purpose. When the criterion is specified on heuristic grounds, or subject to software limitations, for instance as a function of prediction errors (because the program demands it), then this may have little to do with what one is going to use the model for.

Whenever the widest $\mathcal{M}$ structure is too restricted, so that there is no purposive model in $\mathcal{M}$, then it will obviously not be possible to formulate a proper criterion. An attempt to do this using the Bayesian validation criterion, may well result in a set $\mathcal{G}(\mathbf{M})$, where $\mathbf{M}$ is in $\mathcal{M}$, according to the first condition for 'Purpose Equivalence'. But it will never be possible to get the conditional distribution of $\mathbf{M}$ independent of that of the modelling error $\mathbf{E}$. Hence, it will not be possible to satisfy the second condition. Since, that condition cannot be tested within the framework of $\mathcal{M}$, the case is a 'pitfall'.

This means that the designer has *assumed* something about the actual system that is too far from truth. Hence, the 'pitfalls' are unavoidable consequences of the limitations of ones prior information of the object and the experimenter. There is nothing one can do about it by clever analysis. This is simply the point where theory fails the model maker, statistical or other inference and decision theory is not enough, and one must rely also on other, external sources of information. In order to choose $\overline{\mathcal{M}}$ one must know something about the real world in advance, since the answer to whether M is in $\mathcal{G}$ or not depends on what the real world is (if $\mathcal{G}$ depends on **S**). Obviously one cannot deduce that from some universally valid theory operating on data only.

Stopping without a validation rule is always hazardous

When there is no validation rule, the hazard of 'pitfalls' is aggravated, since there are fewer tests available. This would not create more 'pitfalls' *per se*, in the sense that some false *hypotheses* would be accepted, and the procedure therefore come up with a wrong n. In fact, the 'scientist's rule' will not determine a sufficient n, merely reject insufficient n. The stopping (which determines n) will be determined either by the most general structure $\mathcal{M}(\overline{n})$, or by the most extensive experiment $X(\overline{m})$, depending on whether one runs out of hypotheses or data first. *That* creates a 'pitfall'. If the best experiment is still not enough to produce a purposive model, there are no means to diagnose that. Conversely, even if the best model structure cannot be falsified, it still may not contain a purposive model.

Another 'pitfall' is created, if one would 'amend' the 'scientist's rule' by a stopping rule, for instance based on the computed accuracy from $\mathcal{M}(n) \cap \mathcal{B}(m)$. Most identification programs produce estimates of model 'accuracy', that are in fact based on structural assumptions, and are therefore measures of 'precision'. If $\mathcal{M}(n)$ would be narrow but wrong, and simultaneously $\mathcal{B}(m)$ wide enough to have models in common with $\mathcal{M}(n)$, the apparent accuracy would be high, and this would result in a wrong model with a 'high–accuracy' certificate.

Also the experiment is critical

According to the 'scientist's rule', one should not stop, but instead expand the experiment, in order to reduce $\mathcal{B}(m)$. If the expansion is done properly, it will eventually falsify the wrong $\mathcal{M}(n)$. However, there is still a basic problem: If the expansion is done in an *irrelevant* manner, so that the experimenter will never generate stimuli that are representative of those prevailing during the application, then the set of models $\mathcal{M}(n) \cap \mathcal{B}(m)$ may shrink, but never become empty. Formally, the consequences of a too narrow $\mathcal{M}(n)$ and a too wide $\mathcal{B}(m)$ may yield a small enough cut $\mathcal{M}(n) \cap \mathcal{B}(m)$, but in the wrong place. This will yield unfalsified models with ever–increasing precision, but still wrong (see Example 6.8).

Thus, identification without support of a validation rule is more depending on the experiment and apriori information. It will generally be necessary to be more careful, both in the way the experiment is set up, and in the modelling part, using other information than data to design reliable model structures and proper experiments. In other words, there will be more 'assumptions' and less 'hypotheses', since less apriori information can be tested. One has to design more in 'open loop', so to speak.

Theoretically sufficient requirements on a 'proper experiment' were discussed in chapter 3. However, there is a great number of practical obstacles to proper experimentation (see section 1.4 on "The experiment facilities"), and this fact causes a number of 'pitfalls' in practice.

4.6.1 The consequences of unvalidatability

Theoretically, a sufficient requirement for obtaining a purposive model follows immediately, *viz.* $\mathcal{M}(n) \cap \mathcal{B}(m) \subseteq \mathcal{G}$. The requirement is not necessary, however, since a model M need not be in $\mathcal{M}(n) \cap \mathcal{B}(m)$ to be purposive (*Fig. 13*). Neither is the condition directly applicable, since $\mathcal{G}$ depends on the unknown object and is not computable. However, the condition is a useful tool for analysing what can generally go wrong in identification, and what are the conditions one has to satisfy in order to prevent such an unfortunate outcome. The following analysis will assume 'falsifiability', but not 'validatability'.

Depending on $\mathcal{M}(n)$, $\mathcal{B}(m)$, and $\mathcal{G}$ there are several possible outcomes of model making. They may be illustrated graphically:

● If $\mathcal{M}(n) \cap \mathcal{B}(m)$ is empty (*Fig. 19*), then all tentative models will be falsified. Hence, no model in $\mathcal{M}(n)$ will agree with data, which reveals that the structure $\mathcal{M}(n)$ is wrong.

● If $\mathcal{M}(n) \cap \mathcal{B}(m)$ is large (*Fig. 20*), then identification will yield an inaccurate result, but that fact is diagnosable, since one can compute $\mathcal{M}(n) \cap \mathcal{B}(m)$. One may then be able to improve on the situation, either by employing more assumptions, decreasing n and thus narrowing $\mathcal{M}(n)$, or by making a better experiment, increasing m and thus narrowing $\mathcal{B}(m)$.

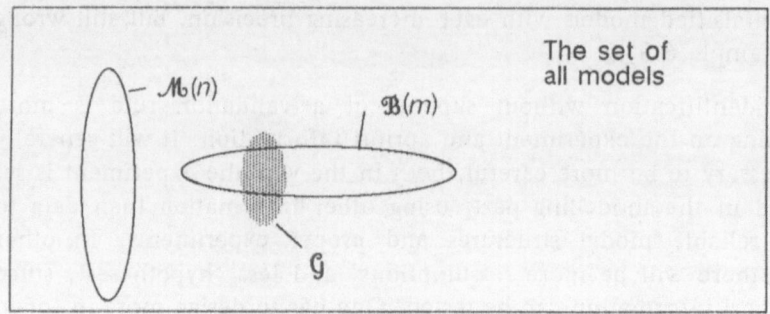

Fig. 19: Illustrating the case of rejected model structure.

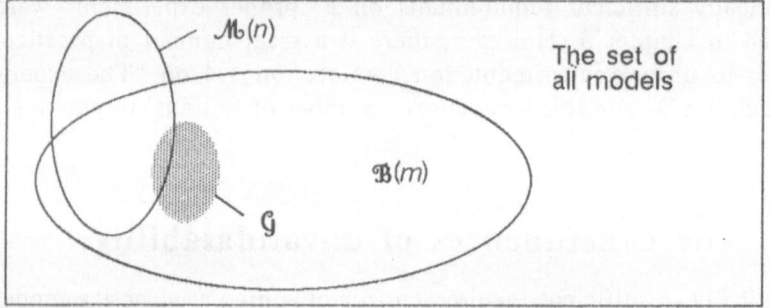

Fig. 20: Illustrating the case of an inaccurate result.

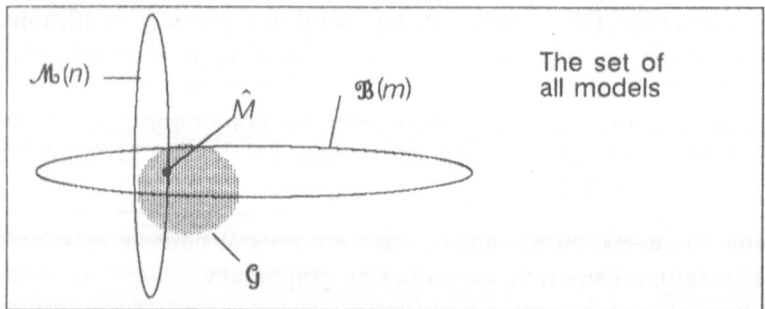

Fig. 21: Illustrating the case of an accurate and good model.

● If $\mathcal{M}(n) \cap \mathcal{B}(m)$ is small, then the accuracy of $\hat{M}$ estimated from the width of $\mathcal{M}(n) \cap \mathcal{B}(m)$ will be high. If, further, $\mathcal{G}$ covers $\hat{M}$ (*Fig. 21*), then identification will succeed. However, if $\mathcal{M}(n) \cap \mathcal{B}(m)$ has elements outside $\mathcal{G}$, then identification may still fail, and if *all* elements are outside (*Fig. 22*), it *will* fail, without any indication that this is the case. If only the purpose is a reasonable one, so that the data has some relevance to the purpose, then there should always be a purposive model in $\mathcal{B}$. Hence, a hazardous situation will occur only if one has

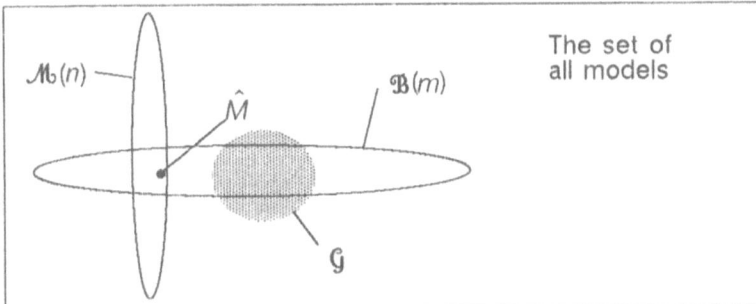

Fig. 22: Illustrating the case of an accurate but bad model.

made false apriori assumptions about the model structure, thus placing $\mathcal{M}(n)$ too far from $\mathcal{G} \cap \mathcal{B}(m)$. In practice this may be the result of trying to amend the case of an inaccurate model by applying an incorrect apriori assumption on the set of tentative models. Notice that this involves assumptions on both the object and the experimenter, as well as assumptions on possible hidden interrelations. The latter is a particularly well concealed 'pitfall' (see Example 6.3).

There are four possible results

In summary, there are four possible results of unvalidatable identification:

● *Failure*: $\mathcal{M}(n) \cap \mathcal{B}(m)$ is empty (*Fig. 19*). All tentative models have been falsified. The proper action is to expand $\mathcal{M}(n)$.

● *Indecision*: $\mathcal{M}(n) \cap \mathcal{B}(m)$ is too wide to be in $\mathcal{G}$ (*Fig. 20*). The computed precision of a model is bad, which gives a warning that the model is not reliable. The proper action is to expand $X(m)$, and thus reduce $\mathcal{B}(m)$.

● *Success*: $\mathcal{M}(n) \cap \mathcal{B}(m)$ is small, and $\hat{M}$ is in $\mathcal{G}$ (*Fig. 21*). A good model has been found.

● *Pitfall*: $\mathcal{M}(n) \cap \mathcal{B}(m)$ is small, but $\hat{M}$ is not in $\mathcal{G}$ (*Fig. 22*). The model cannot be falsified, but it is still bad. The model looks right, since it agrees with ones prior information, its computed precision is high, and no other test one can perform on it using experiment data $(\mathbf{c},\mathbf{d})$ is able to falsify it (since it is inside $\mathcal{B}$). Still, it does not satisfy the purpose.

Hence, without a validation rule one can detect three of the four possible outcomes from data, *viz.* 'no model', 'low–precision model', and

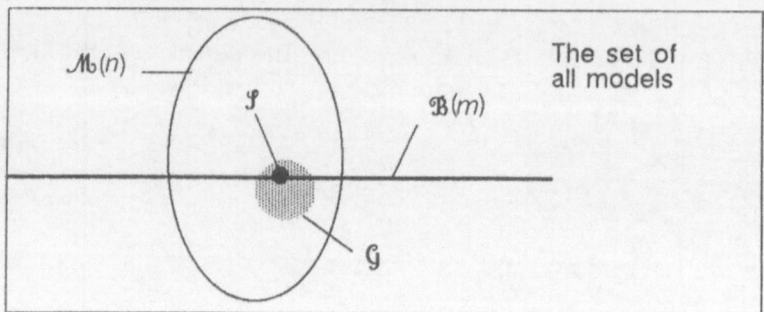

Fig. 23: Illustrating the case of asymptotic ambiguity

'high-precision model', but one cannot say whether the last case is also a 'good' model.

4.6.2 Purposive models without validation

When there is no validation rule to decide whether an unfalsified model is also purposive, there may be other ways to ascertain that $M \in \mathcal{G}$. When $M \in \mathcal{B}(m)$, and since $\mathcal{B}(m)$ depends on the experimenter $X(m)$, one may be able to do an experiment that brings $\mathcal{B}(m)$ inside $\mathcal{G}$. The following are two ways to do that:

Doing an exhaustive search: It may be enough to have a wide structure $\mathcal{M}(n)$, an extensive experimenter $X(m)$, and a purpose such that all models M with the same response distribution as the true data source (X,S) are purposive. The first condition may ensure that $\mathcal{M}(n) \cap \mathcal{B}(m)$ will not become empty, and the third condition $M \in \mathcal{G}$ implies that $\mathcal{G} \cap \mathcal{B}(m) \neq \emptyset$, since $M \in \mathcal{B}(m)$. The second condition may ensure that the set $\mathcal{B}(m)$ of unfalsified models becomes a sufficiently small set to be in $\mathcal{G}$. Hence, conditions for a sufficiently extensive experiment are of interest. For instance, will an infinitely long data sequence imply that $\mathcal{B}(m)$ shrinks to a point (inside $\mathcal{G}$), when $m \to \infty$? That this is not always so, is clear from the fact that, obviously, it also depends on the way the object is stimulated. The case is illustrated in *Fig. 23*. All models in $\mathcal{B}(m)$ produce the same data (c,d), when insufficiently stimulated by an experimenter $X(m)$ (even infinitely long), but not all behave sufficiently similar, when used for the intended purpose, and hence are not in $\mathcal{G}$. However, this negative answer still raises the question of whether there exists a *sufficient experimenter* X such that $\mathcal{B}(m)$ always shrinks to a point, and hence that one can in fact guarantee a purposive model without having to validate it. Asymptotic properties like this are only of partial value in practice, since nothing is infinite (and neither does one always know how fast one approaches infinity).

But they may answer the important question of whether it may be worth while to go on with the experiment and data collection, or whether no data lengths will suffice. Conditions for arriving at good models asymptotically, using only experiment data $(\mathbf{c}, \mathbf{d})$, will be analysed in chapter 6 on "Large–sample theory".

> **Remark:** In essence, the approach is to put so strong demands on the *stimulating quality of the experiment*, that so few models survive the falsification test, that they must certainly be in $\mathbf{G}$. Hence, one does not have to validate the models explicitly. The outer loop is not stopped, but the identification procedure is allowed to go on until one runs out of options. This alternative, if successful, will produce models that are unnecessarily accurate, and is therefore wasteful. But it is sometimes the only possible.

> **Remark:** Preferably one should use both means to ensure a purposive model, *viz.* validation for a stopping rule (that may take care of the accuracy), and designing proper experiments (that may facilitate the validation, and thus make the identification procedure stop sooner, or at all).

The second alternative without validation is:

Doing the 'right' experiment: If the stimulation of an object during the experiment is the same as that when the model is later put to use, then it is enough for the purpose that the model responses mimic the responses of the object. In that case the criterion $\mathcal{M}(n) \cap \mathcal{B}(m) \subseteq \mathbf{G}$ for a purposive model is always satisfied, since the set of unfalsified models $\mathcal{B}(m)$ coincides with the set of good models $\mathbf{G}$. In particular, this is the case in adaptive control.

4.6.3 Sources of information for model design

As argued above, experiment data $(\mathbf{c}, \mathbf{d})$ alone may not be enough information to render it feasible to ascertain a correct outcome of the identification of an object. In that case one needs other information for specifying $\mathcal{M}$ and $\mathbf{X}$ and for computing $\mathcal{B}$. The following information sources are often available to the model maker:

● *Apriori information on* $\mathbf{S}$, for instance physical principles, previously established empirical relations, parameter values measured independently, subjective judgement, or hunches. Also assessments of the reliabilities of the apriori assumptions are useful information. All this may be used to limit the set $\mathcal{M}$ of tentative models.

● *Earlier experience of modelling* ('engineering sense'). May be used to limit the structure 𝓜.

● *Data* (**c,d**) *from apriori experiments, i.e.* experiments carried out for the purpose of generating data for identification. That may be used for computing the set 𝓑 of unfalsified models.

● *Data from aposteriori experiments, i.e.* experiments carried out after identification and for the purpose of validating the usefulness of the model. This includes various kinds of simulation of the designs based on the model in the proper environment.

The apriori information is basically as important as the experiment data (**c,d**); a well designed model depends on both. The fact that in many practical cases one tends to rely more on 'solid data' from experiments on the object than on ones preconception of it, is not that preconceptions are 'biassed' and 'prejudiced', but that one is often *uncertain* of ones apriori information. On the other hand, the apriori information may be based on other experiment data, as 'solid' as the object's responses **d** to the stimulus **c**. And the latter may be contaminated and unreliable. Hence, the *degree of uncertainty* of the apriori information is an important element in model design.

> **Remark:** As argued above, choosing the structure 𝓜 and the experimenter **X** is not an entirely mathematical problem, meaning that it cannot be solved by applying some universally valid theory to data. The choice must be done partly by exercising ones 'engineering sense'. Not even that can, of course, state 'what the real world is', only what it has often turned out to be, based on earlier experience. 'Engineering sense' may therefore be able to indicate how *not* to design experiments, how *not* to model an object with typical properties, and what attempts have typically *failed*, and for what reasons. It is therefore one of the 'external' sources of information one basically needs to deduce anything about 𝓜 and **X**. However, in contrast to the source of 'apriori information about S', the 'engineering sense' is independent of the particular modelling object and should therefore be possible to write down. Some attempts to do that have been published (e.g. Bohlin, 1987a). The recommendations and rules that would thus guide the choice of 𝓜 and **X** are formulated in intuitive terms, not because they cannot logically be formulated in a more precise language, but simply because it seems difficult to do that.

> **Remark:** Research on precise representations of 'information' of other kind than numerical data is a topic in the science of 'artifi-

cial intelligence' (Rada, 1987). If one were to construct a socalled 'expert system' (Sage, 1987b) for engineering modelling, one would first have to solve the problem of formulating precise rules for the choices of $\mathcal{M}$ and $\mathbf{X}$ (so they can be programmed into computers). However, they will first have to be formulated at all and in any way.

If the 'aposteriori experiment' may actually involve the object $\mathbf{S}$, then validation is of course decisive. It means in practice that after a contracted test period, the design must be accepted by the customer, and responsibility removed from the designer (which does not ensure, of course, that nothing can go wrong later). If one is not allowed to experiment on the object, but has to simulate on a model, then aposteriori experiments will, in effect, produce no information that cannot be deduced from the other sources. However, the different stimulations of the model and consequently its different responses may help the model maker too. Generally, one should tap ones sources of apriori information both to set the prerequisites for the modelling *and* to judge the outcome of it. For a discussion and a list of such possibilities see (Gruhl, 1979).

──────────────── **Example 4.5** ────────────────

This is an example of subjective aposteriori judgement.

A number of competent marine ecologists designed models for a number of biological and hydrological processes in a particular lake, based on the best available apriori information and measurement data. However, when the various models were compiled into a total model of the lake, simulations using that model predicted that there ought to be about a half-meter thick layer of dead *chironomidae* insects on the bottom of the lake.

The correct reaction to this preposterous result would be to discard the total model for the purpose of predicting the amount of dead insects on the bottom, but not necessarily any of the submodels for other purposes, or the competence of the model makers, and certainly not the usefulness of model making for ecological purposes.

──

A point to remember is that the model structures are always limited by 'all the designer can think of'. And upon this hinges all the results of model making theory, including all tests defining what is a 'valid', and what is a 'false' model. In fact, and as Nurmi (1978) pointed out, all results of the 'scientific approach' depend on the 'school' the particular scientist belongs to, because this limits the set of models $\overline{\mathcal{M}}$ he/she is willing to try. Of all the 'pitfalls' this may well be the worst one.

5   Modelling

This chapter treats the problem of representing apriori structure information based on facts, assumptions, and hypotheses. It outlines some common ways to bring detail into the descriptions of model structures $\mathcal{M}$. That cannot be done without loss of generality. In fact, the whole purpose of modelling is to restrict the set of all possible models so much that the remaining set has so few degrees of freedom left, that identification will be able to use available data to pick a model from the set. The difficulty is to do the reduction without excluding all valid models.

Information about the structure of the system of experimenter and object is specified by the form one gives the 'probabilistic model' $p[c,d|M]$. The general form allows one to specify causal dependence or independence by moving variables to or excluding variables from the right side of the 'conditional' sign ($|$). This chapter reviews some ways to derive probabilistic models from other structures, that are less general and thus hold more apriori information.

> **Remark:** Efficient modelling requires much theoretical insight into the object to be modelled, as well as practical insight into what problems identification methods can and cannot cope with. Therefore (and in contrast to 'black–box' identification), successful 'grey–box' model making generally requires collaboration between an expert in identification and an engineer or scientist working in the particular 'trade'. Generally, skill in modelling has to be developed from experience of applying some basic principles underlying classes of physical objects. Fasol and Jorgl (1980) give examples of such principles.

5.1   Parametrization

Since one can obviously never infer the values of more than a limited number of unknowns from a limited data record, one has to *parametrize* the model, *i.e.* write $M = (\mathcal{A}, \nu, \theta)$, where ν is an unknown vector of integers, θ is an unknown vector of reals, and $\mathcal{A}$ is the form of the model, which *must be known apriori*. The dimension of ν depends on $\mathcal{A}$ and must be known, but the dimension of θ may depend on the unknown ν.

Normally it has to be much smaller than the dimension of the data record (**c**,**d**). Without loss of generality it is possible to assume that $\theta \in \mathbf{R}$, where **R** is a real space of suitable dimension. Usually, physical parameters do not range over the whole real space, but that is possible to amend by shaping the $\mathcal{A}$–function (see section 5.2 on "The parameter map").

Generally, $\mathcal{A}$ may be envisaged as the formulas or algorithms used to compute the model output, while v are integer parameters used to determine the size of the problem, and θ are real constants. Fundamentally, $\mathcal{A}$ represents the *assumptions,* v the *hypotheses,* and θ the *models.*

After parametrization the probabilistic model of the output of the system of experimenter and object may be written $p[c,d|M] = p[c,d|\mathcal{A},v,\theta]$. Various forms of parametric models are treated in sections 5.3–8. They are the most important means to express apriori information.

> **Remark:** Generally, all physical parameters *a* should be associated with particular components in v, and the given dimension of v is the number of physical parameters in the widest model structure, which of course ought to be known. However, one might be uncertain about what should be counted as 'parameters', for instance, would the acceleration of gravity, $g = 9.82$ m/s², need an entry in the v–vector? The answer is "no", unless the model is to be fitted to data obtained in different gravity fields. The distinction becomes clear, if one thinks of $\mathcal{A}$ as a computer program; if a physical parameter has a value that is known well enough to be entered into the program as a constant, instead of as a variable, then it should not have an entry in v.

> **Remark:** There are 'nonparametric' identification methods in the literature (Wellstead, 1981; see also section 6.2.2 on "The unstructured approach"). This means that the unknowns are real functions or infinite sequences, for instance 'pulse–transfer functions' or 'power spectra'. However, the quantities that can be *inferred* from data are necessarily finite–dimensional approximations of the ideal infinite–dimensional functions.

--- **Example 5.1** ---

This example illustrates the difference between a parametric and a nonparametric model.

A general nonparametric model for deterministic linear discrete–time systems is

$$d(k) = \sum_{l=0}^{k} \theta(l) \, c(k-l)$$

where $\{\theta(l)\}$ is a weighting sequence. The number of unknown elements is the same as the number of data points. If the system is stable, the weights tend to zero. It is also easy to estimate them recursively by the algorithm

$$\theta(k) = [\mathbf{d}(k) - \theta(0) \, \mathbf{c}(k) - \dots - \theta(k-1) \, \mathbf{c}(1)]/\mathbf{c}(0)$$

However, if noise is added to the model, it is necessary to truncate the sequence of weights to finite length, which makes the model parametric. Otherwise the estimates for large lags l will not tend to zero. The difference is that in a 'parametric model' the number of parameters is bounded when the number of output increases. In a 'nonparametric model' the number increases indefinitely.

──────────── **Example 5.2** ────────────

This example illustrates some difficulties of using a nonparametric model.

A general nonparametric model for stationary stochastic time series is defined by its spectral function $\Phi(f)$, $f \in (-\pi, \pi)$. When one wants to estimate the spectrum, one has first to express it by means of a finite number of parameters.

One way is dividing the frequency range into a number of narrow bands and estimating the integrated power density within each band. The variance of each estimate is proportional to (number of bands)/(number of data points). Therefore, the maximum number of bands (and hence the achievable frequency resolution) depends on the sample length, and must be much smaller.

This is not a consequence of the estimation method, but is an inherent property of the mathematical concept of spectrum: the value $\Phi(f)$ at a single point f cannot be estimated. For instance, the raw Fourier transform of the data, the 'periodogram' has infinite variance for all sample lengths. In spectral analysis by an FFT algorithm this is amended by pre– or post–processing, which means that the analysis in fact yields a number of estimates of a *function* of the spectral values in small frequency ranges around a finite number of nominal frequencies. For the theory of spectral estimation see for instance Glover (1987).

5.1.1 Integer parameters

Define a *parametric model structure* as $\mathcal{M}_\nu \triangleq (\mathcal{A},\nu)$. With a slight abuse of notation, this can be interpreted as the set of all models of given form and parametrization. In order to define the 'sequence of expanding structures' postulated in section 4.4 on "Basic identification procedures" one has first to define a partial sequencing of the various indexed structures $\mathcal{M}_\nu$. This means that one must determine a sequence $\mathcal{N} = \{\mathcal{N}_1, \mathcal{N}_2, \ldots, \mathcal{N}_{\bar{n}}\}$ of sets $\mathcal{N}_n$ of structure indices ν, as part of the modelling task. Then the model sets defined by

$$\mathcal{M}(n) = \mathcal{M}(n-1) \bigcup_{\nu \in \mathcal{N}_n} \mathcal{M}_\nu, \quad \mathcal{M}(-1) = \emptyset \tag{5.1}$$

clearly defines an expanding structure, since $\mathcal{M}(n) \supset \mathcal{M}(n-1)$.

It would not make sense to sequence the model structures in any other way than towards increased complexity. If ν is a scalar, there is a natural sequencing. If ν is a vector, there are several possible. However, the integer n may still be used as 'complexity number' in the following way:

Partition the set of all structure indices as $\mathcal{N}_1 \cup \mathcal{N}_2 \cup \ldots \cup \mathcal{N}_{\bar{n}}$, so that $\mathcal{N}_n$ has complexity n, *e.g.* the total number $|\nu_n|$ of parameters is r_n for all $\nu_n \in \mathcal{N}_n$. Then apply the principle of Parsimony to order the sets $\mathcal{N}_n$ after increasing n.

It helps falsification to have 'nested' structures

Some efficient statistical methods to reject a model structure $\mathcal{M}(n)$ as being too small to hold a model that data does not contradict (the 'inner stopping rule'), require that one can define an alternative wider structure to compare with. The condition is satisfied, if each model structure in $\mathcal{N}_n$ has a wider structure in some more complex $\mathcal{N}_{n'}$, $n' > n$. The condition will be referred to as *'nesting'*.

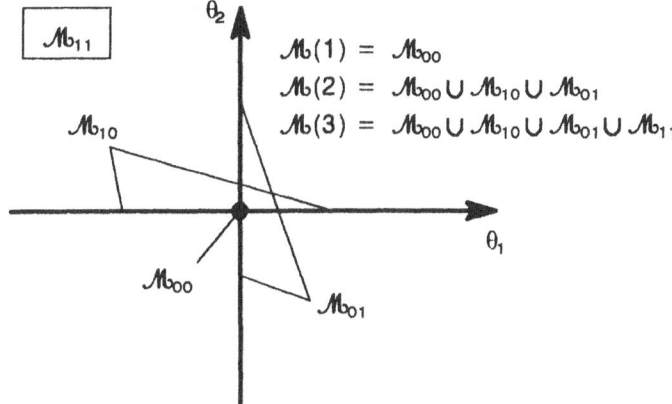

$$\mathcal{M}(1) = \mathcal{M}_{00}$$
$$\mathcal{M}(2) = \mathcal{M}_{00} \cup \mathcal{M}_{10} \cup \mathcal{M}_{01}$$
$$\mathcal{M}(3) = \mathcal{M}_{00} \cup \mathcal{M}_{10} \cup \mathcal{M}_{01} \cup \mathcal{M}_{11}$$

Fig. 24: Illustrating nested composite structures.

More precisely, the condition is the following: Two parametric structures $(\mathcal{A}, v)$ and $(\mathcal{A}, v')$ are *nested*, if $v' > v$, and for every θ there is some θ' such that $p[c, d | \mathcal{A}, v, \theta] = p[c, d | \mathcal{A}, v', \theta']$ for all (c, d).

Nesting can be obtained in a number of ways. For instance, write the model in such a way that increasing any component v_i means that old components in θ retain their places and meanings, when new components are added. Then determine the partitioning of the set of possible v-values in such a way that no structure component in $\mathcal{N}_{n-1}$ is larger than its counterpart in $\mathcal{N}_n$, and at least one is smaller. In this way all models in $(\mathcal{A}, \mathcal{N}_{n-1})$ become equivalent to some models in $(\mathcal{A}, \mathcal{N}_n)$ with some zero parameter values (if zeroes are admissible values), and the complexity counted in number of (non–zero) parameters will always increase with n. No other way of specifying the set of 'expanding structures' will be considered in the sequel.

A structure $(\mathcal{A}, \mathcal{N}_n)$ of given complexity n is *composite*, if $\mathcal{N}_n$ contains more than one structure index v_n.

Remark: *Fig. 24.* Illustrates the concept of (composite) nested structures. The partial sequencing of possible indices $\{(0,0), (1,0), (0,1), (1,1)\}$ is $\mathcal{N}_1 = \{(0,0)\}$, $\mathcal{N}_2 = \{(1,0), (0,1)\}$, $\mathcal{N}_3 = \{(1,1)\}$. The structure $(\mathcal{A}, \mathcal{N}_2) = \mathcal{M}_{10} \cup \mathcal{M}_{01}$ of complexity 2 is composite and cannot be described by a single parametric structure. The same holds for $\mathcal{M}(2)$. The structure $(\mathcal{A}, \mathcal{N}_3) = \mathcal{M}_{11}$ is not composite. The structure $\mathcal{M}(3)$ is composite, but can be described by the single parametric structure $(\mathcal{A},\mathcal{N}_3)$, because the structures are nested.

Example: A simple nested structure is a rational transfer function, where ν is the common order number, and θ are the $2\nu+1$ coefficients. A lower–order transfer function is a special case of a higher–order function, where some coefficients are zero.

Example: Often one has partial information about the structure; there are components that are conceivable but uncertain. Then one can assign parameters to the uncertain parts in such a way that zero values exclude those parts. Thus the lower–complexity models are special cases of the higher–complexity models.

Remark: A systematic way to define nested structures is to use the concept of 'parameter map' in section 5.2.

Nesting also simplifies the search

An important consequence of nesting is that for all kinds of loss functions Q holds that
$$\min\{Q(\nu,\theta|c,d)|\nu \in \mathcal{N}_1 \cup \ldots \cup \mathcal{N}_n, \theta \in R\}$$
$$= \min\{Q(\nu,\theta|c,d)|\nu \in \mathcal{N}_n, \theta \in R\}$$
Hence. nesting has the advantage that one will in effect have to cope with only the set $(\mathcal{A},\mathcal{N}_n)$ instead of the whole $\mathcal{M}(n)$, when searching for a model.

However, $(\mathcal{A},\mathcal{N}_n)$ is still a 'composite' structure; it contains models with different indices ν, even if all have the same number of parameters. This means that the search strategy may be a serious problem in complicated cases. It will therefore be a great advantage, if the apriori information allows the designer to have a complete sequencing of the structures, such that every set $\mathcal{N}_n$ contains only one value of the structure index: $\mathcal{N} - \{\nu_1,\nu_2,\ldots,\nu_{\bar{n}}\}$. This is called *'simple nesting'*. The search will then have to be carried out only over the simple structure $(\mathcal{A},\nu_n)$, which means an ordinary search over the real parameters θ.

However, this has also a disadvantage. Since it is required that $\nu_{n-1} < \nu_n$, there is generally no sequencing such that all ν will be included in the sequence $\mathcal{N}$. This does not necessarily mean that any structure $(\mathcal{A},\nu)$ will be excluded from the competition, since it may still be included in a higher–complexity structure tried later. But it may result in a model of unnecessarily high complexity. Because of this one may have to carry out so called 'model reduction' on the first model that passed all tests — there might still be a simpler one that would also pass the tests.

The concept of *'full nesting'* means that there is a common widest structure $\mathcal{M}(\bar{n})$, *i.e.* $\mathcal{N}_{\bar{n}} = \nu_{\bar{n}}$, while the less complex structures may still be composite. Full nesting simplifies the problem of determining ν.

Both the combined search over integer and real parameters and the model reduction are generally intricate problems (see chapter 9 on "Structure identification").

———————————————— **Example 5.3** ————————————————

Illustrates nesting in the case of vector–valued structure indices.

Let the model structure be that of the ARMA series.

M: $\quad d(k) + \ldots + a_{\nu_1}\, d(k{-}\nu_1) = \lambda\, [\omega(k) + \ldots + c_{\nu_2}\, \omega(k{-}\nu_2)]$

$\theta = \{a_1, \ldots, a_{\nu_1}, c_1, \ldots, c_{\nu_2}, \lambda\}$

The set of possible ν–values is

$\{(0,0),(1,0),(0,1),(2,0),(1,1),(0,2),(3,0),(2,1),(1,2),(0,3),\ldots\}$

There are several reasonable ways to sequence this:

I. Partially, after number of parameters:

$\mathcal{N}_1 = (0,0)$
$\mathcal{N}_2 = \{(1,0),(0,1)\}$
$\mathcal{N}_3 = \{(2,0),(1,1),(0,2)\}$
. . . .

This implies the partial nesting

$\mathcal{M}(1) = (0,0)$
$\mathcal{M}(2) = (0,0) \cup (1,0) \cup (0,1)$
$\mathcal{M}(3) = (0,0) \cup (1,0) \cup (0,1) \cup (2,0) \cup (1,1) \cup (0,2)$
. . . .

This has the disadvantage that one will have to deal with 'composite' model structures in each set $\mathcal{M}(n)$, *e.g.* $(2,0) \cup (1,1) \cup (0,2)$ in $\mathcal{M}(3)$. On the other hand, the end result will have the simplest structure.

II. Simple nesting, assuming that the numerator has the same order as the denominator, or one lower:

$\mathcal{N}_1 = (0,0)$
$\mathcal{N}_2 = (1,0)$
$\mathcal{N}_3 = (1,1)$
$\mathcal{N}_4 = (2,1)$
$\mathcal{N}_5 = (2,2)$
. . . .

This implies the simple nesting

$\mathcal{M}(1) = (0,0)$
$\mathcal{M}(2) = (0,0) \cup (1,0)$
$\mathcal{M}(3) = (0,0) \cup (1,0) \cup (1,1)$
$\mathcal{M}(4) = (0,0) \cup (1,0) \cup (1,1) \cup (2,1)$
. . . .

The simple nesting has the advantage that one will have to search over only one structure, *e.g.* $(2,1)$ in $\mathcal{M}(4)$. On the other hand, if the best structure would be $(0,1)$, then the first model to be accepted would be in $(1,1)$, and one would have to carry out a 'model reduction' procedure to arrive at the simplest model in $(0,1)$.

III. A full nesting is feasible, if both numerator and denominator orders are bounded. Assume the largest order is 2. Then

$\mathcal{N}_1 = (0,0)$
$\mathcal{N}_2 = \{(1,0),(0,1)\}$
$\mathcal{N}_3 = \{(2,0),(1,1),(0,2)\}$
$\mathcal{N}_4 = \{(2,1),(1,2)\}$
$\mathcal{N}_5 = \{(2,2\}$

This implies the full nesting

$\mathcal{M}(1) = (0,0)$
$\mathcal{M}(2) = (0,0) \cup (1,0) \cup (0,1)$
$\mathcal{M}(3) = (0,0) \cup (1,0) \cup (0,1) \cup (2,0) \cup (1,1) \cup (0,2)$
$\mathcal{M}(4) = (0,0) \cup (1,0) \cup (0,1) \cup (2,0) \cup (1,1) \cup (0,2)$
$$\cup (2,1) \cup (1,2))$$

$\overline{\mathcal{M}} = \mathcal{M}(5) = (0,0) \cup (1,0) \cup (0,1)$
$$\cup (2,0) \cup (1,1) \cup (0,2) \cup (2,1) \cup (1,2) \cup (2,2)$$

_____

5.1.2 The parameter distribution

The last factor in Bayes' rule (and the second way to express apriori information) is $dP[M] = dP[\mathcal{A},\nu,\theta]$. To be used for design it must first be defined mathematically.

Factorize $dP[\mathcal{A},\nu,\theta] = dP[\theta|\mathcal{A},\nu]\ dP[\nu|\mathcal{A}]\ dP[\mathcal{A}]$. Since $\mathcal{A}$ is postulated, the last factor is trivial. Since $\mathcal{M}_\nu$ are disjoint, and ν are integers, it follows that $\int dP[\nu|\mathcal{A}] = \sum_\nu p[\nu|\mathcal{A}] = 1$, where $p[\nu|\mathcal{A}]$ are apriori probabilities for the structure indices ν.

Since θ is a real vector of the dimension required to define a point in $\mathcal{M}_\nu$, it is possible to write $dP[\theta|\mathcal{A},\nu] = p[\theta|\mathcal{A},\nu]\ d\theta$, where $p[\theta|\mathcal{A},n]$ is a density (possibly including Dirac δ-functions). This defines the 'density' of the system by

$$dP[\mathcal{A},\nu,\theta] = p[\nu|\mathcal{A}]\ p[\theta|\mathcal{A},\nu]\ d\theta \qquad (5.2)$$

It is now possible to define a L
likelihood function for parametric models of a given structure:

$$L(\nu,\theta|\mathcal{A},\mathbf{c},\mathbf{d}) = p[\mathbf{c},\mathbf{d}|\mathcal{A},\nu,\theta]\ p[\nu|\mathcal{A}]\ p[\theta|\mathcal{A},\nu] \qquad (5.3)$$

The factors in (5.3) must be specified apriori, but notice that $p[\theta|\mathcal{A},\nu]$ are ordinary probability density functions, and $p[\nu|\mathcal{A}]$ are nonnegative real numbers depending only on ν.

> **Remark:** In analogy with the reasoning in section 2.2 on 'Likelihood', the factor $dP[\nu,\theta|\mathcal{A}]$ would be the probability density of 'true' parameters, when nothing has been observed. If the existence of a 'true model' **S** may be doubted, that doubt is even more motivated for the 'true parameters'. However, it is clearly permitted to use $dP[\nu,\theta|\mathcal{A}]$ as a *design parameter*. This means that it is no longer regarded as a description of some 'truth', but as a means to influence subjectively the result of the identification. It is called *'subjective probability'*. In order to understand *how* this design parameter, the subjective probability, will influence the result, it helps to reason in terms of a 'true' model and 'true' parameters. Often one may assume that nothing is known apriori about the values of the θ-parameter (or specify that one does not apriori prefer some parameter value before any other), and hence assign a subjective probability that does not depend on θ. However, if one

knows some, it will help the identification. Notice again, that one usually cannot assign a constant value to $p[v|\mathcal{A}]$.

Regardless of how one chooses to interpret the factors $p[v|\mathcal{A}]$ and $p[\theta|\mathcal{A},v]$, the following are three practical ways to utilize these possibilities to provide apriori information on the model:

● *Reducing complexity*: One can specify a bias towards simple models by letting $p[v|\mathcal{A}]$ have the values of arbitrary weights, decreasing with increasing number of parameters (see chapter 9 on "Structure identification"). The end result is that low—complexity models are weighted more, and are therefore more likely to come out as result of the identification, unless data is so heavily against, that the result will still be more complex than one would like.

● *Improving a prior estimate*: In this case a first estimate of parameters has been calculated based on prior data (c^o, d^o). If the estimate is not satisfactory, because the prior data is insufficient or measurements were not well chosen, one can try to improve it by observing more and/or taking other measurements for (c, d). The apriori density will then be $p[\theta|\mathcal{A},v,c^o,d^o]$ and the aposteriori density will be
$p[\theta|\mathcal{A},v,c^o,c,d^o,d] =$

$$p[c,d|\mathcal{A},v,\theta,c^o,d^o]\ p[\theta|\mathcal{A},v,c^o,d^o]/p[c,d|\mathcal{A},v,c^o,d^o].$$

The point of using this relation is that one can calculate $p[\theta|\mathcal{A},v,c^o,c,d^o,d]$ without having to repeat the calculation of $p[\theta|\mathcal{A},v,c^o,d^o]$ based on the previous data, and thus one can save time. If the procedure is repeated regularly, each time using a few more measurements, this yields a procedure for identifying recursively (Peterka, 1981).

● *Excluding inadmissible models*: It is often known apriori that parameters outside a certain admissible domain, $\theta \notin \Theta$, cannot appear in the physical object. For instance, it might be known that a linear system is stable, so that θ—values yielding a pole in the right half—plane are outside Θ. This information can be utilized by putting $p[\theta|\mathcal{A},v] = 1$, if $\theta \in \Theta$, and zero otherwise. Normally, this has no effect on the ML (peak—value) estimate, since normally the peak of $p[c,d|\mathcal{A},v,\theta]$ is already in Θ, but if data are poor, it may fall outside. Then the weighted ML estimate will fall on the boundary of Θ, while the unweighted estimate will fall outside and be inadmissible. In practice this happens not too rarely, since practical problems are often ill—conditioned. In such cases the likelihood is distributed along a 'ridge' that may well extend into the inadmissible domain, and where the maximum on the ridge happens to occur depends on chance. Then the weight $p[\theta|\mathcal{A},\nu]$ may mean all the difference between a useful solution and a ridiculous one.

5.2 The parameter map

The following are two ways to represent apriori information and uncertainty in connection with parametric model structures:

● *Specifying apriori parameter distribution*: The factor $p[\theta|\mathcal{A},\nu]$ in the likelihood function can be tailored to express both admissible ranges and apriori knowledge or preference of the set of admissible parameter values θ. The possibility was pointed out in section 5.1 on "Parametrization".

● *Mapping*: In case one knows the physical meanings of a number of parameters a in the model structure, it may also be possible to express apriori information on the values of these parameters by a function: $a = \mathcal{P}(\nu,\theta)$, where $\dim(\theta) \leq \dim(a)$. Apriori parameter uncertainty is expressed by assuming that θ has a standard distribution, for instance gaussian with zero mean and a unit covariance matrix. In this model structure the free parameter vector θ is interpreted as a 'primitive random variable', not necessarily with a physical meaning, while the function $\mathcal{P}$ represents the apriori information about the physical parameters. The approach is similar to the way 'randomness' is expressed in algorithmic stochastic models, where ω are the primitives, and the structure information is expressed by the model structure Λ (section 5.3 on "Algorithmic models").

> **Remark:** A priori uniform distribution between 0 and 1 (for example) is modelled by the mapping $a = \Phi(\theta)$, where Φ is the cumulative gaussian distribution function.

It will be possible to express a number of object properties known apriori by shaping the function $\mathcal{P}(\bullet,\bullet)$:

● *Nominal values* (best apriori guesses): There is often some idea apriori of what the value of a physical parameter might be, whether obtained from literature, measured independently, estimated in some crude way, or just the result of an 'educated guess'. Denote that value a°, and form $\mathcal{P}$ in such a way that $\mathcal{P}(\nu,0) = a^{\circ}$.

● *Parameter uncertainty*: Together with a nominal value one has usually an opinion of the uncertainty of that value. This can be expressed by forming $\mathcal{P}$ in such a way that $\mathrm{grad}_\theta\mathcal{P}(\nu,0) = a_\theta$, where a_θ is a diagonal matrix of given *scales*. The scales are then determined by the apriori uncertainty.

● *Parameter dependence*: In case one suspects that the physical parameters are related, it is possible to express this relation by replacing the diagonal a_θ matrix with a full matrix that specifies both relation and scale.

> **Remark:** The following is a hint to the interpretation of the matrix a_θ in this case: In an environment around the nominal values the physical parameters are regarded as stochastic variables with mean a° and covariance matrix $a_\theta\, a_\theta^T$. By the well–known 'orthogonal factorization' or by a 'singular value decomposition', one can write this matrix as $a_\theta\, a_\theta^T = K \wedge K^T$, where $\wedge$ is a diagonal matrix of squared 'scales' and K is a matrix that expresses dependence between the components in a.

● *Structure uncertainty*: A scalar integer parameter n can be used to express the following type of structure uncertainty: Often the model structure is written naturally as a collection of submodels, with varying apriori credibility. Some parts, based on basic physical laws, or conservation laws, such as the conservation of material, energy, or chemical balances have the highest credibility. Others, based on empirical relations, may be well established, but it may still be uncertain whether the requirements for their validity are satisfied in the particular case. Some are just suspected to play a role in the phenomenon one wants to model. Some have to be modelled as 'black boxes', because one knows nothing about them, or indeed, if they have any relevance at all. Notice that a submodel may describe a phenomenon that is very well known physically, while it is still uncertain whether it has any influence on the object's response. What matters here is not only how much one *knows* apriori, but also how much of that knowledge is *needed* for the purpose. One can express this type of uncertainty in the following way: Sequence the submodels (or collections of submodels) after increasing uncertainty. Let $a = \{a_n | n=1,2,\ldots\}$, where a_n is the vector of parameters in submodel #n. Create a sequence of increasingly complex (and increasingly uncertain) model structures $\mathcal{M}(n)$, each one formed by including one more substructure (and hence one more vector of parameters a_n). Assign 'free parameters' θ (by designing $\mathcal{P}$) in such a way that $\mathrm{grad}_\theta\mathcal{P}(n,0)$ will be block–wise right–triangular. Then the dimension of θ will depend on n, one will in this way have created the set of 'expanding structures' $\{\mathcal{M}(n)\}$ required for the sequential identification schemes outlined in section **4.4**, and it will be possible to decide how many of the uncertain substructures that are needed.

● *Parameter range*: Physical parameters usually do not range from minus to plus infinity (as do the 'free parameters' θ). Also this type of constraints can be expressed by shaping the $\mathcal{P}$-function. For instance, a positive scalar parameter may be expressed as $a = \exp(\theta)$, a parameter in the range $(-1,1)$ as $a = 2 \; \text{arctg}(\theta)/\pi$.

> **Remark:** A practical advantage of expressing parameter ranges in this way is that it is usually easier to search for the values of unconstrained parameters θ, than for those of range–constrained parameters a. However, this may be a double–edged tool; if the relation $\theta \mapsto a \mapsto d$ curves too much, many search algorithms will still have difficulties.

All cases of expressing apriori information can be combined into the form

$$a = a^\circ + \mathcal{P}[K(\nu) \; a_\theta(\nu) \; f(\theta)] \tag{5.4}$$

where $K(\nu)$ is upper–right block triangular and expresses parameter interdependence, $a_\theta(\nu)$ is the diagonal matrix of scaling factors (parameter uncertainty), f expresses range, and $\mathcal{P}$ is free for the designer to express other things.

The 'parameter map' is all specification required to compute a given θ. Besides defining the 'free parameter range' as the set of points $\{a|\theta\}$ reachable by transferring any real vector θ into $\{a\}$, it also defines a distribution of a around a° by mapping gaussian vectors θ with covariance $\Gamma_\theta^T \Gamma_\theta$ into $\{a\}$. The apriori covariance is $\Gamma_\theta^T \Gamma_\theta = I$, but the aposteriori covariance should be much smaller and usually not diagonal.

─────────────────── **Example 5.4** ───────────────────

Illustrating the scaling with rated values and using dimensionless variables — Modelling a permanent–magnet servo motor:

From basic mechanical and electromechanical principles follows immediately

$d\phi/dt = \omega, \; J \, d\omega/dt = \Psi \, I, \; L \, dI/dt + R \, I = V$

where ϕ = rotor position, ω = rotor speed, I = stator current, V = stator voltage, J = moment of inertia of rotor and load, Ψ = linked stator and rotor field, L = stator inductance, and R = stator resistance.

Assume that one can control V freely and precisely using some drive circuit, and that one can measure ϕ, ω, and I, using some transducers with not negligible errors. Then

$$c = V, \quad d_1 = \phi + v_1, \quad d_2 = \omega + v_2, \quad d_3 = I + v_3$$

where v_i are 'white noises' with zero means and standard deviations σ_i. Assume also that the parameters J, Ψ, L, R are constant but unknown. Obviously, only $\Psi/J, L, R$ affect the motor response, so that all models with the same values of Ψ/J are Object–Equivalent. Hence, define the parameters $a_1 = \Psi/J$, $a_2 = L$, $a_3 = R$.

There should be no problem in simulating the model on a computer, and even searching for the unknowns automatically, given some sufficient data. The modelling could stop here. However, if one wants the *designer* to take more active part in the identification or simulation (for instance because the model structure may be in doubt), then it is necessary to have a model structure with a better interface to the designer. It is necessary to define the parameter map $\mathscr{P}$ and the model structure $\mathscr{M}$ in such a way that the variations of parameters and variables are possible to interpret intuitively. For instance, what would be suitable start values of $\Psi/J, L, R$, and how much or how little should the simulated variables vary to be considered 'significant' or 'negligible' respectively? Both are questions of *scaling*.

A systematic scaling can sometimes be based on '*rated values*': Assign normative values to input, output, and the time variable, and use them as units in a scaling operation, as follows:

$$\phi = \phi_0 \, x_1, \quad \omega = \omega_0 \, x_2, \quad I = I_0 \, x_3, \quad V = V_0 \, u, \quad t = t_0 \, \tau$$

where x_1, x_2, x_3, u, τ are dimension–less variables. The rated values $\phi_0, \omega_0, I_0, V_0, t_0$ are set from consideration of how the object is supposed to operate, or can be estimated from a look at the experiment data. Reasonable choices are
$\phi_0 = 360°$ (alternatively, the operating range),
ω_0 = the maximum speed of the motor,
I_0 = the maximum stator current allowed,
V_0 = the maximum input voltage.

The maximum speed may be replaced by the 'cruising speed', etc. The related time may be defined as the time it takes to reach ϕ_0 with the maximum speed (then $t_0 = \phi_0/\omega_0$) or with maximum current or voltage.

This yields the dynamic equations $(x_i' = dx_i/d\tau)$

$$\begin{bmatrix} x_1' = t_0\ \phi_0^{-1}\ \omega_0\ x_2 \\ x_2' = t_0\ \omega_0^{-1}\ I_0\ a_1\ x_3 \\ x_3' = -\ t_0\ a_3\ a_2^{-1}\ x_3 + t_0\ I_0^{-1}\ V_0\ a_2^{-1}\ u \end{bmatrix}$$

In order to define $\mathcal{P}$ assume that the motor is designed so that the three states have time constants of the same order of magnitude (if not, one should remove some of the states). With rated time such that $t_0\ \phi_0^{-1}\ \omega_0 = 1$, this yields the 'origin' a_i^0 of the free parameter range from the equations:

$$\begin{bmatrix} t_0\ \omega_0^{-1}\ I_0\ a_1^0 = 1 \\ t_0\ I_0^{-1}\ V_0\ (a_2^0)^{-1} = 1 \\ t_0\ a_3^0\ (a_2^0)^{-1} = 1 \end{bmatrix}$$

This might do as start values in a search, unless one has a better value from other sources. For instance, in large motors the resistance is often negligible, and one could choose $a_3^0 = 0$ instead.

The model structure will be

$$\mathcal{M}: \begin{bmatrix} x_1' = x_2 \\ x_2' = (1 + \theta_1)\ x_3 \\ x_3' = -\ (1 + \theta_3)\ x_3 + (1 + \theta_2)\ u \end{bmatrix}$$
$$\begin{bmatrix} d_1 = \phi_0\ x_1 + \sigma_1\ \omega_1 \\ d_2 = \omega_0\ x_2 + \sigma_2\ \omega_2 \\ d_3 = I_0\ x_3 + \sigma_3\ \omega_3 \end{bmatrix}$$

Hence, the parameter map is defined by

$$\mathcal{P}: \begin{bmatrix} a_1 = a_1^0\ (1 + \theta_1) \\ a_2 = a_2^0/(1 + \theta_2) \\ a_3 = a_3^0\ (1 + \theta_3)/(1 + \theta_2) \end{bmatrix}$$

The free coordinates θ_i are defined in that way in order to retain linearity in the parameters, which is generally favourable for search. All simulations and search variables are dimension–free, and there will be no problem with too fast or too slow responses with too large or too small amplitudes (unless the initial guess is much out of range).

The parameter map is a design tool

The parameter map may be used for several purposes:

● To specify apriori restrictions on a for estimation; only $a \in \{a|\theta\}$ are free to fit to data, and θ are the parameters to be estimated.

● To describe the aposteriori distribution of a after estimation. In that case the origin is at the estimate, $a^\circ = \hat{a}$, and its error distribution (asymptotically) is given by $\mathcal{P}$ and (7.15)

$$\Gamma_\theta^T \Gamma_\theta = [-\text{grad}_\theta \, \text{grad}_\theta \, \log L(0|\mathcal{A}, \nu, \mathbf{c}_N, \mathbf{d}_N)]^{-1}$$

This requires a structure such that the ML–estimate $\hat{a}$ will be asymptotically normal.

● To generate random variation in a for simulation in order to appraise uncertainty, robustness, etc. In that case random a are generated via $\theta = \Gamma_\theta^T \omega$, where ω is gaussian with zero mean and unit covariance matrix.

● To specify an alternative hypothesis $a \in \{a|\theta\}$ to the *null* hypothesis $a = \hat{a}$ in the falsification of the model: The coordinate θ is then implicit and does not appear in the hypothesis (see section 8.3 on "Conditional falsification of models").

● To determine parameter identifiability prior to fitting: The covariance factor Γ_θ^T computed from $(\mathcal{A}, \nu, \mathbf{c}_N, \mathbf{d}_N)$ reveals which components of θ that are not identifiable.

> **Remark:** The parameter map is an important structure element, since it allows the designer to change model complexity at 'run time', in essence by manipulating the dimension of the θ–vector. That makes it possible to imbed differently complex models in a 'chinese–box' pattern, and to find the simplest model compatible with data in the following way: One or more sequences of increasingly complex tentative model structures is required from the designer and processed by a sequence of falsifications. In this way the designer is offered means to affect complexity and to trade this for a computed confidence in the model in a way that is possible to assess subjectively (see section 9.2 on "Sequential falsification").

5.3   Algorithmic models

A probability distribution can be defined by means of an *algorithm d ←*
A(ω) that generates random vectors *d* from other random vectors ω with
a given distribution *p*[ω]. Generally, the notation *y ← F*(*u*) means that
the algorithm/routine/procedure *F* operates on the input *u* to produce
the output *y*.

> **Remark:** The left–arrow symbol (←) corresponds to the 'assign-
> ment' symbol used in algorithmic programming languages
> (FORTRAN uses =, Pascal uses :=). It can often be replaced by an
> equality sign — it is legitimate to write the model *d = A*(ω) — but
> not always. For instance, the expression 0 ← *A*(ω) has no meaning,
> while 0 = *A*(ω) may have. The equality sign signifies an equation,
> which defines a set of ω–values. Conversely, *n ← n* + 1 has a
> meaning, while *n = n* + 1 has not. A right–arrow symbol (→) will
> also be used, and means that the left–hand side tends to the right-
> hand side in the limit. The right–arrow also symbolizes a mapping
> from one set of variable values to another.

The density that corresponds to the algorithm *A* — the probabilistic
model — can be expressed as

$$p[d|A] = \int \delta[d - A(\omega)] \, p[\omega] \, d\omega \qquad (5.5)$$

where δ is the Dirac distribution, *i.e.* one whose mass is all con-
centrated at the origin. If ω has a standard distribution, *e.g.* its components are independent gaussian variables with zero means and unit
variances, then the algorithm *A* alone determines the distribution of *d*.
It is called a *stochastic model*, and ω are *primitive random variables* driving the model.

> **Remark:** The concept of a stochastic model raises the question of
> whether it is possible to model all conceivable distributions {*p*[*d*]}
> in this way, with *p*[ω] gaussian. But that is a mathematical
> problem. The following is the outline of a proof of the existence of
> an *A* for every distribution:

Proof:
Let $d_k = \{d(1),...,d(k)\}$ be the vector of the first *k* components of
d. Define a series of scalar functions $\{\Omega_i(d_i)|i=1,...,k\}$ such that

$$\Phi[\Omega_i(d_i)] = \int_{-\infty}^{d(i)} p[z(i)|d_{i-1}] \, dz(i)$$

where Φ is the cumulative gaussian distribution function. Since Φ is monotone, this defines all Ω_i uniquely. Define $\omega(i) = \Omega_i(d_i)$. Hence, the existence of a unique ω has been shown by construction. In addition, it follows that

1) ω is gaussian, with zero mean and unit covariance matrix
2) There is a bijective (one–to–one and onto) correspondence between d_i and ω_i,
3) $\text{grad}_{d(i)}\Omega_i[d_i]$ may be chosen left–triangular.   ∎

Remark: The idea of a stochastic model is that it allows a kind of *separation* in the modelling of two important qualities of reality: Physical objects have *properties* that should influence any design based on models, and there is also an element of *randomness*, which affects response data, but which should not be allowed to influence the designs too much (if it would, then the design would depend too much on chance data to be of any use in the 'application phase'). In a stochastic model the properties of the distribution are embodied entirely in A, and that is a deterministic element and known when the model has been found. The randomness is embodied in ω, but since it has a standard distribution, it does not affect any design.

Remark: A probabilistic model is a mapping: $M: R^c \times R^d \to R^+$, where R^c and R^d are the real spaces of the sequences c and d respectively, and R^+ is the nonnegative real axis. Willems (1986) defines a *deterministic* model as the set of pairs (c,d) not mapped to zero. Thus, a deterministic model is basically a restriction of the set of all possible stimulus–response pairs. If no other variables affect the response, then the distribution defined by M would be a point, *i.e.* both c and d would be given apriori. Since that would be a trivial case, also other variables ω affect the response. Then the deterministic model does not correspond to a point, but is the set of all possible stimulus–response pairs generated by all possible ω. The set is defined by the algorithmic model $(c,d) \leftarrow A(\omega)$. A case when this description is suitable is that when ω is the initial state. If, further, ω can be assigned a distribution P, then one can define a stochastic model by

$$P[c,d] = \int \text{Ind}[(c,d) \leq A(\omega)] \, dP(\omega).$$

This covers the case when ω is a persistent disturbance as well as that when ω is an unknown initial state. The only formal difference between deterministic and stochastic algorithmic models is that the latter hypothesizes a distribution of ω.

After parametrization the probabilistic model of the system of experimenter and object was written $p[c,d|\mathcal{A},\nu,\theta]$. The corresponding stochastic model will be

$$(c,d) \leftarrow \mathcal{A}(\omega|\nu,\theta) \qquad\qquad (5.6)$$

where $\mathcal{A}$ must be known apriori (it determines $\mathcal{M}_b$) and ω,ν,θ are unknown. By giving different forms to $\mathcal{A}$, based on more or less detailed apriori information, one can bring more or less information into $p[\bullet,\bullet|\mathcal{A},\nu,\theta]$.

> **Remark:** In the simplest case the algorithm $\mathcal{A}$ is a formula, but more often it is a computer program that can be used for simulation in the following way: The characteristics of the model are entered as parameters (ν,θ), and ω is obtained from a pseudo–random–number generator. When executed the program produces the output (c,d).

> **Remark:** The three unknowns in (5.6) have different roles: The parameters (ν,θ) carry information that is invariant in time and hence may be of use for the application phase, the primitives ω do not. Their information is only valid for the data $(\mathbf{c},\mathbf{d})$. This means that ω takes into account the possibility that the same experiment on the same object may produce different data $(\mathbf{c},\mathbf{d})$ on different occasions. If one were to repeat the experiment at another time, the values of ν and θ would be the same, but the value of ω would not. In other words, (ν,θ) is the 'information' and ω is the 'noise'. The parameters (ν,θ) define the information–carrying distribution $\{p[c,d|\mathcal{A},\nu,\theta]\}$, but ω does not.

─────────────── **Example 5.5** ───────────────

This example illustrates the apriori separation into invariant and random properties inherent in a stochastic model.

When one wants to design transmission equipment for random signals, like speech, any optimal filter and channel design will depend on the frequency spectrum of the signal. Thus, the spectrum is the invariant 'property' that will influence the design, and the 'randomness' is the actual message that will be transmitted. If, on the other hand, one would want the transmission equipment to be tuned even more, for instance to a few standard messages (thus saving channel capacity and increasing noise tolerance), then other properties than the spectrum would affect the design, and hence be regarded as 'invariant' and 'information carrying'.

Hence it depends on the design purpose what phenomena of the object that should be regarded as 'invariant' and 'random' respectively, and that affects the modelling.

An *'equivalent filter'* is a stochastic model for random processes

$$d_k \leftarrow \mathcal{A}[\omega_k|\theta] \quad (k=1,2,\ldots)$$

where ω_k is a 'white noise' sequence, and θ is a parameter vector that characterizes the filter. When the filter is fed by 'white noise', it will produce signals with the same statistical properties (for instance the same spectrum) as the signals it is meant to describe. It is 'equivalent' to the object in that respect. The model will not be used to produce any actual signals, but only be used as a design tool. The characteristics θ are supposed to be invariant, and are the only properties of the actual object that are allowed to affect any design.

A convenient equivalent filter is the ARMA model defined by

$$d(k) = -a_1 d(k-1) - \ldots - a_n d(k-n) + c_0 \omega(k) + \ldots + c_n \omega(k-n)$$

Its spectrum is (Porat, 1987)

$$\Phi(f) = |[c_0 + \ldots + c_n \exp(2\pi inf)]/[1 + \ldots + a_n \exp(2\pi inf)]|^2$$

It depends only on parameters $\theta = (a_1,\ldots,a_n,c_0,\ldots,c_n)$, and can be used for design. The random part is the ω-sequence. It holds no information, since the signals produced by the equivalent filter do not correspond to any actual output from an object (such as a message).

5.4   The modelling of dynamic systems

The proper interpretation of the c and d variables and the system $(\mathbf{X,S})$ is fundamental for a correct modelling. When one has decided what physical variables that are to be recognized as stimulus c and response d, then both the experimenter and the object have in fact been *defined* by this decision (see section 2.1 on "Bayes' idea"). The requirement that

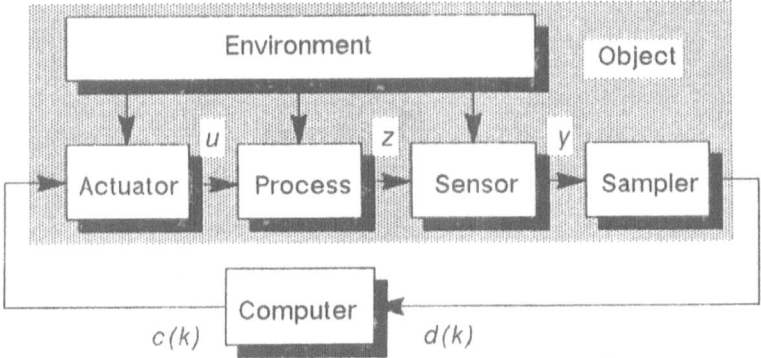

Fig. 25: Modelling of a system for experiment with computer–generated test signals or feedback control.

the values of c and d must be *known exactly* may seem restrictive, for instance in cases when one has only got contaminated measurements of the object's input and output. In cases of closed loop it may even be difficult to distinguish between input and output. However, the requirement of exactness is a matter of *interpretation* and not an actual restriction. In order to illustrate this, four common ways to experiment on dynamic systems will be modelled:

● **Identification using test signals** (*Fig. 25*): When the stimulus to the object is provided by the same computer as is logging the responses, the actuator (final controlling element) model will be able to extrapolate from c_k to produce the process input u for the consecutive control interval, usually by zero–order hold. Some process variables z have sensors (measuring transducers), that provide measured values y, which are sampled by a sampling device (data acquisition system). All uncertainty is allocated to within the object.

● **Identification under computer feedback control:** When the object is computer controlled, the same model of the object is valid, but the algorithm generating the stimulus $c(k)$ may use previous response data d_k as input.

● **Identification using spontaneous stimulation** (*Fig. 26*): When stimulus is provided by an external source and measured with errors, then u is input to the sensor, its noise–contaminated value will be in the sensor output y, and the sampled values will be in the data $d(k)$. In order to model the case correctly one should specify what may be known about the external source and include that into the model structure $\mathcal{M}$.

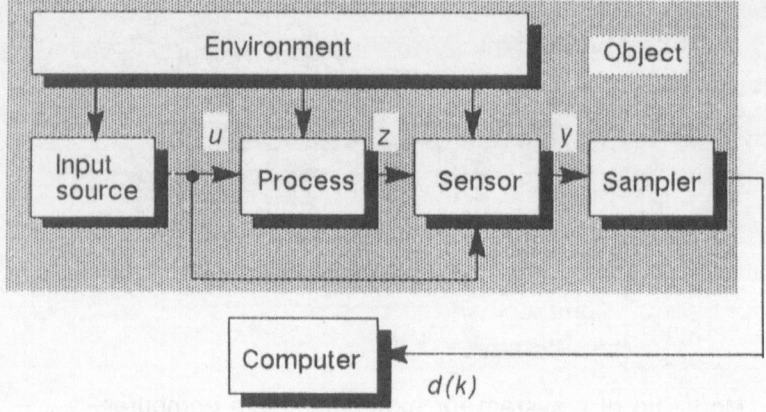

Fig. 26: Modelling of a system for identification with "natural input".

● **Identification of closed systems** (*Fig. 27*): When stimulus is generated by feedback through other channels than the data–logging computer, then the model structure 𝔐 must include also the controller, and the object is the closed–loop system. In the case that the computer can change the set point of the controller, there is a reference input $c(k)$ to the controller.

5.4.1 Linear *vs* nonlinear models

Most physical objects are no doubt nonlinear — doubling the stimulus signal does not usually yield twice the response. Nonlinear models, on

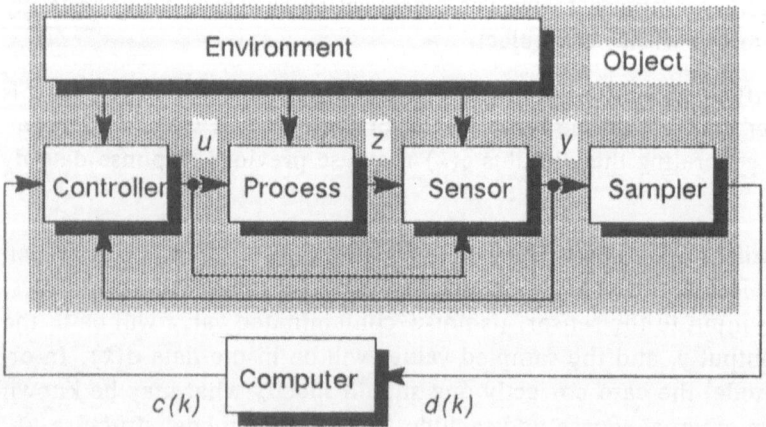

Fig. 27: Modelling for identification of a closed system.

the other hand, frequently exhibit unexpected behaviour, sometimes even chaotic, which may or may not mimic the normal behaviour of the object. Because of this, nonlinear modelling is generally tricky, and one tends to adhere to a linear description as far as possible.

This is often a successful approach, when the application is such that the model needs to be valid only in a vicinity of some nominal trajectory (solution), in particular for the purpose of controlling the object around the same trajectory. Even nonlinear objects are often approximately linear for small *deviations* in the stimulus, and if control is sucessful, the behaviour of the object will never deviate far from the nominal trajectory. Hence, the linear model will be a sufficiently accurate description for the purpose.

If, as is also common, there are several nominal trajectories, it would be conceivable to make a linear model for each trajectory (*local* model). It would even be practical, provided the trajectories are not too many. This would be the case, for instance, when the object is to be run in only a few different ways, or used to manufacture a few different products with well–defined specifications.

However, in this case a nonlinear model may still be preferable, provided it is *linearizable*. It may be possible to utilize apriori information about the (nonlinear) structure, and hence arrive at one and the same model for all trajectories (*global* model). This is clearly to prefer when the nominal trajectories are unknown apriori, or many. This is the case when 'grey–box' identification has a good chance of success.

However again, linearizability is the crucial point. Two necessary conditions for this are the following:

● The nonlinear model must be continuously differentiable, so that it is possible to do a Taylor expansion around the nominal trajectory.

> **Remark:** This generally excludes models with discontinuities like dry friction, hysteresis, bumps, or backlash, all describing important real phenomena, but difficult to analyse using general methods.

● The linear model must be stable around the trajectory — small deviations in initial state and stimulus signal should not make the response deviate much.

> **Remark:** For autonomous state–vector models the mathematical condition is that the eigenvalues of the linearized model should have strictly negative real parts. If some eigenvalues have strictly positive parts, the trajectory is unstable, and if some are zero, the

linear model will not necessarily behave as the nonlinear around the nominal trajectory. The set of state values for which this is the case is called 'the center manifold', and a linearizable model will not be a successful modelling attempt in such cases.

In general, linearizable parametric models cause no unsurmountable identification problems (and are also quite common in practice), but other nonlinearities require sophisticated specialized techniques (Billings, 1980; Mehra, 1979). They are also the ones that exhibit the 'strange' behaviour, but so do many real objects when stimulated strongly.

The problems of strong nonlinearities will not show up explicitly in the sequel. They are hidden in the likelihood function $L(\theta|c,d)$. For instance, if a model has a chaotic behaviour for some parameter values, then $L(\theta|c,d)$ will be a complicated function of θ, and so will the B and V functions, making fitting, validation, and falsification difficult or unmanageable. For long samples the properties of the limiting log likelihood function $Q(\theta|c_\infty,d_\infty)$ determine the identifiability of the case (Chapter 6 on "Large–sample theory"). Even when the limit exists, it may not be a continuous function of θ. The general *analysis* is still valid for all nonlinear models, but the identification *techniques* may get into trouble.

> **Remark:** It has been argued that chaotic deterministic systems can be modelled as stochastic for control purposes (Fowler, 1989). This would clearly facilitate the use of the Likelihood function.

Linear modelling tends to be 'black box' oriented, even if some apriori information is needed to decide on the range of time constants in the model. Even non–linear modelling is often based on little apriori information. Billings (1980) has given a survey of nonlinear model structures that are feasible for identification. The most commonly used of the 'black–box' models are the Volterra model, the block–oriented models composed of linear dynamic blocks and static nonlinear blocks, the bilinear, and the linear–in–the–parameters models, simply because estimation methods have been developed for their parameters (see also Billings and Leontaritis, 1982). However, fitting nonlinear 'black–box' models is normally a very demanding task and requires at best much computing. The structure selection part (see section 9.6) is even more difficult, and often well nigh impossible. The reason is the prohibitive number of possible combinations of nonlinear elements within a reasonably general nonlinear model class. The basic idea of 'grey–box' modelling is that apriori information about the object should reduce this multitude of possibilities to a manageable size.

Frequently, identification requires much computing also in the case of linear but multivariable objects. It may therefore be necessary to take into account the algorithms available to carry out the identification step already in the modelling step. In particular, linear multivariable structures are often interchangeable from the point of view of describing input–output behaviour, and selected instead on account of how convenient their associated identification algorithms are to handle. El-Sherief and Sinha (1979) has written a survey of such structures.

Linearity yields a choice between time– and frequency descriptions

When a linear model will be adequate, the modelling is frequently affected by other aspects than the properties of the object, such as the convenience of computing, and how transparent the theory is to intuition. Most notably, linear time–invariant models allow a convenient shortcut in the analysis of dynamical systems, which affects the identification techniques as well as an intuitive design. It is well known that if the Fourier transform of the input signal is multiplied by the transfer function of the model, the result is the Fourier transform of the output signal. Secondly, the signals can be interpreted as superpositions of weighted sinusoidal signals, so that the kernel in the transform can be interpreted as the distribution of the frequency contents of the signals. This appeal to intuition is the basis of the popularity of this way of modelling (Wellstead, 1981). It implies the following: *i*) It is easy to assess how the model properties affect the spectrum of the response (since it is easy to predict the consequences of the multiplication operation simultaneously for many frequences), and hence to design input signal and model structure to suit a given purpose. *ii*) If the input signal is chosen in particularly convenient ways (*e.g.* sinusoids or PRBS), then the otherwise cumbersome identification procedure will be simple. It is also possible to express estimation accuracy in formulas that are easy to interpret. These advantages are however acompanied by disadvantages, and the alternative of modelling in the time domain has gradually become more popular. However again, frequency–domain modelling still has its proper applications. Ljung and Glover (1981) discuss the pros and cons of the two alternatives, as well as relations between them.

5.5   Internal and external models

The terms 'internal' and 'external' models were introduced briefly in section 1.1 on "The terminology". The distinction has to do with computing the information in the data, *i.e.* the unbiassed likelihood $L(M|c,d) = p[c,d|M]$. With a dynamic model the likelihood is from (3.13)

$$L(M|\mathbf{c}_N,\mathbf{d}_N) = \prod_{k=1}^{N} p[\mathbf{c}(k)|\mathbf{c}_{k-1},\mathbf{d}_k,k,M]\ p[\mathbf{d}(k)|\mathbf{c}_{k-1},\mathbf{d}_{k-1},k,M] \quad (5.7)$$

and one needs to compute $p[\mathbf{d}(k)|\mathbf{c}_{k-1},\mathbf{d}_{k-1},k,M]$ for all $\mathbf{c}_{k-1},\mathbf{d}_k$ and all M in the particular structure $\mathcal{M}$ one has arrived at by modelling.

5.5.1   External models

An *external* model is one that describes only the relation between the *observable* variables c and d. If the model also has the form of a 'probabilistic model' $p[c,d|M]$, then one can compute its (unbiassed) likelihood immediately by inserting the values of $(\mathbf{c},\mathbf{d})$.

Algorithmic external models:
If the model has the form of an *algorithmic* external model $(c,d) \leftarrow A(\omega)$, then it is also easy to compute the likelihood, provided the model is *invertible, i.e.* one can solve for ω in the equality $A(\omega) = (c,d)$ by an algorithm $\omega \leftarrow W(c,d)$. Since the density $p[\omega]$ of the primitive random variables is given, one gets immediately that

$$p[c,d|M] = p[W(c,d)]\ \mathfrak{J}(c,d) \quad (5.8)$$

where $\mathfrak{J}(c,d) \triangleq |\det \mathrm{grad}_{cd} W(c,d)|$ is the Jacobian of the transformation from (c,d) to ω. A necessary condition is obviously that the model A use the minimum dimension of the primitive random vector ω needed to describe the density $p[c,d|M]$, which is the dimension of the data vector (c,d) (see the proof in section 5.3 on "Algorithmic models").

   Remark: In most cases one is only interested in modelling S. In that case $d \leftarrow A^d(c,\omega^d)$ has the inverse $\omega^d \leftarrow W^d(c,d)$, and (5.8) corresponds to $p[d|c,M] = p[W^d(c,d)] \mathcal{J}^d(c,d)$ where $\mathcal{J}^d(c,d)$ $= |\det \text{grad}_d W^d(c,d)|$

———————————— **Example 5.6** ————————————

Illustrating the modelling of a nonlinear stochastic object — The probabilistic model of a saturating amplifier.

Let a nonlinear amplifier have the following relation between input and output signals u and y:

$y = \mathbf{a} \, u \, (\mathbf{b}^2 + u^2)^{-1/2}$

where $\mathbf{a}$ and $\mathbf{b}$ are unknown.

Assume the amplifier is fed with unknown gaussian noise $u = \lambda \omega$, and one observes the response $\mathbf{d} = y$. Then the algorithmic model is

$A(\omega) = \lambda \, \omega \, a \, (b^2 + \lambda^2 \, \omega^2)^{-1/2}$

In order to compute the probabilistic model needed for the identification use (5.8). The inverse model is

$W(d) = \lambda^{-1} \, b \, d \, (a^2 - d^2)^{-1/2}$

and its Jacobian is

$\mathcal{J}(d) = \text{grad}_d W(d) = \lambda^{-1} \, b \, d^2 \, (a^2 - d^2)^{-3/2}$

From (5.8)

$p[d|a,b] = \lambda^{-1} \, b \, (a^2 - d^2)^{-3/2} \, \phi[\lambda^{-1} \, b \, d \, (a^2 - d^2)^{-1/2}]$

where ϕ is the gaussian density function. One may now estimate $(\mathbf{a},\mathbf{b}/\lambda)$ by fitting p to the observed amplitude distribution of the output $\mathbf{d}$.

_____

Dynamic external models:
A *probabilistic dynamic* external model would have the form $p[c(k),d(k)|c_{k-1},d_{k-1},k,M]$. The corresponding *algorithmic* dynamic external models may be derived from the algorithmic model for (c_N,d_N), as follows:

According to section 5.3 on "Algorithmic models" the probabilistic model $p[c_N, d_N | M]$ corresponds to

$$(c_N, d_N) \leftarrow A[\omega^c{}_N, \omega^d{}_N | M] \tag{5.9}$$

where $\omega^c{}_N = \{\omega^c(1), \ldots, \omega^c(N)\}$ and $\omega^d{}_N = \{\omega^d(1), \ldots, \omega^d(N)\}$ are now independent sequences of independent random vectors with independent components with a common known distribution, usually gaussian with zero mean and unit variance.

The form (5.9) is not convenient to work with, however, since the complexity increases indefinitely with N. A simpler, equivalent form can be derived as follows:

It was shown in section 5.3 that A is a triangular function, which means that there are equalities of the form

$$A^d[\omega^c{}_{k-1}, \omega^d{}_k | M] = d(k)$$
$$A^c[\omega^c{}_k, \omega^d{}_k | M] = c(k) \tag{5.10}$$

for $k = 1, \ldots, N$. Secondly, since the relation $(\omega^c{}_k, \omega^d{}_k)$ to (c_k, d_k) is bijective, there is an inverse (triangular)

$$W^d[c_{k-1}, d_k | M] = \omega^d(k)$$
$$W^c[c_k, d_k | M] = \omega^c(k) \tag{5.11}$$

By using these relations recursively one can compute $c_{k-1}, d_k, \omega^c(k), k, M$ $\mapsto c(k)$ and $c_{k-1}, d_{k-1}, \omega^d(k), k, M \mapsto d(k)$ in the order $d(1), c(1), d(2), c(2),$ $\ldots, d(N), c(N)$. Hence, there are algorithms

$$d(k) \leftarrow A^d[c_{k-1}, d_{k-1}, \omega^d(k), k | M]$$
$$c(k) \leftarrow A^c[c_{k-1}, d_k, \omega^c(k), k | M] \tag{5.12}$$

that generate any density $p[c(k), d(k) | c_{k-1}, d_{k-1}, k, M]$.

Now, if one may assume that the experiment system is 'separable', for instance the experimenter is *Isolated* from the object (see section 3.5), then $M = (M^c, M^d)$, and there are algorithms that correspond to the probabilistic dynamic external models $p[d(k) | c_{k-1}, d_{k-1}, k, M^d]$, $p[c(k) | c_{k-1}, d_k, k, M^c]$:

External algorithmic dynamic model:

$$d(k) \leftarrow A^d[c_{k-1}, d_{k-1}, \omega^d(k), k | M^d]$$
$$c(k) \leftarrow A^c[c_{k-1}, d_k, \omega^c(k), k | M^c] \qquad (5.13)$$

Inverse model:

$$\omega^d(k) \leftarrow W^d[c_{k-1}, d_k, k | M^d]$$
$$\omega^c(k) \leftarrow W^c[c_k, d_k, k | M^c] \qquad (5.14)$$

Remark: The second algorithm in (5.13) can be used directly by the experimenting computer to generate stimuli for an actual experiment; $\omega^c(k)$ is then a pseudo–random vector.

Remark: The algorithmic dynamic external model M^d is also called *predictor model*, because it can be used for predicting the output $d(k+1)$ based on data (c_k, d_k). Since the model determines a conditional distribution of $d(k+1)$, it also determines a conditional estimate. In particular, the minimum–prediction–error estimate is defined by $E\{d(k+1)|c_k, d_k\}$, which can be computed as $E\{d(k+1)|c_k, d_k\} = \int A^d[c_k, d_k, \omega, k+1|M^d] \; p[\omega] \; d\omega$. A simpler estimate is $A^d[c_k, d_k, 0, k+1|M^d]$, which is the *median* of $d(k+1)$, if A^d is monotone in ω.

5.5.2 Innovations

If one applies the inverse algorithm (5.14) of the external model to any model M and to the data (c_N, d_N) it produces, the resulting sequences will be $\{\omega^d(k)\}$ and $\{\omega^c(k)\}$, and hence have a known, independent, standard distribution, usually gaussian. However, if one applies the same algorithm to the actual data $(\mathbf{c}_N, \mathbf{d}_N)$, and uses an arbitrary model $M = (M^c, M^d)$, one gets the sequences

$$w^d(k|M^d) = W^d[\mathbf{c}_{k-1}, \mathbf{d}_k, k|M^d]$$
$$w^c(k|M^c) = W^c[\mathbf{c}_k, \mathbf{d}_k, k|M^c] \qquad (5.15)$$

which normally do not have the standard distribution. This distinguishes the particular model $(\mathbf{X},\mathbf{S})$ from the majority of other models M (even if not all), and makes the variables $\mathbf{w}^d(k|M^d)$ useful enough to merit a particular name, *'innovations'*. In analogy, $\mathbf{w}^c(k|M^c)$ are the innovations for the experimenter.

The innovations play an important role in the evaluation of the likelihood: If one can invert the model (5.13) to obtain (5.15), and compute the gradients

$$\nabla_c \mathbf{w}^c(k|M^c) \triangleq \text{grad}_{c(k)} W^c[\mathbf{c}_k,\mathbf{d}_k,k|M^c],$$

$$\nabla_d \mathbf{w}^d(k|M^d) \triangleq \text{grad}_{d(k)} W^d[\mathbf{c}_{k-1},\mathbf{d}_k,k|M^d],$$

then

$$\log L(M|\mathbf{c}_N,\mathbf{d}_N)$$

$$= \sum_{k=1}^{N} \log p[\mathbf{c}(k)|\mathbf{c}_{k-1},\mathbf{d}_k,k,M^c] + \sum_{k=1}^{N} \log p[\mathbf{d}(k)|\mathbf{c}_{k-1},\mathbf{d}_{k-1},k,M^d]$$

$$= \sum_{k=1}^{N} \{\log \phi[\mathbf{w}^c(k|M^c)] + \log \det[\nabla_c \mathbf{w}^c(k|M^c)]\}$$

$$+ \sum_{k=1}^{N} \{\log \phi[\mathbf{w}^d(k|M^d)] + \log \det[\nabla_d \mathbf{w}^d(k|M^d)]\} \qquad (5.16)$$

where $\phi(\omega)$ is the standard distribution of the primitive random variables (usually gaussian).

An external model has the fewest primitive random variables

An *external model* uses the minimum dimension of the primitive random vector ω needed to describe the distribution $\{p[c,d|M]\}$ by an algorithmic model $(c,d) \leftarrow A[\omega]$, since it must be *invertible*.

> **Remark:** The name 'innovation' originates in the following consideration: Given a model M, the innovations sequence $\mathbf{w}_k$ can be computed from the data sequence $(\mathbf{c}_k,\mathbf{d}_k)$ for any k, and vice versa. Hence, $\mathbf{c}(k)$ and $\mathbf{d}(k)$ can be computed from $(\mathbf{c}_{k-1},\mathbf{d}_{k-1})$ and $\mathbf{w}(k)$, *i.e.* the innovation $\mathbf{w}(k)$ carries the same information as the current data pair, given previous data (or previous innovations). Hence, the 'innovation' represents the information that is 'novel' in the data sequence, that is, what cannot be computed

from previous data. The point is that innovations can usually be represented by fewer bits than the raw data (see Example 1.2).

Remark: The related quantity, the *residual*, $e(k) \triangleq [\nabla_d w^d(k|M^d)]^{-1} w^d(k)$ also plays an important role. In connection with certain model structures it may be interpreted as the 'one–step prediction error'. This makes it the prime target for minimization in the Minimum Prediction Error (MPE) identification method (Ljung, 1987a). Notice however, that it is not always correct to interpret it as a 'modelling error'. Its value depends on how wrong the model is, but also on the size of disturbances from the environment, and the latter may well dominate over the modelling error, causing large residuals even from a useful model.

Remark: In addition to being instrumental in the evaluation of the likelihood, the innovations sequence is generally a very practical device for revealing all sorts of misrepresentations in a model. It can be used to falsify models in a way described in section 8.2 on "Unconditional falsification".

Remark: The 'external algorithmic dynamic model' $c(k), d(k) \leftarrow A[c_{k-1}, d_{k-1}, \omega(k), k]$ offers a way to *test* reproducibility: It is clear that if one *defines* a model by a given algorithm A *and* the innovation sequence w_N computed from the equations $A[c_{k-1}, d_{k-1}, w(k), k] = (c(k), d(k))$, $(k = 1, \ldots, N)$, then one can also define the set of unfalsified models $\mathcal{B}(c, d) = (A, w_N)$, *i.e.* as containing only the algorithm A and the innovation sequence w_N. Repeating the experiment would yield a set $\mathcal{B}(c, d)$ consisting of several models $\{A, w_N\}$. If the models in the set would eventually aggregate according to a distribution, then one would be able to *predict* in which domain the next innovation sequence will be with a given probability. Any design based on the model would be able to use the invariant A (including the invariant distribution of the innovations), but not the variable w_N.

──────────────── **Example 5.7** ────────────────

Illustrating the concept of 'residuals'.

A useful algorithmic model for a large class of random signals is the ARMA–model (see also Example 5.3):

A: $d(k) = - a_1 d(k-1) - \cdots - a_n d(k-n)$
$\qquad + \lambda [\omega(k) + c_1 \omega(k-1) + \cdots + c_n \omega(k-n)]$, $d(k) = 0$ for $k < 1$

The inverse of A is

W: $\omega(k) = - b_1 \omega(k-1) - \cdots - b_n \omega(k-n)$
$\qquad + \lambda^{-1} [d(k) + a_1 d(k-1) + \cdots + a_n d(k-n)]$, $\omega(k) = 0$ for $k < 1$

The residuals and innovations are generated by the recursion

$\mathbf{e}(k|a,b) = - b_1 \mathbf{e}(k-1|a,b) - \cdots - b_n \mathbf{e}(k-n|a,b)$
$\qquad + \mathbf{d}(k) + a_1 \mathbf{d}(k-1) + \cdots + a_n \mathbf{d}(k-n)$, $\mathbf{e}(k) = 0$ for $k < 1$,
$\mathbf{w}(k|a,b,\lambda) = \mathbf{e}(k|a,b)/\lambda$

They depend on (a,b). If $(a,b) = (\mathbf{a},\mathbf{b})$, then $\mathbf{w}(k|a,b) = \omega(k)$ for all k. It is easy to prove that for all other values of (a,b):

$$E\{\mathbf{e}^2(k|a,b)\} \geq E\{\mathbf{e}^2(k|\mathbf{a},\mathbf{b})\} = \lambda^2$$

Hence, one can determine $(\mathbf{a},\mathbf{b})$ from a long sample by minimizing the residual variance $\dfrac{1}{N} \sum\limits_{k=1}^{N} \mathbf{e}^2(k|a,b)$. Furthermore, the minimum value is a consistent estimate of λ^2.

The one–step predictor is

$\hat{d}(k|k-1,a,b) = E\{\mathbf{d}(k)|d_{k-1}\}$
$= E\{- a_1 \mathbf{d}(k-1) - \cdots - a_n \mathbf{d}(k-n)$
$\qquad\qquad + \lambda [\omega(k) + b_1 \omega(k-1) + \cdots + b_n \omega(k-n)]|d_{k-1}\}$

But $\omega(k-1),\ldots,\omega(k-n)$ are given by d_{k-1} and the inverse function W, and $\omega(k)$ is independent of d_{k-1}. Hence,

$\hat{d}(k|k-1,a,b) = - a_1 \mathbf{d}(k-1) - \cdots - a_n \mathbf{d}(k-n)$
$\qquad\qquad + b_1 \mathbf{e}(k-1|a,b) + \cdots + b_n \mathbf{e}(k-n|a,b)]$.

The prediction error is $\mathbf{d}(k) - \hat{d}(k|k-1,a,b) = \mathbf{e}(k|a,b)$. Hence, $(\mathbf{a},\mathbf{b})$ minimizes the prediction–error variance

$$E\{\frac{1}{N} \sum\limits_{k=1}^{N}[\mathbf{d}(k) - \hat{d}(k|k-1,a,b)]^2\}$$

which is an application of the Minimum–Prediction–Error criterion (see section 4.3 on "Fitting").

5.5.3 Internal models

An *internal* model is one that describes also the relations between other variables than those that can be observed. An algorithmic model for the unobservable variables z would be $z \leftarrow A^z(c, \omega^z)$. Relating unobservable z to the observations d will require a second model $d \leftarrow A^d(z, \omega^d)$. The second model may be interpreted as that of the instrumentation; it specifies what components or functions of z that are measured and with what errors ω^d. When put together the two models will define the algorithmic model $d \leftarrow A(c, \omega)$. However the primitive random variables ω will now be the pair (ω^d, ω^z) and will have *more* than the minimum number of components $\dim(d)$. The point is that one may still be able to determine both A^d and A^z. Since the model has to be *parametric* in practice, it must be written $A^z = \mathcal{A}^z(c, \omega^z | \theta)$, $A^d = \mathcal{A}^d(z, \omega^d | \theta)$, and the possibility of estimating θ from data depends on $\dim(\theta)$ and not on $\dim(\omega)$. This means that the more detail one builds into the model, the more of that detail must be known apriori. That is not surprising, since given data can obviously supply only a certain amount of information, which means, loosely, a certain number of accurate parameter estimates.

> **Remark:** In particular, one may know the sources of randomness in the responses, and locate them to a number of *disturbances* v. If one can also model the density $p[v|\theta]$ of those disturbances, then the latter will have an algorithmic model $v \leftarrow \mathcal{A}^v(\omega^v | \theta)$.

——————————————— **Example 5.8** ———————————————

Illustrating the difference between internal and external models.

Let the object be given by the internal algorithmic model

S: $z \leftarrow \theta c + \lambda \omega^z$, $d \leftarrow z + \omega^d$

where ω^z and ω^d are gaussian $(0, 1)$. The corresponding input–output model is

$$d \leftarrow \theta c + \begin{bmatrix} \lambda & 1 \end{bmatrix} \begin{bmatrix} \omega^z \\ \omega^d \end{bmatrix}$$

It is not invertible, since $\left[\omega^z \ \omega^d \right]$ has more components than d, and is therefore not an external model. However, its output distribution is equal to that of

$$d \leftarrow \theta\, c + \omega\,\sqrt{1 + \lambda^2}, \quad \omega \text{ gaussian } (0,1)$$

which is an external algorithmic model

Dynamic algorithmic internal model:

Dynamic internal models appear naturally, when the unobservable variables z depend on time t. Since the variables (c,d) available to the computer depend on the reading k of the sampling counter, set by the computer's real–time clock, the computer and object together make a *hybrid* system. The setup in *Fig. 28* is common:

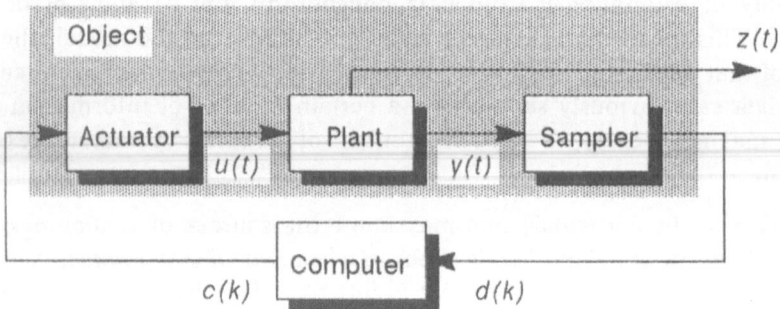

Fig. 28: Internal model of the hybrid system of clock– and time–driven dynamic processes.

Let time t be discrete (for simplicity), and divide the 'object' into the three parts '*Plant*', '*Actuator*', and '*Sampler*', where 'Plant' is the part of the object that is described by variables in time t. The two other parts are needed as interfaces between the clock–driven 'computer' and the time–driven 'plant'. The labels reflect the interpretation of the physical object as a computer controlled production process, but the model is general. Collect the time–dependent physical variables of interest for the purpose of the modelling into three vectors

$u(t)$ = exogenous input to the plant,

$z(t)$ = dependent output (objects for prediction/control, if any),

$y(t)$ = sensor output.

Let subscript t mean the past history, *e.g.* $u_t = \{u(t), u(t-1), \ldots, u(t_0)\}$.

Then an *algorithmic dynamic internal model of the Plant* may be written in the form

Plant model: $y(t), z(t) \leftarrow F[u_t, \omega_t, t | t \in T]$ (5.17)

where T is the time interval for which the model is defined, and ω_t is a sequence of uncorrelated gaussian vectors with zero means and unit covariance matrices. The sequence models unpredictable random 'disturbances'. Physically, the latter may have their sources inside the plant or in an 'environment'. In the latter case a model of the environment is included in F. In any case, modelling disturbances that are not white and gaussian requires that shaping filters be also included in F.

> **Remark:** With reference to *Fig. 30* the model describes the environment, process proper, and sensors. An alternative that emphasizes this would be
>
> *Environment model*: $v(t) \leftarrow F^v[\omega_t, t]$
>
> *Process model*: $z(t) \leftarrow F^z[v_t, u_t, \omega_t, t]$
>
> *Sensor model*: $y(t) \leftarrow F^y[z_t, \omega_t, t]$
>
> where $v(t)$ means a second input to the process, coming from the environment. The three submodels would normally include different components of ω as arguments. This alternative is less general (and more informative), since it excludes the possibility that, for instance, the sensor output depend on input u and v in other ways than through the output z.

One has also to model the two interfaces Sampler and Actuator, *i.e.* how the variables in time are transformed into data sequences, and vice versa. In order to define the Sampler let $\mathbf{d}(k)$ be the data obtained at time t_k and let $\mathbf{d}_k = \{\mathbf{d}(k), \mathbf{d}(k-1), ..., \mathbf{d}(1)\}$. Define the process sampling the sensor output for data

Sampler: $d(k) \leftarrow D[y(t_k), k | 1 \leq k \leq N]$ (5.18)

In order to take into account that not all sensor output $y(t_k)$ may be in the data sample, even at the sampling instants t_k, let $D[\bullet, k]$ pick the components that are recorded for the purpose of identification at t_k. This covers the case of different sampling rates as well as occasionally missing data.

> **Remark:** The model structure implicitly assumes that there is a sensor that produces output $y(t)$ at the same (nearly continuous) rate that the process produces output $z(t)$. It models *sampling* by means of the sampler model. In this way one can change the sam-

pling interval without having to change F. This means no loss of applicability, but it has the slight inconvenience that one will have to conceive a fictitious nearly continuous 'sensor', in case there is no real one. An alternative would be to include the sensor in the Sampler model, thus eliminating the y–variables.

The Actuator model computes the exogenous input $u(t)$ within sampling intervals and has the form:

$$Actuator: \; u(t) \leftarrow C[c_{k-1}, \omega_t, k, t | t_{k-1} < t \leq t_k] \qquad (5.19)$$

where $c_k = \{c(k), c(k-1), \ldots, c(1)\}$ is a sequence of control input. Notice that the control sequence c_k is available input to an identification algorithm, while the plant input sequence u_t is not. The rule is similar to that for the output variables; the data sequence d_k is available input to the algorithm, while the plant output (z_t, y_t) are not.

> **Remark:** The meaning of C depends on how one does the experiment. In the particular case of a computer carrying out the experiment in open–loop (see section 1.3 on "The purpose"), C includes a description of the sample–and–hold mechanism and DA conversion. Similarly, D includes a model of the sampling mechanism and the AD–conversion. However, several other ways to experiment are common practice and have been discussed in section 5.4 on "The modelling of dynamic systems".

The indexing must preserve causality

Since there are two sets of sequences involved in an internal dynamic model, indexed by time t and by the sampler clock reading k respectively, and since there are causality relations between those variables, it is important how the variables are indexed. The indexing applied in the Actuator, Plant, and Sampler models is illustrated in *Fig. 29*. It follows from the Actuator model (5.19) and the figure that at three consecutive times around t_k it holds that

$$u(t = t_k - 1) = C[c_{k-1}, \omega_t, k, t | t_{k-1} < t \leq t_k]$$
$$u(t = t_k) = C[c_{k-1}, \omega_t, k, t | t_{k-1} < t \leq t_k]$$
$$u(t = t_k + 1) = C[c_k, \omega_t, k+1, t | t_k < t \leq t_{k+1}] \qquad (5.20)$$

The indexing guarantees that the Actuator model will always respond *after* the control signal c. The Plant model may respond faster than the time quantum, which then appears to be instantaneous in the discrete-

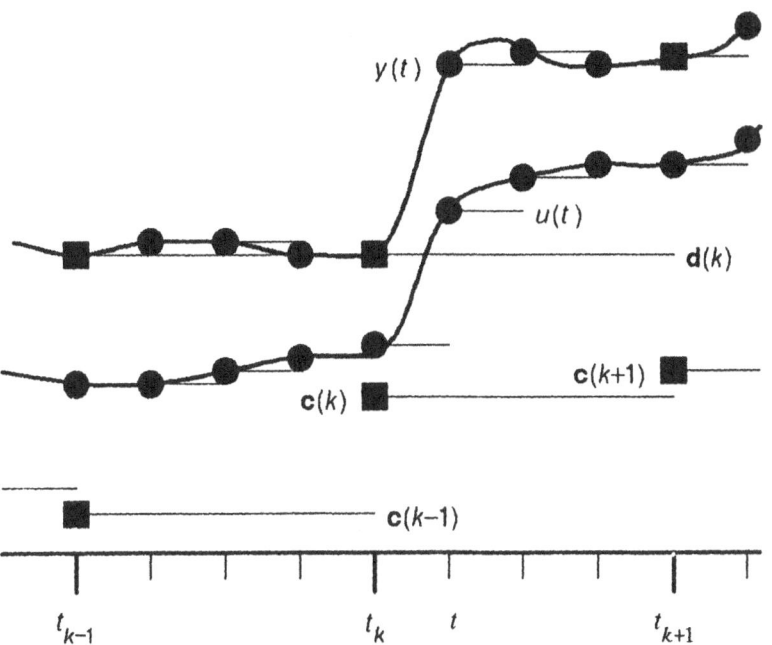

Fig. 29: Illustrates the causal dependence of stimulus $c(k)$, plant input $u(t)$, sensor output $y(t)$, and response data $d(k)$. Data are marked ■ and discrete–time variables ●. The thin horizontal lines mark the time intervals when variables are unchanged in the computer memory.

time model. That means that the model allows so–called 'direct terms'. However, there can still be no 'direct term' in the relation $c \mapsto u \mapsto y \mapsto d$. The reason is that there may be a 'direct feedback term' in the algorithm computing the stimulus $d \mapsto c$, and one cannot have both at the same time without introducing an algebraic loop.

Remark: Normally the Plant model output $y(t)$ is computed by integration of differential equations over the interval $(t-1, t)$. With a direct term the continuous–time model structure may include algebraic relations between $u(t)$ and $y(t)$. Notice that with a sufficiently short time quantum, all algebraic input–output relations must vanish in a causal system, since nothing reacts instantaneously. However, if the reaction is very fast compared to the sampling interval, the number of evaluations of the model between the sampling points may be prohibitive. A direct term will approximate the fast response, and the time quantum will have to be adapted only to the moderately fast responses. This works well when the time constants of the fast responses are well separated

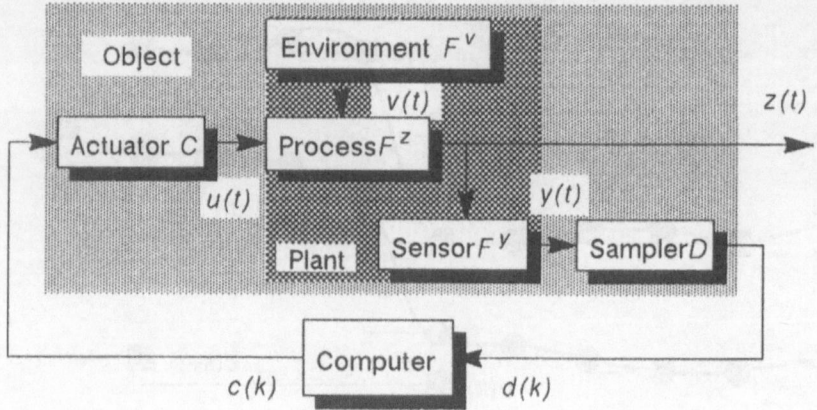

Fig. 30: Structure of an internal algorithmic dynamic model.

from the other. That is the simplest case of a so-called 'stiff' dynamic system. In more complex cases one will have to use a solution algorithm for stiff differential equations to integrate between the discrete times t (see also section 3.4 on "Dynamic systems" and 5.6 on "Implicit and explicit models").

An *algorithmic dynamic internal model* structure is illustrated in *Fig. 30*.

Sampling an internal model makes a hybrid system

The two time variables involved in identification using internal models, *viz.* physical time t and the sampling counter k ('real–time clock') make an analysis unnecessarily complicated. Since the two time variables are related by the sampling instants $\{t_k\}$, it is possible to eliminate the explicit physical time. If all y–variables are not sampled and all u–variables are not changed at the same times, then $\{t_k\}$ are the instants when *any* variable is sampled or changed. Introduce
$\omega(k) = \{\omega(t)|t=t_{k-1}+1,\ldots,t_k\}$, $u(k) = \{u(t)|t=t_{k-1}+1,\ldots,t_k\}$,
$z(k) = \{z(t)|t=t_{k-1}+1,\ldots,t_k\}$, $y(k) = \{y(t)|t=t_{k-1}+1,\ldots,t_k\}$.
It follows that $\omega_k = \{\omega(t)|t=t_0,\ldots t_k\}$, etc. Then the internal model may be simplified as

Plant model: $y(k), z(k) \leftarrow F[u_k, \omega_k, k]$ (5.21)

Sampler: $d(k) \leftarrow D[y(k), k]$ (5.22)

Actuator: $u(k) \leftarrow C[c_{k-1}, \omega_k, k]$ (5.23)

This is only an apparent simplification, since F and C are now more complicated functions. It serves only to facilitate the theory, while an application will in most cases be better off using the forms in (5.17–19).

The plant and actuator models (5.21,5.23) define together a probability density $p[y_N, z_N, u_N | c_N, C, F]$. By applying Bayes' chain rule:

$$p[y_N, z_N, u_N | c_N, C, F]$$

$$= \prod_{k=1}^{N} p[y(k), z(k), u(k) | c_N, k, C, F]$$

$$= \prod_{k=1}^{N} p[y(k), z(k) | c_N, y_{k-1}, z_{k-1}, u_k, k, C, F]$$

$$\cdot \prod_{k=1}^{N} p[u(k) | c_N, y_{k-1}, z_{k-1}, u_{k-1}, k, C, F] \quad\quad (5.24)$$

As before, the relation holds generally, and hence yields no structure information so far. All structure information (for instance causality) is contained in the definitions of C and F. The latter yield for the factors in (5.24)

$$p[y(k), z(k) | c_N, y_{k-1}, z_{k-1}, u_k, k, C, F] = p[y(k), z(k) | y_{k-1}, z_{k-1}, u_k, k, F]$$

$$p[u(k) | c_N, y_{k-1}, z_{k-1}, u_{k-1}, k, C, F] = p[u(k) | c_{k-1}, k, C] \quad\quad (5.25)$$

This means that the factors are interpreted as *probabilistic dynamic models*, the first one for the object and the second one for the input.

The unbiassed likelihood of the model (C, F) will be from (5.16)

$$L(C, F | c_N, d_N, M^c, D) = p[c_N, d_N | M^c, D, C, F]$$

$$= \prod_{k=1}^{N} p[c(k) | c_{k-1}, d_k, k, M^c] \, p[d(k) | c_{k-1}, d_{k-1}, k, D, C, F] \quad\quad (5.26)$$

That expression is much more difficult to compute than the likelihood (5.16) of an *external* dynamic model. A closed form can be derived thus:

     Derivation: The joint probability density of $(c_N, d_N, y_N, z_N, u_N)$ is

$$p[c_N, d_N, y_N, z_N, u_N | M^c, D, F, C]$$

$$= \prod_{k=1}^{N} p[c(k)|c_{k-1},d_k,y_k,z_k,u_k,k,M^c,D,F,C]$$

$$p[d(k)|c_{k-1},d_{k-1},y_k,z_k,u_k,k,M^c,D,F,C]$$
$$p[y(k),z(k),u(k)|c_{k-1},d_{k-1},y_{k-1},z_{k-1},u_{k-1},k,M^c,D,F,C]$$

$$= \prod_{k=1}^{N} p[c(k)|c_{k-1},d_k,k,M^c] \; p[d(k)|y_k,k,D]$$

$$p[y(k),z(k),u(k)|c_{k-1},y_{k-1},z_{k-1},u_{k-1},k,F,C]$$

The data distribution is the marginal distribution. Hence,
$$p[c_N,d_N|M^c,D,F,C]$$

$$= \prod_{k=1}^{N} p[c(k)|c_{k-1},d_k,k,M^c] \int\int\int p[d(t)|y_k,k,D]$$
$$p[y(k),z(k),u(k)|c_{k-1},y_{k-1},z_{k-1},u_{k-1},k,F,C] \; dy(k) \; dz(k) \; du(k)$$

Inserting the *algorithmic* models (5.21–25) yields
$$p[c_N,d_N|M^c,D,F,C]$$

$$= \prod_{k=1}^{N} p[c(k)|c_{k-1},d_k,k,M^c] \int\int\int\int \delta\{d(k) - D[y(k),k]\}$$
$$\delta\{[y(k),z(k)] - F[u_k,\omega_k,k]\} \; \delta\{u(k) - C[c_{k-1},\omega_k,k]\}$$
$$p[\omega(k)] \; dy(k) \; dz(k) \; du(k) \; d\omega(k)$$

$$= \prod_{k=1}^{N} p[c(k)|c_{k-1},d_k,k,M^c]$$

$$\int \delta\{d(k) - A[c_{k-1},\omega_k,k]\} \; p[\omega(k)] \; d\omega(k)$$

where $d(k) \leftarrow A[c_{k-1},\omega_k,k]$ is defined by
$$u(k) \leftarrow C[c_{k-1},\omega_k,k]; \quad y(k),z(k) \leftarrow F[u_k,\omega_k,k];$$
$$d(k) \leftarrow D[y(k),k].$$

Evaluating the expression is feasible only for certain classes of models (Bohlin, 1987b). Approximations have been suggested for nonlinear state–vector models (Kramer and Sorenson, 1988b).

Remark: In case one has more structure information, one will also obtain more structure in the probabilistic model. With a separation of the plant into environment, process proper, and sensor the probabilistic dynamic model will be
$$p[y(k),z(k),v(k),u(k)|c_N,y_{k-1},z_{k-1},v_{k-1},u_{k-1},k,F^y,F^z,F^v,C]$$

$$= p[y(k)|z_k,k,F^y] \quad p[z(k)|z_{k-1},v_k,u_k,k,F^z]$$
$$p[v(k)|v_{k-1},k,F^v] \quad p[u(k)|c_N,u_{k-1},k,C]$$

5.5.4 Discrete–time *vs* continuous–time models

The assumption that time is discrete simplifies the theory considerably, and would seem to be no restriction in practice, since one may choose the time unit arbitrarily short (*e.g.* 1 s, or 1 µs, or one 'time quantum' for that matter, if a physicist would claim there is one). If the model structure would originally be defined by a continuous–time model, *e.g.* a system of differential equations, then one would also have to supply an integration routine to compute the solution defining F. And that integration routine has a minimum step length, which may be chosen as time unit. When integration may be carried out apriori over the time unit, then what is important in practice is how sparsely data is collected counted in number of time units.

This does not mean, however, that it would always be pointless to investigate the continuous–time case. Even if the fundamental *theory* would become more complicated and the data sets infinite (in theory), the resulting identification procedure may still be simpler. There are two reasons for this:

Fast responses require short time quanta

Calculations will be cumbersome, if the time unit is much shorter than the sampling interval. This may be the case in particular if the differential equations describing a continuous–time system are stiff, or the integration routine for this or other reasons would change the step length much.

Numerical integration with stochastic input is tricky

A problem with discrete–time modelling is also that of choosing the time unit, and that may cause more difficulties than the previous paragraphs might suggest. In order to get a good approximation of a continuous–time model one would want a discrete–time model that tends to a limit in some sense, when the time unit tends to zero. This means in the first place, that the results of some interpolation between the discrete times of all internal and external discrete–time variables should approximate the continuous–time variables for all small time units. The latter would be no problem for the control input u, since that can be as-

sumed (piece–wise) continuous, and hence to be approximated arbitrarily well, for instance by a step function. But a continuous–time
driving noise ω cannot be approximated by a step function in the same
way, since it is 'white noise' and not continuous. It becomes a problem
even to understand what should be the meaning of a 'white–noise' input.

A way out is to let the nonlinear continuous–time plant model have a
continuous 'disturbance' variable v as input, instead of the 'white noise'
ω, and model the disturbance as the output of a *linear*, possible time
varying process, whose input is 'white noise'. This restriction will
remove much theoretical difficulty with the solution of stochastic differential equations, and still model most physical disturbances encoun-
tered in practice. However, the parameters in the discrete–time model
will have to depend on or include the time unit, in order for the discrete–time model to be invariant for small time units. The following example illustrates this. (See also Wahlberg, 1988).

Example 5.9

Illustrating a difficulty with discretizing continuous–time disturbance
models.

Let the continuous–time disturbance $v^c(t)$ be a Wiener process
(Åström, 1970), *i.e.* $v^c(t)$ is gaussian with independent increments

$E\{[v^c(t+\tau) - v^c(t)]\ v^c(t)\} = 0$, for $\tau > 0$
$E[v^c(t+\tau) - v^c(t)]^2 = \mu^2\ \tau,$

where μ is the parameter characterizing the average rate of
change of the disturbance. The process can be modelled as the
result of 'white noise' through an integrator, which is zeroed at t =
0. The output is continuous and known as 'Brownian motion'.

To approximate this by a discrete–time model let

$v(k+1) = v(k) + \lambda\ \omega(k)$, $\omega(k)$ Gaussian$(0,1)$, $k = 1,2,...$

Obviously, this satisfies the requirement of having independent
increments. The variance of the increment after v steps is

$E[v(k+v) - v(k)]^2 = \lambda^2\ E[\omega(k+1) + \cdots + \omega(k+v)]^2 = v\ \lambda^2$

Now, in order to define what one should mean by an approximation
of $v^c(t)$ it is reasonable to require the following from the discrete–
time approximation:
1) The joint distributions of the values of $v(k)$ should approximate

those of $v^c(t)$ at $t = kh$ with arbitrary accuracy, when $h \to 0$.
2) Interpolations of $v(k)$ to points t, $kh < t < (k+1)h$, should approximate $v^c(t)$ with arbitrary accuracy, when $h \to 0$.
3) The parameters in the discrete–time model should be invariant when $h \to 0$.

The first requirement is obviously satisfied if the increment variances agree at $t = kh$, i.e. $k \lambda^2 = \mu^2 \tau = \mu^2 kh$. Hence, if the discrete–time model is defined as

$$v(k+1) = v(k) + \mu \sqrt{h} \; \omega(k)$$

the third requirement is also satisfied, since μ has the same value for all h. However, the 'time unit' h appears explicitly as an additional parameter. The second requirement is satisfied, in particular by a linear interpolation, since $v^c(t)$ is continuous.

This means that a suitable approximation of 'continuous white noise' is a sequence of step functions with independent amplitudes proportional to $1/\sqrt{h}$.

It is possible to describe all objects driven by random signals in this way. However, this costs at least one additional state variable for each signal. If the driving signal has an approximately flat spectrum over the frequency range where the objects can respond, then one can avoid the extra state variable by using the concept of idealized 'white noise'. This means however that the ordinary interpretation of the resulting differential equation no longer holds. Mainly, the Itoh differentiation rule applies to (nonlinear) systems driven by 'white noise', and this has implications on the interpretation and solution of stochastic differential equations (Åström, 1970; Marcus, 1987). For instance, they cannot be solved directly by algorithms designed for ordinary deterministic differential equations. In essence, a 'correction term' that Itoh's differentiation rule introduces has first to be added. See Graebe (1990) for an extensive discussion of the problem of modelling continuous–time objects with stochastic input.

5.6   Implicit and explicit models

The terms 'implicit' and 'explicit' models have to do with the *form* of the apriori information available to the model maker.

An *explicit* model is an algorithmic model $d \leftarrow A(c,\omega)$, *i.e.* one that can be used directly for computing d, when arguments ω and c are given. The system

$$u(t) \leftarrow C[c_{k-1},\omega_t,k,t|t_{k-1} < t \le t_k]$$
$$y(t),z(t) \leftarrow F[u_t,\omega_t,t|t \in T]$$
$$d(k) \leftarrow D[y(t_k),k|1 \le k \le N]$$

is a *dynamic explicit* model, which can be used sequentially to compute the data sequence $d(1),\dots,d(N)$.

Often the apriori structure information is available *implicitly*, *i.e.* in the form of *equations*, for instance differential or integral equations, between variables z related to c,ω, and d. Such equations take the form $A(\mathcal{O}c,\mathcal{O}d,\mathcal{O}\omega,k) = 0$, or in the case of internal dynamic models $A(\mathcal{O}y,\mathcal{O}z,\mathcal{O}u,\mathcal{O}\omega,t) = 0$, where $\mathcal{O}$ is an operator, for instance a differential or a difference operator (notice that one has to use the *equality* sign $=$ instead of the *assignment* symbol $\leftarrow$). In order to get equations into computable form, *i.e.* assignment statements, one has to supply a *solution algorithm*. Thus, the combination of an implicit model and a solution method makes an explicit model.

There are basically two problems involved here:

● One has to solve the equations for the output d or (z,y). In case of a differential operator $\mathcal{O}$ acting on the output, one has to solve for the highest derivative. This is a problem of solving a system of nonlinear equations, which can be quite nasty if the variables are many. The fundamental problem is causality, *i.e.* to determine which variables cause what others to vary. In an explicit model the causality is established by the algorithm; it is given by the order in which the statements appear in the algorithm. What is on the right side of the assignment symbol $\leftarrow$ *causes* what is on the left side. In an implicit model one has to *decide* which variables should be placed on the right side of $\leftarrow$ and which should be on the left side, *i.e.* be picked out as 'dependent'. Then one has to solve for the dependent variables. The problem of *algebraic loops* appears when the causality is not clear, but one has relations of the type $z_i = f_i(z_j,\dots)$, $z_j = f_j(z_i,\dots)$, which would mean that z_j causes z_i, and z_i

causes z_j. One may interpret this as if the object has reached an *equilibrium*, *i.e.* has established a *balance* between the variables. The cause–and–effect relationships determine the dynamic process of reaching the balance, but is lost when only the equilibrium is described. Also, it may not be of value for the purpose to recover the actual causal relation. Many common models, like much of the theory of electrical networks, are based on balances and cause algebraic loops. So one may in fact have to deal with them. Often there is also a problem of *uniqueness*: If one collects a large number of equations from what one happens to know about the object, one may well end up with too many or too few of them. If they are too few, the solution will not be unique. If they are too many, some may contradict each other, and there is no solution.

● In case some dependent variables have an operator **O** attached to them, one has to solve a system of differential, integral, or other dynamic equations, possibly with delayed variables. In principle that is a smaller problem than selecting and solving for the dependent variables. One can always transform operator equations to difference equations, if one can afford a short enough integration interval. But often one cannot. In case of 'stiff' differential equations, the integration interval would be prohibitively short, and one has to use special numerical methods for solving such equations.

Generally, transforming implicit models into explicit models are problems that concern developers of simulation software. For an investigation of the problems see Elmqvist (1980).

> **Remark:** A convenient structure is the NARMAX model (Nonlinear Auto–Regressive Moving–Average with eXogeneous input):
> $$d(k) = A[d(k-1),\ldots,d(k-n),c(k),\ldots,c(k-n),e(k-1),\ldots,e(k-n)] + e(k)$$
> $$e(k) = \Lambda\, \omega(k)$$
> suggested by Leontaritis and Billings (1985).

5.7 Finite–memory models

The restriction of *finite memory* implies that the model structures can be evaluated efficiently by computers. An algorithmic dynamic model $d(k) \leftarrow A(c_k,\omega_k)$ is a function of the whole input history (c_k,ω_k) and the latter grows unboundedly as time progresses. Even if the computer would have a sufficiently large primary memory to store all the histories of all the variables involved, the time of evaluating the algorithms would grow unboundedly with k. The algorithms would also grow in complexity, since the number of variables grows, and the computations would get increasingly sluggish. This is the price one would have to pay for allowing the freedom of an increasingly complex model, and it is normally neither acceptable nor necessary.

If the admissible algorithms have to be restricted so that they can be evaluated using a number of operations that for each k does not grow over a limit, even when k does, it is important that the restriction be done in such a way that most of the *information* the input carries (and which grows with k) is not lost in the evaluation. That is the case if the algorithm has a finite–dimensional *state* $x(k)$, *i.e.* it can be expressed as:

External state–vector model:

$$x(k+1), d(k) \leftarrow A^d[c(k), \omega(k), x(k), k] \qquad (5.27)$$

The restriction is one of 'finite memory'. However, this does not mean that some information in past data will necessarily be discarded, only that the number of parameters x that *carry* this information must be limited. When data increases, the information these parameters hold may be better (without limit), but must not require more places to store it. Hence, the condition is a constraint on the complexity.

> **Remark:** An algorithm having a state is a *recursive relation* and is well suited to the way computers operate. In fact, all recursive algorithms involved in identification, not only those that describe the model structures, must have three kinds of variables:
> *Input*: Variables that must be set prior to evaluation.
> *Output:* Variables that are set by the evaluation.
> *State:* Variables that must both be set before, and are set again by the evaluation, and hence must be saved between. The whole identification program is defined by such algorithms, which use each others output.

Also phase–variable models have finite memory

An alternative finite–memory model is the 'phase–variable' representation, which is obtained by limiting the dependence of the past histories of input *and* output variables to data no older than a fixed age. This actually discards past data, but notice that the recent past output still contain older *information*. Hence, this restriction does not necessarily lose more information than a general state–vector model (and none of them loses much). In fact, the two forms are equivalent for linear systems (Bose, 1987).

In all systems a phase–variable model is a special case of a state–variable model: $d(k) \leftarrow A^d[c_k, d_{k-1}, \omega_k, k]$, where A^d does not use the parts of c_k, d_{k-1}, ω_k that are older than $k-n$, can conveniently be written

$$d(k) \leftarrow A^d[c(k), \omega(k), x(k), k]$$
$$x(k+1) \leftarrow \mathfrak{J}[c(k), d(k), \omega(k), x(k), k] \tag{5.28}$$

with the state

$$x(k) \triangleq \{c(k-1), \ldots, c(k-n), d(k-1), \ldots, d(k-n), \omega(k-1), \ldots, \omega(k-n)\},$$

and the state–transfer algorithm $\mathfrak{J}$ is the shift operation.

Phase–variables are typically used in 'black–box' models, where little is known apriori about the internal structure. The latter may be linear or nonlinear (Ljung, 1987a).

5.7.1 Internal state–vector models

The *internal* model defining a Plant will have the following forms for the state–vector and the phase–variable cases:

Finite–memory Plant models:

State–vector: $x(t+1), y(t), z(t) \leftarrow F[u(t), \omega(t), x(t), t]$ (5.29)

Phase–variable: $y(t), z(t) \leftarrow F[y_{t-1}, z_{t-1}, u_t, \omega_t, t]$ (5.30)

The state–vector forms for the Actuator and Sampler models are more complicated, because of the two counters involved, *viz.* physical time t and the real–time clock counter k. Using the notations in (5.21–23) changes the hybrid model (5.19) into

$$u(k) \leftarrow C[c_{k-1}, \omega_k, k] \tag{5.31}$$

which has the finite–memory forms

Finite-memory Actuator models:

State-vector: $x^c(k+1), u(k) \leftarrow C[c(k), \omega^c(k), x^c(k), k]$ (5.32)

Phase-variable: $u(k) \leftarrow C[c_{k-1}, u_{k-1}, \omega_k^c, k]$ (5.33)

Obviously, the sampler model (5.19) is not dynamic and does not need a state vector.

--- **Example 5.10** ---

Phase-variable model for zero-order hold:

$$\{u(t_{k-1}+1), u(t_{k-1}+2), \ldots, u(t_k)\}$$
$$\leftarrow C[c_{k-1}, u_{k-1}, \omega_k^c, k] \triangleq \{c(k-1), c(k-1), \ldots, c(k-1)\}$$

Evaluating the likelihood requires an information state

Also a probabilistic model must normally have a finite state to be possible to evaluate. Thus, it is required that

$$p[d(k)|c_{k-1}, k] = p[d(k)|x(k), k]$$
$$x(k+1) \leftarrow \mathfrak{I}[c(k), x(k), k] \tag{5.34}$$

which means that the distribution of $d(k)$ conditional on c_{k-1} is the same as that conditional on $x(k)$. If so, the state vector $x(k)$ carries as much information about $d(k)$ as does the whole past history c_{k-1}. For that reason a state vector for probabilistic models is called 'information state' (Ljung, 1987a). Another name is 'sufficient statistic', in particular in the field of Mathematical Statistics (Kendall and Stuart, 1967). That name derives from the fact that $x(k)$ is a 'statistic', *i.e.* a function of data c_{k-1}, and sufficient to compute the distribution of $d(k)$. The condition of finite memory is also called the 'Markov property', and $\{x(k)\}$ is a 'Markov process' (Marcus, 1987).

Notice that the evaluation of the likelihood of a model M always requires an information state: From (5.7)

$$L(M|\mathbf{c}_N,\mathbf{d}_N) = \prod_{k=1}^{N} p[\mathbf{c}(k)|\mathbf{c}_{k-1},\mathbf{d}_k,k,M]\ p[\mathbf{d}(k)|\mathbf{c}_{k-1},\mathbf{d}_{k-1},k,M]$$

$$= \prod_{k=1}^{N} p[\mathbf{c}(k)|\mathbf{d}(k),\mathbf{x}(k),k,M]\ p[\mathbf{d}(k)|\mathbf{x}(k),k,M]$$

$$x(k+1) \leftarrow \mathfrak{I}[\mathbf{c}(k),\mathbf{d}(k),\mathbf{x}(k),k] \tag{5.35}$$

and the factors in the product are now functions of a fixed number of variables.

─────────────────── **Example 5.11** ───────────────────

Illustrating the concept of information state.

Assume one wants to estimate the value of a constant θ, and for that purpose takes repeated measurements with the results $\mathbf{d}(k) = \Theta + \lambda\,\omega(k)$, where $\omega(k)$ are gaussian errors with zero means and unit variances. One wants to stop the measuring when the estimation error is $< \sigma$.

The log likelihood of (θ,λ) is after N measurements

$$\log L(\theta,\lambda|\mathbf{d}_N) \triangleq \log p[\mathbf{d}_N|\theta,\lambda] = \sum_{k=1}^{N} \log p[\mathbf{d}(k)|\mathbf{d}_{k-1},\theta,\lambda]$$

$$= -\frac{1}{2}\sum_{k=1}^{N} [\mathbf{d}(k) - \theta]^2\,\lambda^{-2} - \frac{1}{2}N\log(2\pi\,\lambda^2)$$

Since the estimation accuracy of the ML–estimate of θ is

$$D^2[\hat{\theta}(N)] \rightarrow [-\nabla_\theta^2 \log L(\hat{\theta},\hat{\lambda}|\mathbf{d}_N)]^{-1}$$

where ∇_θ is the gradient operator, one could choose to evaluate the likelihood function, and then maximize it with respect to (θ,λ) for increasing N, until $D^2[\hat{\theta}(N)] < \sigma^2$. This would yield

$$\log L(\theta,\lambda|\mathbf{d}_N)$$

$$= -\frac{1}{2}\{N\,\theta^2 - 2\,\theta\sum_{k=1}^{N}\mathbf{d}(k) + \sum_{k=1}^{N}\mathbf{d}(k)^2\}\,\lambda^{-2} - \frac{1}{2}N\log(2\pi\,\lambda^2)$$

$$= -\frac{1}{2}N\{[\theta - \frac{1}{N}\sum_{k=1}^{N}\mathbf{d}(k)]^2\,\lambda^{-2} + \frac{1}{N}\sum_{k=1}^{N}[\mathbf{d}(k) - \frac{1}{N}\sum_{k=1}^{N}\mathbf{d}(k)]^2\,\lambda^{-2}\}$$

$$- \frac{1}{2}N\log(2\pi\,\lambda^2)$$

It is obviously maximized by

$$\hat{\theta}(N) = \frac{1}{N} \sum_{k=1}^{N} \mathbf{d}(k)$$

$$\hat{\lambda}^2(N) = \frac{1}{N} \sum_{k=1}^{N} [\mathbf{d}(k) - \hat{\theta}(N)]^2 = \frac{1}{N} \sum_{k=1}^{N} \mathbf{d}(k)^2 - \hat{\theta}(N)^2$$

and the likelihood function becomes

$$\log L(\theta, \lambda | \mathbf{d}_N)$$
$$= - \frac{1}{2} N [\theta - \hat{\theta}(N)]^2 \lambda^{-2} - \frac{1}{2} N \hat{\lambda}^2(N) \lambda^{-2} - \frac{1}{2} N \log(2\pi \lambda^2)$$

The estimation error variance is $D^2[\hat{\theta}(N)] = \hat{\lambda}^2(N)/N$. Hence, a solution to the problem would be to evaluate $\hat{\lambda}^2(N)$ and $\hat{\theta}(N)$ for increasing values of N and stop when $\hat{\lambda}^2(N)/N < \sigma^2$.

However, this would mean having to save all $\{\mathbf{d}(k)|k=1,\ldots,N\}$ for every N, and also evaluating increasingly longer sums. It is better to introduce state variables. One has immediately the recursive equations for evaluating the sums

$$k \hat{\theta}(k) = (k-1) \hat{\theta}(k-1) + \mathbf{d}(k)$$
$$k [\hat{\lambda}^2(k) + \hat{\theta}(k)^2] = (k-1) [\hat{\lambda}^2(k-1) + \hat{\theta}(k-1)^2] + \mathbf{d}(k)^2$$

which yields

$$\hat{\theta}(k) = \hat{\theta}(k-1) + [\mathbf{d}(k) - \hat{\theta}(k-1)]/k$$
$$\hat{\lambda}^2(k) = \hat{\lambda}^2(k-1) + \{[\mathbf{d}(k) - \hat{\theta}(k-1)]^2 (k-1)/k - \hat{\lambda}^2(k-1)\}/k$$

This defines recursive relations that will need the saving only of the latest measurement $\mathbf{d}(k)$ and the *information state* $x(k) = \{\hat{\theta}(k), \hat{\lambda}^2(k)\}$. The likelihood function can be expressed as a function of that state:

$$\log L(\theta, \lambda | \mathbf{d}_N) = \log L(\theta, \lambda | \hat{\theta}(N), \hat{\lambda}^2(N))$$
$$= - \frac{1}{2} N [\theta - \hat{\theta}(N)]^2 \lambda^{-2} - \frac{1}{2} N \hat{\lambda}^2(N) \lambda^{-2} - \frac{1}{2} N \log(2\pi \lambda^2)$$

5.8   Classification of models by purpose

The information contents of a model is at best equal to the union of those of the apriori information and of the data. Hence, the apriori information about the object and the quality of the experiment determine the place in a 'hierarchy' of models (ordered after information contents) that one can obtain under the circumstances. Since different purposes have different claims on the information contents, the purpose determines the lowest level of model that is required. It is possible to link a number of general purposes to model type, defined by what information it requires (in the form of assumptions and specifications) and what information it will provide:

● The *Data Description* takes the lowest place in the hierarchy of models with increasing information contents. It yields a model that is able to reproduce the experiment data $(\mathbf{c},\mathbf{d})$. It may be used to *compress data* for transmission.

> **Remark:** An external algorithmic model A is a Data Description, if one includes the 'innovations' $\{\mathbf{w}(k)\}$ computed from the equation
> $A[\mathbf{c}_{k-1},\mathbf{d}_{k-1},\mathbf{w}(k),k] = (\mathbf{c}(k),\mathbf{d}(k))$
> (see section 5.5.2). It will then be possible to reconstruct the original data sequence. Thus, encoding that sequence into model and innovations, transmitting both through a channel, and finally decoding innovations using the model, will make an efficient way of transmitting data.

● The *Source Description* comes next. It yields a probability distribution of data $(\mathbf{c},\mathbf{d})$ produced by the particular experiment system $(\mathbf{X},\mathbf{S})$. The model will hold also if the experiment is repeated. Hence, it may serve the purpose of *forecasting* the outcome of experiments using the same system.

> **Remark:** An external algorithmic model
> $c(k),d(k) \leftarrow A[c_{k-1},d_{k-1},\omega(k),k]$
> 'innovations' excluded, is a Source Description. It yields the probability density $p[c(k),d(k)|c_{k-1},d_{k-1},k,\mathbf{X},\mathbf{S}]$ of the data pair $(c(k),d(k))$, when previous data (c_{k-1},d_{k-1}) have been observed.
> It may be used, in particular, for predicting random processes and signals, provided one has reasons to believe (from apriori knowledge) that the experiment is *Reproducible*. It is conceivable

that Source Descriptions would also be useful for design of channels for transmission of data. However, there is no guarantee that the model will hold when subject to other stimuli. Source Descriptions are normally quite useless for control purposes, and also hazardous, since they are the source of most of the 'pitfalls' in identification, which were discussed in section 4.6 on "The origin of 'pitfalls'".

● The *External Object Description* comes next. It describes the object's responses to stimuli also during an application phase. Hence, it serves the purpose of allowing design of *feedback control* of d.

   Remark: A probabilistic external model describes the object response density $p[d(k)|c_{k-1},d_{k-1},k,\mathbf{S}]$ when subject to any stimulus c_{k-1} coming from any experimenter X.

● The *Internal Object Description* comes next. It describes the responses of object output z that are not measured. It serves the purpose of allowing design of *feedforward control* of those unmeasured variables z as well as feedback control from d.

   Remark: 'Feed–forward control' is also called 'inferential control', since in the absence of direct measurement of the controlled variable one has to 'infer' its value from other measurements.

● The *Physical Object Description* requires most information apriori, and may satisfy other purposes than control. A Physical Object Description contains information of relations between parameters a that have physical meanings and the physical variables defined by the object **S**. Thus, it may serve the purpose of *system design*, *i.e.* of modifying parameters to obtain a desired response.

   Remark: The first model type in the hierarchy requires no apriori structure information. By satisfying various conditions (see Chapter 6 on "Large–sample theory") one gets statistics for estimating the density $p[c(k),d(k)|c_{k-1},d_{k-1},k,\mathbf{X},\mathbf{S}]$ of the output of the experiment system from data $(\mathbf{c}_N,\mathbf{d}_N)$. By also satisfying other conditions (see Chapter 3 on "The experiment") one gets statistics for estimating the object density $p[d(k)|c_{k-1},d_{k-1},k,\mathbf{S}]$ defining an External Object Description. If one needs one of the last two model types, then the additional information must come from structural restrictions $\mathcal{M}$. They serve the purpose of reducing the set of models M satisfying the equation

$$p[c(k),d(k)|c_{k-1},d_{k-1},k,M] = p[c(k),d(k)|c_{k-1},d_{k-1},k,\mathbf{X},\mathbf{S}]$$

so that a model of the particular type is determined.

Remark: The Physical Object Description must necessarily be parametric. In practice, the Internal Object Description must also be parametric, and often also the External Object Description. A difference between the 'physical' and the 'internal' models is then that the latter does not need to have all parameters a determined. The values of a are not interesting for the purpose.

Remark: Mathematically more precise would be to define the various model types as different mappings:

Data Description: $\omega, a \rightarrow c, d$

Source Description: $\Omega, a \rightarrow C, D$

External Object Description: $C, \Omega, a \rightarrow D$

Internal Object Description: $C, \Omega, a \rightarrow Z, D$

Physical Object Description: $C, \Omega, A \rightarrow Z, D$

where $c \in C$ is stimulus, $d \in D$ is observed response, $z \in Z$ is unobserved response, $a \in A$ is constant object parameters, and $\omega \in \Omega$ is the primitive random vector making the variables (c,d,z) stochastic. For instance, the difference between a Data Description and a Source Description is that the former is defined only for one point ω, while the latter is defined for all points in Ω.

Remark: The mathematical definition of a Physical Object Description actually means that if one would redesign the object using different a–parameters, the model would still hold. Usually one cannot redesign the object for different experiments, and this means that one cannot in fact get data from more than one point a. Hence, the information for expanding the validity of the model to other points in A must come from other sources than experiment data. This is known as the 'scaling problem' and will not be considered in this book. The problem of scaling appears, for instance, when one has made a model based on experiments on a laboratory plant, and from that wants to derive a model for a full scale production unit. That might work, but one should be aware that phenomena that were negligible in the laboratory plant, and that therefore have not been modelled, might become important when the scale changes.

Remark: Gevers and Bastin (1982) have written an enlightening short tutorial on modelling problems. The authors stress the importance of the purpose and the prior knowledge, and discuss the consequences on the modelling of the three purposes of simulation, prediction, and control. In particular, modelling for prediction requires much less apriori information than for simulation, and applying the former type of models for the latter purpose may be disastrous.

—————————————— **Example 5.12**——————————————

This example (Gevers and Bastin, 1982) illustrates the possible consequences of using a models for a different purpose than that for which it was designed.

Let the true model be

$$d(k) + \mathbf{a}\, d(k-1) = \mathbf{b}\, c(k-1) + \mathbf{c}\, \omega(k)$$

where $\{c(k)\}$ and $\{\omega(k)\}$ are independent random variables of unit variances (see the figure).

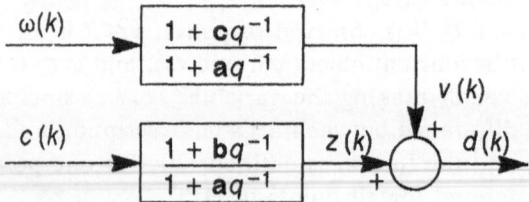

Consider first the 'prediction model':

$\mathcal{M}_1: \hat{d}(k+1|a,b) = -a\, d(k) + b\, c(k)$

Then the best predictor will yield $a = \mathbf{a} - \mathbf{c}/r$, $b = \mathbf{b}$, where $r = E\{d(k)^2\}$.

Consider now a 'simulation model':

$\mathcal{M}_2: \hat{d}(k+1|a,b) = -a\, \hat{d}(k|a,b)\, d(k) + b\, c(k)$
$= b\, q^{-1}\, (1 + a\, q^{-1})^{-1}\, c(k+1)$.

Within model set $\mathcal{M}_2$, the best predictor will yield $a = \mathbf{a}$, $b = \mathbf{b}$.

Now suppose that the steady-state gain of the system must be computed. Then the simulation model will give the correct answer, $b/(1+a)$, while the prediction model will yield a completely erroneous answer $b/(1 + a - c/r)$.

——

5.9   'Black–box' and 'grey–box' models

'Black–box', 'white–box', and 'grey–box' models were introduced in intuitive terms in section 1.2 on "The software". In essence, a 'black box' uses no apriori structure information, a 'grey box' some, and a 'white box' only such information. All can be written as probabilistic or algorithmic models, or in any other general form, but the difference is how much *known detail* one writes into these forms, and how much is left to determine by identification. Less 'white' boxes have more to determine from data. The trade–off depends on how much one can rely on each one of the two information sources *apriori knowledge* and *data*. It also depends on how much information one needs.

> **Remark:** For the purpose of feedback control one often needs very little information, since feedback control is rather insensitive to model errors. That is the main advantage of using feedback. That is also the reason why 'black–box' models work well in adaptive control, a situation where one wants a 'universal' controller, without built–in specifications of what object it is going to control.

Generally, the purposes of 'feedback control' and 'feedforward control' require External and Internal models respectively. They have also close relations to the 'black–box' and 'grey–box' models, but this may not be obvious. The following will therefore motivate the distinction made between the two model types.

5.9.1   The requirements of the control purpose

The following analysis will show that *for the purposes of forecasting and control*, one will (generally) need more information than derived from data, *i.e.* one will need a 'grey–box' model.

Assume the plant one wants to forecast or control responds as a dynamic system

$$z(t), y(t) \leftarrow F[u_t, \omega^s_t, t] \tag{5.36}$$

where u is the control input

z is the output to be controlled/forecast

y is the output from sensors (measuring transducers)

ω^s is the primitive random variables (see section 5.3) and adding subscripts t denotes the whole time series up to and including time t. The formalism means that the plant F operates on the past histories of the control and random input and produces instantaneous values of the controlled and measured output.

Introduce a feedback controller from data to control sequence

$$c(k) \leftarrow X[d_k, \omega^c_k, k] \qquad (5.37)$$

where ω^c is another primitive random variable, independent of ω^s. Notice that X defines an experimenter, since it produces a control variable $c(k)$; that is why it is denoted by X.

> **Remark:** Most often the controller is not a stochastic mechanism and does not use any ω–component at all. The reason why there is still a random element in the controller is that this simplifies the analysis; the conditional density $p[c(k)|c_{k-1}, d_k, X]$ will not be singular.

The purpose of this analysis is to advocate that a 'black box' is not enough for all control purposes (Bohlin, 1986). Assume therefore the most favourable condition for making an efficient 'black box', *viz.* that y be measured at all times t, and that the actuator be a perfect zero–order hold, so that $k = t$, $d(k) = y(t)$, $u(t) = c(k)$, and $S = F$ (see *Fig. 28* in section 5.5.3). Make also the still more favourable (and for other purposes preposterous) assumption that $S \in \mathcal{M}$.

When the object S is closed by the controller X, the system produces the three sequences c_T, d_T, z_T in the time interval $t \in [1,T]$. They have the joint probability density

$$p[c_T, d_T, z_T | X, S] \qquad (5.38)$$

The controller–design problem is to choose X so that the density (5.38) will be to ones liking, for instance minimizing the expected value of the loss function $\mathcal{L}[c_T, d_T, z_T]$ (any reference trajectory is included in $\mathcal{L}$). For that purpose one has to know $p[c_T, d_T, z_T | X, S]$ for all admissible X.

Bayes' chain rule yields

$$p[c_T, d_T, z_T | X, S] = \prod_{t=1}^{T} p[c(t), d(t), z(t) | c_{t-1}, d_{t-1}, z_{t-1}, t, X, S]$$

$$= \prod_{t=1}^{T} p[c(t)|c_{t-1}, d_t, t, X] \, p[d(t), z(t) | c_{t-1}, d_{t-1}, z_{t-1}, t, S] \qquad (5.39)$$

The second equality follows from the definitions of X and S. In order to design control of S one has to have a sufficiently good model of

$$p[d(t),z(t)|c_{t-1},d_{t-1},z_{t-1},t,S] \qquad (5.40)$$

but not necessarily of the whole S.

The factorization (5.39) for control design cannot be used for the purpose of identification, since z is not observed. The data density can be used, and it factorizes as

$$p[c_T,d_T|X,S] = \prod_{t=1}^{T} p[c(t)|c_{t-1},d_t,t,X]\ p[d(t)|c_{t-1},d_{t-1},t,S] \qquad (5.41)$$

Now, assume a class of models $M \in \mathcal{M}$. Assume also, for the sake of argument, that the whole data distribution can be evaluated experimentally so that one has a satisfactory approximation of $p[c_T,d_T|X,S]$. As pointed out before, that distribution is the maximum one can ever get out of an experiment. There are conditions for getting even that (see chapter 6 on "Large–sample theory"), but it is clear that one can never get more. Hence, one can never do better than fitting densities in the class $\{p[c_T,d_T|M]\,|\,M \in \mathcal{M}\}$ to the data densities $p[c_T,d_T|X,S]$ for all points (c_T,d_T), *i.e.* solving for $\hat{M}$ in the equation.

$$p[c_T,d_T|\hat{M}] = p[c_T,d_T|X,S] \qquad (5.42)$$

This means (in the best case)

$$p[d(t)|c_{t-1},d_{t-1},t,\hat{M}^d] = p[d(t)|c_{t-1},d_{t-1},t,S] \quad (t=1,\ldots,T) \qquad (5.43)$$

where $\hat{M}$ is a model that agrees with the data distribution. This gives the relation between the object S and any model $\hat{M}^d$ with the same response pattern.

However, there may well be several, even much different such models, since equation (5.43) probably does not have a unique solution. Hence, one may not get enough of S to determine the desired distribution $\{p[d(t),z(t)|c_{t-1},d_{t-1},z_{t-1},t,S]\}$, not even if the object S would be in $\mathcal{M}$, and not even if the experimenter X would be the best one can conceive. Determining the distribution of $\{d(t),z(t)|c_{t-1},d_{t-1},z_{t-1},t,S\}$ may require more of S than determining the distribution of $[d(t)|c_{t-1},d_{t-1},t,S]$ by (5.43).

That should not be surprising. What one wants to control may in the worst case have little to do with what one can measure, and then one would obviously get problems. In fact, Bayes' rule gives immediately

$$p[c_T,d_T,z_T|\mathbf{X},\mathbf{S}] = p[z_T|c_T,d_T,\mathbf{X},\mathbf{S}]\ p[c_T,d_T|\mathbf{X},\mathbf{S}] \tag{5.44}$$

and only the last factor on the right side has been determined. Determining the first factor is still unresolved. That cannot be done without further *structural assumptions*. Three such assumptions will be analysed here.

5.9.2 The 'black–box' approach

Assume that *an observer is known for z*, *i.e.* an algorithm

$$z(t) \leftarrow O[c_{t-1},d_t,t] \tag{5.45}$$

> **Remark:** Normally, one has to know **S** to find an exact observer, but working approximate observers, *e.g.* 'Luenberger observers' (see for instance Anderson and Moore, 1971), may be feasible on much less apriori information than the whole **S**.

> **Remark:** A common special case is that when $s = d$, *i.e.* only measured output is to be controlled.

The assumption of an observer implies that the unknown density can be determined by

$$p[d(t),z(t)|c_{t-1},d_{t-1},z_{t-1},t,\mathbf{S}] = p[d(t),z(t)|c_{t-1},d_{t-1},t,O,\mathbf{S}]$$
$$= \delta\{z(t) - O[c_{t-1},d_t,t]\}\ p[d(t)|c_{t-1},d_{t-1},t,\mathbf{S}] \tag{5.46}$$

since $\widetilde{z_{t-1}}$ is determined by (c_{t-1},d_t,O), and the last factor of (5.46) is identifiable from data.

Since the first factor in (5.46) is given by the observer, the 'black–box' approach puts no restriction on the model structure $\mathcal{M}$, and *for the purpose of control* the form of $\mathcal{M}$ does not matter much, as long as it is sufficiently rich to describe the conditional distribution of d. Control may in this way be satisfactory, even if the model $\hat{M}$ arrived at by identification bears little resemblance with **S**. One actually gives up trying to learn something about the object **S**, but one does not give up trying to control it. Control leans heavily on feedback, as the main information on the actual state of the object. Having a closed loop is vital.

5.9.3 The approach of separating disturbances

This is the 'white–box' approach. One assumes that the random variables, if any, that affect the process output z_T and data d_T are *statistically independent*. With an algorithmic model

$$y(t),z(t) \leftarrow F[u_t,\omega_t,t|t \in T]$$

the assumption means that $\omega_t = (\omega^y{}_t,\omega^z{}_t)$, and

$$z(t) \leftarrow F^z[u_t,\omega^z{}_t,t|t \in T]$$
$$y(t) \leftarrow F^y[u_t,\omega^y{}_t,t|t \in T] \tag{5.47}$$

The assumption is generally motivated by a desire to make a model where the control input c is the dominating cause of variation in z and d (with $u = c$, $y = d$). One wants an *internal* description, *i.e.* to know about S.

Since the assumption implies that $z(t)$ does not depend on d_t and $d(t)$ does not depend on z_{t-1}, it follows that the wanted density factorizes as

$$p[d(t),z(t)|c_{t-1},d_{t-1},z_{t-1},t,\mathbf{S}]$$
$$= p[z(t)|c_{t-1},d_t,z_{t-1},t,\mathbf{S}] \; p[d(t)|c_{t-1},d_{t-1},z_{t-1},t,\mathbf{S}]$$
$$= p[z(t)|c_{t-1},z_{t-1},t,\mathbf{S}] \; p[d(t)|c_{t-1},d_{t-1},t,\mathbf{S}] \tag{5.48}$$

Since the first factor does not depend on the response data, it follows that it can never be determined from data, unless enough of S can be determined from data. Hence the equation

$$p[d(t)|c_{t-1},d_{t-1},t,M^d] = p[d(t)|c_{t-1},d_{t-1},t,\mathbf{S}] \tag{5.49}$$

that determines the model must have a solution $\hat{M}^d$ such that $p[z(t)|c_{t-1},z_{t-1},t,M^d] = p[z(t)|c_{t-1},z_{t-1},t,\mathbf{S}]$. And in order to achieve that, one must simply know sufficiently much in advance about the structure $\mathcal{M}$.

The 'white box' yields open–loop control

A consequence is that controllers designed on the basis of 'white–box' models will not use feedback! If the design criterion $\mathcal{L}[c_T,d_T,z_T]$ involves only (c_T,z_T) (which is reasonable), the density $p[z_T|c_T,X,\hat{M}^d] = p[z_T|c_T,X,\mathbf{S}]$ will be enough information for design of a controller X.

The fact that X will not have data d_T as an argument (*i.e.* feedback), may be explained as follows: If d is determined only by c and by disturbances that do not also affect z, then d does not carry more information about z than do c and $\hat{M}^d$. Consequently, there is no logical need for measuring d and feeding it back. Control will be 'open loop' and rely only on forecasts computed from the first factor of the last member of (5.48). The second factor is used only to determine the model.

> **Remark:** A great number of conventional model structures belong to this category, for instance do all deterministic dynamic models measured with independent measurement errors:
>
> $z(t) \leftarrow F^z[u_t, t]$
>
> $d(t) \leftarrow z(t) + \omega^y(t)$
>
> They are not suitable for design of control. This means that *fitting the output of a deterministic model to experiment data is a hazardous way to design models for the purpose of feedback control.* Fitting a *predictor* model of the form $z(t) \leftarrow F^z[d_{t-1}, z_{t-1}, u_t, t]$ is safer, since that model uses on-line data d_{t-1}.

In summary, the use of 'black boxes' is limited by the requirement of observability of z, which may not be satisfied for an actual object. Control decisions are based mainly on measurements in real time and feedback. The use of 'white boxes' is limited by the requirement that a model structure $\mathcal{M}$ be known to such detail that when its parameters have been determined, there should be no more information to gain by real time measurements. Often none of those requirements are satisfied in actual practice. Experience tells us, that elements of open and closed-loop control will often be needed together to make efficient control. This is motivation for widening the class of models to such that exploits both apriori structure information $\mathcal{M}$ for open-loop control and real-time measurements d for closed-loop control.

5.9.4 The 'grey-box' approach

Assume that so much apriori knowledge about the object S is available that this allows supplementing an observer of z with predictions of unobservable z based on such apriori knowledge. This means that the condition (5.43)

$$p[d(t)|c_{t-1}, d_{t-1}, t, \hat{M}^d] = p[d(t)|c_{t-1}, d_{t-1}, t, S] \tag{5.43}$$

should also yield the equality

$$p[z(t)|c_{t-1},d_t,z_{t-1},t,S] = p[z(t)|c_{t-1},d_t,z_{t-1},t,\hat{M}^d] \qquad (5.50)$$

Remark: Notice that this is a weaker assumption than that behind the 'white box', which is that the same fitting condition should imply that

$$p[z(t)|c_{t-1},d_t,z_{t-1},t,S] = p[z(t)|c_{t-1},z_{t-1},t,\hat{M}^d]$$

In comparison, the assumption of a 'black box' is that

$$p[z(t)|c_{t-1},d_t,z_{t-1},t,S] = p[z(t)|c_{t-1},d_t,t,O] = \delta\{z(t) - O[c_{t-1},d_t,t]\}$$

As in the 'white–box' case the model $\hat{M}^d$ is used to predict the controlled variable z, instead of the observer O; the difference is that since $\hat{M}^d$ is normally less detailed, the right member of (5.50) normally has a more complicated dependence on the other arguments instead, and this makes both the identification and the estimation of z more difficult. Notice that in (5.50) $z(t)$ depends on d_t, which means that control will use feedback from d_t as well as apriori knowledge from $\hat{M}^d$.

Remark: The limit of the 'grey–box' approach is reached when no apriori knowledge is available about unobservable output. But then there is obviously nothing else to do than either to get such information, or to add more real–time measurements.

──────────────── **Example 5.13** ────────────────

This example illustrates the fundamental difference between deterministic and stochastic control design.

Let the control object be

S: $z(k) = z(k-1) + c(k-1) + \omega_1(k)$
    $d(k) = z(k) + \omega_2(k)$

If the disturbance $\omega_1(k)$ is small, it is tempting to 'separate disturbances' and assume a deterministic model

M: $z(k) = z(k-1) + c(k-1)$
    $d(k) = z(k) + \omega_2(k)$

In order to control this to a reference $r(k)$ one should obviously choose $c(k) = r(k+1) - r(k)$, $c(0) = r(1) - d(0)$. The theoretical control error (assuming no ω_1) will be small and equal to $-\omega_2(0)$ for all k. However the actual control error, assuming the 'true' object S, will tend to infinity with k, irrespective of how small $\omega_1(k)$ are.

Open–loop control obviously would not work in this case, as every control engineer would realize quickly. The problem is that the 'reasonable' modelling yields an 'unreasonable' control. Recognizing the fact that the disturbances on the unobservable and the observable output z and d are in fact coupled, however small an ω_1, yields the closed–loop controller $c(k) = r(k+1) - d(k)$ with the error $\omega_1(k) - \omega_2(k-1)$.

A pocket reference for modelling

The various model forms outlined in this chapter have been introduced for different purposes and with different motivations. The classification was either done for convenience of analysis and computing, or based on general assumptions on the object, or with the purpose of the modelling in mind. Table 1 contains the most important of the motivations for each one of the model types. A model structure may have several of the properties, when several motivations are relevant.

Table 1: Model types and their motives

Model type	Motif
	Mathematical properties:
Probabilistic	Convenient for general analysis
Algorithmic	Useful for simulation
	Separates structure from statistics
Parametric	Convenient description of structure
External	Easy evaluation of likelihood
Internal	Describes unobservables
Finite memory	Computing does not grow unboundedly
Phase–variable	Simplifies interpretation
	(states are physical variables)
Explicit	Simplifies identification
	(allows recursive solution)
Implicit	Simplifies modelling
	(but requires a solution algorithm)
	Physical conditions:
Black box	There is an exact observer
Grey box	There is no exact observer
White box	Disturbances on observables and
	unobservables separate
	Purpose:
Data description	Allows reproduction of data
Source descriptions	Allows forecasting
External object description	Allows feedback control
Internal object description	Allows also feedforward control
Physical model	Allows modification of object

6   Large–sample theory

This chapter continues the topic of modelling by stating rules for proper choice of the model structure $\mathcal{M}$. The results are based on the *limiting* properties of data distributions from dynamic systems, which means that one does not assume that there is a validation rule for the model to stop the sequential procedure of identification. Instead one may envisage the procedure to go on indefinitely with the lengths of the data samples tending to infinity. The most important limiting property of an identification problem is *identifiability*. It means, loosely, the existence of suitable statistics, *i.e.* functions of data, that converge to given qualities of the object or experiment system. Thus, identifiability answers the question of whether it may help at all to increase the sample length. Only if it does, is it meaningful to ask how long a sample has to be to yield a sufficient *accuracy* for the purpose, and to hope for success in validation tests. The recommendations on 'proper' experiments and structures that will ascertain identifiability (and hence a purposive model) may not be necessary in all cases, but they provide an ideal to aim at. Anyhow, if identifiability does not hold in theory, something is likely to go wrong in practice (for instance in the form of run–time error messages from the computer).

The rules for proper design of $\mathcal{M}$ do of course depend on the purpose $\mathcal{G}$. Since the set of conceivable purposes is immense, it is necessary to limit the analysis to a number of typical and general purposes. In particular, identifiability conditions for three of the five types defined in section 5.8, will be studied, *viz.* the Source Description, the External Object Description, and the Physical Object Description. Each will put some general requirements on the model structure.

Generally, the relations between purpose, model type, and the information each will require are not obvious. Neither are the conditions one has to satisfy to achieve identifiability. Since there are several types of models, each for a different purpose and with different information contents, there are also several identifiability definitions. Before they can be formulated, there is a number of more basic conditions for identification that have to be sorted out.

6.1 Equivalent dynamic models

The first problem is to decide what one wants to estimate, in case one has no well–defined purpose $\mathcal{G}$, or has a purpose one cannot validate. Without a stopping rule, the identification procedure goes on indefinitely (in theory) and one aims att shrinking the set $\mathcal{B}$ of unfalsified models, preferably until $\mathcal{M} \cap \mathcal{B}$ contains only one model. But what should that model mean? It cannot be the 'true' model $(\mathbf{X},\mathbf{S})$, since that is not in any mathematically feasible set. So, what should substitute 'truth'? Another problem is that if one would carry the experimentation too far (the 'inner' loop), one would probably falsify all models in any $\mathcal{M}$, and would have no result at all.

'True' must be replaced by 'equivalent'

It is evident, that if there would be a 'source–equivalent' model, that is, one yielding the same distribution of data (c,d) as the true source, then one could never separate it from the true source only on the basis of data from a given experiment, however long. Neither would one be able to do so, if the same experiment would be repeated, whatever number of times. This makes 'source–equivalent' models the best substitute for the 'true source' one can get for a given experiment. For *dynamic* systems the condition would be:

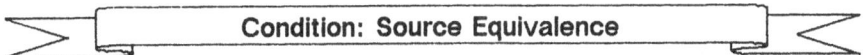

Condition: Source Equivalence

The source $(\mathbf{X},\mathbf{S})$ and the widest model structure $\bar{\mathcal{M}}$ must be such that $\mathcal{G} \cap \bar{\mathcal{M}} \neq \emptyset$, where

$$\mathcal{G} = \{X,S \mid p[c_K,d_K|X,S] = p[c_K,d_K|\mathbf{X},\mathbf{S}], \text{ all } (c_K,d_K) \in \mathcal{R}_K(\mathbf{X})\}$$

––––––– •••••• –––––––

Remark: This means that the set of models that will satisfy the definition depends on the degree of stimulation by the experimenter. If the experimenter would yield no stimulation at all, and there is no other stimulation from an environment, then all models in $\bar{\mathcal{M}}$ would be Source Equivalent. For this extreme case to yield a useful answer it is required that $\overline{\mathcal{M}}$ be a sufficiently restricted set for the purpose. The case means designing the model from apriori information only.

It follows that all Source–Equivalent models $\mathbf{M}$ will be in $\mathcal{B}(\gamma|\mathbf{c}_N,\mathbf{d}_N)$, whatever the values of γ and N, and, in particular, $\mathbf{M} \in \mathcal{B}(0|\mathbf{c}_\infty,\mathbf{d}_\infty)$. If $\mathcal{M} \cap \mathcal{B}(0|\mathbf{c}_\infty,\mathbf{d}_\infty)$ would shrink to a single point, that point would be a unique Source–Equivalent model, and an obvious candidate for identification.

In general, Source–Equivalent models will not be unique however. In particular, the experimenter may be stimulating enough to ascertain uniqueness of the probabilistic source model $p[c(k),d(k)|c_{k-1},d_{k-1},k,\mathbf{M}]$, but not to allow a separation of the equivalent experimenter $\mathbf{M}^c$ from the equivalent object $\mathbf{M}^d$, or even a separation between the corresponding probabilistic models $p[c(k)|c_{k-1},d_k,k,\mathbf{M}^c]$ and $p[d(k)|c_{k-1},d_{k-1},k,\mathbf{M}^d]$. In that case $\mathbf{M} = (\mathbf{M}^c,\mathbf{M}^d)$ is not unique. Depending on the purpose different cases of 'identifiability' are relevant. In particular, the identifiability of an External Object Description requires the existence of an Object–Equivalent model:

Condition: Object Equivalence

The object $\mathbf{S}$ and the corresponding widest model structure $\overline{\mathcal{M}}^d$ must be such that $\mathcal{S}^d \cap \overline{\mathcal{M}}^d \neq \emptyset$, where

$$\mathcal{S}^d = \{S|p[d(k)|c_{k-1},d_{k-1},k,S] = p[d(k)|c_{k-1},d_{k-1},k,S],$$

$$\text{all } k, \text{ all } (c_k,d_k) \in \mathcal{R}_k(\mathcal{X})\}$$

••••••

Object Equivalence is a stronger condition than Source Equivalence. It can however be ascertained from other conditions:

Lemma of Object Equivalence: If the system $(\mathbf{X},\mathbf{S})$ has a *Source Equivalent* model $(\mathbf{M}^c,\mathbf{M}^d)$, and the experimenter $\mathbf{X}$ is *Isolated* and *Sufficiently Stimulating*, then $\mathbf{M}^d$ is *Object–Equivalent*.

Proof:
From the condition of an Isolated Experimenter:

$p[c_{k-1},d_k|\mathbf{X},\mathbf{S}]$
$= p[d(k)|c_{k-1},d_{k-1},k,\mathbf{X},\mathbf{S}]\, p[c_{k-1},d_{k-1}|\mathbf{X},\mathbf{S}],\ (c_k,d_k) \in \mathfrak{R}_k(\mathbf{X})$
and
$p[c_{k-1},d_k|\mathbf{M}]$
$= p[d(k)|c_{k-1},d_{k-1},k,\mathbf{M}^d]\, p[c_{k-1},d_{k-1}|\mathbf{M}],\ \mathbf{M} \in \bar{\mathcal{M}},\ (c_k,d_k) \in$
$\mathfrak{R}_k(\mathbf{X})$

From the condition of Source Equivalence:
$p[c_{k-1},d_k|\mathbf{X},\mathbf{S}] = p[c_{k-1},d_k|\mathbf{M}],\ (c_k,d_k) \in \mathfrak{R}_k(\mathbf{X})$
From the condition of Sufficient Stimulation:
and $p[c_k,d_k|\mathbf{X},\mathbf{S}] > 0,\ (c_k,d_k) \in \mathfrak{R}_k(\mathbf{X})$.

Hence, all factors in the product are positive, and
$p[d(k)|c_{k-1},d_{k-1},k,\mathbf{X},\mathbf{S}]$
$= p[c_{k-1},d_k|\mathbf{X},\mathbf{S}]/p[c_{k-1},d_{k-1}|\mathbf{X},\mathbf{S}]$
$= p[c_{k-1},d_k|\mathbf{M}]/p[c_{k-1},d_{k-1}|\mathbf{M}] = p[d(k)|c_{k-1},d_{k-1},k,\mathbf{M}^d]$,
all k, all $(c_k,d_k) \in \mathfrak{R}_k(\mathfrak{X}) \subseteq \mathfrak{R}_k(\mathbf{X})$.

This is the definition of Object Equivalence. ∎

The Source–Equivalent and Object–Equivalent models solve the conceptual problem of defining a 'purpose' mathematically, when the actual purpose does indeed depend on a 'true source' that is not mathematical in nature. It does this by suggesting that one would not be interested in other purposes than those defined by $\mathcal{G}(\mathbf{M})$, where $\mathbf{M}$ is 'equivalent' in some sense, suitable to the purpose (see section 4.5 on "Validatability conditions"). In case one has no mathematically well-defined purpose at all, one has still a way to define the set of 'good' models as $\mathcal{G}(\mathbf{X},\mathbf{S}) = \mathbf{M}$ or $\mathbf{M}^d$. This means taking the 'scientist's approach'. Investigating 'identifiability' means investigating under what conditions $\mathcal{M} \cap \mathfrak{B}(\mathbf{c}_N,\mathbf{d}_N) \to \mathbf{M}$ as $N \to \infty$.

> **Remark:** In the strict sense, there is probably no Source–Equivalent model. However, there may be *approximately* equivalent models satisfying $p[c,d|\mathbf{M}] \approx p[c,d|\mathbf{X},\mathbf{S}]$. The point of the definition is that it provides a natural way of defining approximations. Notice that approximations in probability densities must be measured in terms of the closeness of the corresponding *cumulative* distributions, but there is no conceptual difficulty in doing that. Hence, in practice $\mathbf{M}$ will have to be replaced by all models that satisfy the approximate equivalence within a prescribed error. This means that the 'accuracy' of the model will in fact be defined by the given approximation level of the response distribution, which usually does not suit the primary purpose of the modelling. And that may result in a more accurate model than is actually needed

for that purpose. One would preferrably want to analyse the conditions under which a sequence of models for increasing sample lengths would tend to any model inside an arbitrary set $\mathcal{G}(X,S)$ of purposive models. However, that seems a prohibitively difficult problem, so the analysis will aim at reaching **M** asymptotically.

Remark: One may well choose to disregard the problems connected with 'equivalent' models entirely — most text books evade it by assuming that (X,S) is a feasible model, and hence that one should aim at estimating **S**. In order to understand identifiability theory one may safely do this, nothing essential changes, if one would replace **M** with (X,S).

A principle of Parsimony

Generally, there are many Source–Equivalents **M**, in particular if $\overline{\mathcal{M}}$ is a wide set. In order to define the 'best' of those models as a target for identification, one may apply the principle of *Parsimony*. It says that if several models accord with data, one should prefer the simplest one. One may apply this principle here, if one has defined what is 'a simpler model'. But the basic sequential validation and falsification procedure outlined in section 4.4 already requires the necessary information in the form of the sequence of expanding structures $\mathcal{M}(1),\mathcal{M}(2),\dots,\mathcal{M}(\overline{n})$. If the structure $\mathcal{M}(n)$ is *parametric*, and a model in $\mathcal{M}(n)$ requires more parameters than one in $\mathcal{M}(n-1)$, the definition of Object–Equivalent models may be supplemented by

The principle of Parsimony: A 'simplest' Source–Equivalent model **M** is defined as one in $\mathcal{M}(\mathbf{n})$, where $\mathbf{n} = \min\{n \mid \mathcal{S} \cap \mathcal{M}(n) \neq \emptyset\}$

> **Remark:** It is an inconvenient consequence of the principle of Parsimony, that **M** depends on the somewhat arbitrary ordering of the hypotheses that define the sequence $\mathcal{M}(1),\mathcal{M}(2),\dots,\mathcal{M}(\overline{n})$. Furthermore, **M** may still not be unique. Some of the analysis in the sequence will assume that one has a way to define a unique **M**. This could be achieved, either by a suitable choice of the sequence of expanding structures, or by any other predefined rule for selection, including that of random selection.

6.2 Consistency

Bayes' rule (2.5) for computing aposteriori probability of the unknown source is the basis of validation, falsification, and estimation of models from data. It can also be used for determining identifiability by investigating the limiting properties when $N \rightarrow \infty$ (see section 6.2.1 on "The structured approach").

However, for long samples the principle of *fitting data distributions* is a theoretical alternative. This means that one would first estimate the data distribution $p[c_K, d_K | \mathbf{X}, \mathbf{S}]$ by some statistic $\hat{p}[c_K, d_K | \mathcal{M}, \mathbf{c}_N, \mathbf{d}_N]$ based on data $(\mathbf{c}_N, \mathbf{d}_N)$ and the model structure $\mathcal{M}$, and then determine a model $\hat{M}$ from the equation

$$p[c_K, d_K | \hat{M}] = \hat{p}[c_K, d_K | \mathcal{M}, \mathbf{c}_N, \mathbf{d}_N] \tag{6.1}$$

Notice that fitting data distributions does not answer any questions of the properties of $\hat{M}$ for finite N (as does Bayes' rule); the statistical properties depend on how $\hat{p}$ is computed. However, the fitting equation (6.1) for infinite N is useful for investigating *identifiability* properties.

> **Remark:** Evaluating data distributions is basically done by repeating the experiment so many times that for all possible coordinates (c, d) in the distribution one gets the desired densities $\hat{p}[c, d | \mathcal{M}, \{\mathbf{c}, \mathbf{d}\}]$ from counting the frequences of occurrences of data $(\mathbf{c}, \mathbf{d})$ at each point (c, d). The set of data $\{\mathbf{c}, \mathbf{d}\}$ one gets by repeating the same experiment on the same object is called 'ensemble', and the resulting frequency distribution is called 'histogram'. Thus, one reduces the effects of randomness by taking averages for each point (c, d). Even so, one has to be cautious by how to define experimental density, since if the space of $\{c, d\}$ is continuous, each point $(\mathbf{c}, \mathbf{d})$ in the experiment data will probably occur only once in the data record. However, if one has a dynamic system $(\mathbf{X}, \mathbf{S})$ to identify (or if one cannot repeat the experiment for some reason) this way of evaluation is not feasible. Data from a dynamic system does not yield a large number of points $\{\mathbf{c}(k), \mathbf{d}(k)\}$ with a common distribution. It yields *a single* point $(\mathbf{c}_N, \mathbf{d}_N)$ of very high dimension. One experiment yields one sample, and it may not be possible to repeat it, at least not so many times that one can take averages over the samples. Hence, evaluating

$\hat{p}[c_K, d_K | \mathcal{M}, \mathbf{c}_N, \mathbf{d}_N]$ is not possible, unless one restrict the model structure $\mathcal{M}$.

Remark: Notice that the dimensionality of the distribution, determined by K, is kept constant while the sample length N tends to infinity. In practice the maximum value of K for which the model is to hold is determined by the length of the 'application phase' (see section 1.3 on "The purpose"), and that is often assumed to be much longer than the 'identification phase' determined by N. In fact, one would want K to tend to infinity, while N is fixed. Since, this is not possible without constraining the set of possible distributions $\hat{p}[c_K, d_K | \mathcal{M}, \mathbf{c}_N, \mathbf{d}_N]$ — one cannot determine more free parameters than there is data, and usually much less — the necessary predictability has to be based on some *invariance* condition (see section 1.3 on "The purpose"). First, $K/N \to 0$, to ensure identifiability, and then $K/N \to \infty$, on account of the source invariance. Two invariance conditions will be formulated below, *viz.* the 'structured' and the 'unstructured approaches'

Remark: Notice that there are two types of approximation error involved here, one *random*, which depends on finite data length, and one *systematic*, which depends on the definition of Source–Equivalent models. Generally, errors are due to chance effects *and* to the practical impossibility of modelling reality exactly. Compare with the case of measuring a real variable, like length: The achievable accuracy is limited by 1) random measurement errors and 2) systematic errors due to the fact that one must use a finite number of bits to represent the measurement.

Hence, a basic requirement for all types of identifiability, except that of the Data Description, is that experiment data be consistent enough to yield an arbitrarily accurate Source Description, if the sample is long enough. A Data Description can be identified unconditionally (Rissanen, 1976, 1978), and the models higher up in the hierarchy require more conditions to be satisfied.

'Consistency' is generally a limiting property, and would mean in this case that $\hat{p}[c_K, d_K | \mathcal{M}, \mathbf{c}_\infty, \mathbf{d}_\infty] = p[c_K, d_K | \mathbf{X}, \mathbf{S}]$ for all (c_K, d_K). With a consistent data distribution the unknown dynamic system $(\mathbf{X}, \mathbf{S})$ would satisfy the following equation for all (c_K, d_K):

$$\hat{p}[c_K, d_K | \mathcal{M}, \mathbf{c}_\infty, \mathbf{d}_\infty] = \prod_{k=1}^{N} p[c(k), d(k) | c_{k-1}, d_{k-1}, k, \mathbf{X}, \mathbf{S}] \qquad (6.2)$$

However, if one would want to compute a probabilistic dynamic model from the marginal distribution, *i.e.* by the formula

$$p[c(k),d(k)|c_{k-1},d_{k-1},k,\mathbf{X},\mathbf{S}]$$

$$= \hat{p}[c_k,d_k|\mathcal{M},\mathbf{c}_\infty,\mathbf{d}_\infty]/\hat{p}[c_{k-1},d_{k-1}|\mathcal{M},\mathbf{c}_\infty,\mathbf{d}_\infty] \qquad (6.3)$$

this cannot be done even for infinite samples, unless the denominator is positive for all (c_{k-1},d_{k-1}). However again, there may be other ways to compute the model. The feasibility of doing this depends on how much constraint the designer has placed on the set $\mathcal{M}$ of tentative models. For instance, under feedback control with a deterministic controller the density $\hat{p}[c_K,d_K|\mathcal{M},\mathbf{c}_N,\mathbf{d}_N] = 0$ for all data pairs (c_K,d_K) that do not satisfy the control law. Still the case may or may not be identifiable.

This motivates the following condition on source and model structure:

Condition: Consistency

The source $(\mathbf{X},\mathbf{S})$ and the model structure $\mathcal{M}$ are such that one can compute an asymptotically unbiassed data distribution

$$\hat{p}[c(k),d(k)|c_{k-1},d_{k-1},k,\mathcal{M},\mathbf{c}_N,\mathbf{d}_N] \rightarrow p[c(k),d(k)|c_{k-1},d_{k-1},k,\mathbf{X},\mathbf{S}]$$
all k, all $(c_k,d_k) \in \mathfrak{R}_k(\mathbf{X})$, and almost all $(\mathbf{c}_N,\mathbf{d}_N)$.

––––––––– •••••• –––––––––

> **Remark:** The attribute 'almost' is a precisely defined term in mathematical statistics (the set of data for which the condition does not hold has probability zero). It means that one excludes cases that are mathematically possible but unlikely, such as constant data sequences, and indeed all those that exhibit an exactly predictable 'pattern'.

When does consistency hold?

The next two sections will investigate more closely the condition of Consistency, *viz.* that an arbitrarily accurate Source Description be computable from a sufficiently long data sample. It is a fundamental requirement for attaining identifiability of all model types except the Data Description. As argued above, a single data record is never enough for evaluating a data distribution without some restricting conditions on the source that produced the data. Since a completely unrestricted source requires much more specification than the data it produces, most of the possible models must be excluded apriori by structural restrictions. Hence, consistency depends on the experimenter X and on the model structures $\mathcal{M}$ (and indirectly also on the object S, since there must be an equivalent model in $\mathcal{M}$). A number of ways to represent $\mathcal{M}$ were outlined in chapter 5 on "Modelling", and the choice between alternative structures will depend on how much one knows apriori about the object, and also on what the purpose will be. However, not even this does necessarily ensure consistency without further ado. Two principles that introduce *necessary* restrictions are the 'structured' and the 'unstructured approaches' treated below.

6.2.1 The structured approach

If it is possible to limit the set of models $\mathcal{M}$ enough, then one can use the (unbiassed) Likelihood function introduced in section 2.2:

$$L(M|\mathbf{c}_N,\mathbf{d}_N) = p[\mathbf{c}_N,\mathbf{d}_N|M] = \prod_{k=1}^{N} p[\mathbf{c}(k),\mathbf{d}(k)|\mathbf{c}_{k-1},\mathbf{d}_{k-1},k,M] \quad (6.4)$$

The approach to computing the data distribution for large samples $\hat{p}[c_K,d_K|\mathcal{M},\mathbf{c}_N,\mathbf{d}_N]$ is based on the following

Lemma of the Maximum Likelihood: The 'true model' (X,S) maximizes the ensemble average of the logarithm of the unbiassed Likelihood function $E\{\log L(X,S|\mathbf{c}_N,\mathbf{d}_N)|X,S\}$.

> **Proof:** (Kendall and Stuart, 1967)
> $E\{\log L(X,S|\mathbf{c}_N,\mathbf{d}_N)|X,S\}$
> $= \int \log p[c_N,d_N|X,S] \; p[c_N,d_N|X,S] \; d(c_N,d_N)$
> $= \int \log p[c_N,d_N|X,S] \; p[c_N,d_N|X,S] \; d(c_N,d_N)$
> $\quad + \int \log\{p[c_N,d_N|X,S]/p[c_N,d_N|X,S]\} \; p[c_N,d_N|X,S] \; d(c_N,d_N)$
> $\leq \int \log p[c_N,d_N|X,S] \; p[c_N,d_N|X,S] \; d(c_N,d_N)$
> $\quad + \int \{p[c_N,d_N|X,S]/p[c_N,d_N|X,S] - 1\} \; p[c_N,d_N|X,S] \; d(c_N,d_N)$
> $= \int \log p[c_N,d_N|X,S] \; p[c_N,d_N|X,S] \; d(c_N,d_N)$
> $\quad\quad\quad + \int \{p[c_N,d_N|X,S] - p[c_N,d_N|X,S]\} \; d(c_N,d_N)$
> $= \int \log p[c_N,d_N|X,S] \; p[c_N,d_N|X,S] \; d(c_N,d_N) + 1 - 1$
> $= E\{\log L(X,S|\mathbf{c}_N,\mathbf{d}_N)|X,S\}$ ∎

This indicates that one should aim at maximizing the expectation of the logarithm of the Likelihood function. Since one usually does not have an ensemble of samples to evaluate the ensemble average, but only one (long) sample $(\mathbf{c}_N, \mathbf{d}_N)$, the lemma has to be supplemented by a condition that allows one to replace ensemble averages of the Likelihood function with time averages, and to aim at maximizing those instead. This motivates a closer look at the Likelihood function for long samples.

Remark: Åström and Söderström (1974) have shown that the ARMA structure yields only a global maximum of the Likelihood function, when the sample gets sufficiently long. However, other structures may yield several local maxima. For such cases Bohlin (1971) devised a method to test aposteriori whether the search has reached a false local maximum. Generally, the problem is solved by the falsification techniques in Chapter 8.

Remark: It follows from the Lemma of the Maximum Likelihood that for long samples the global maximum is also the correct one. However, this fact hides a potential 'pitfall': If the structure does not hold an Equivalent model, the best approximation one would want instead may not correspond to the global maximum. In particular, in the case of identification in closed loop the global maximum may correspond to an approximate inverse controller model (see Example 3.5). To avoid such a calamity one should always try and falsify a local maximum before accepting the corresponding model.

There must be a 'law of large numbers'

Introduce the loss function defined by (3.3)

$$Q(M|\mathbf{c}_N, \mathbf{d}_N) \triangleq - \log L(M|\mathbf{c}_N, \mathbf{d}_N)/N$$

$$= - \frac{1}{N} \sum_{k=1}^{N} \log p[\mathbf{c}(k), \mathbf{d}(k)|\mathbf{c}_{k-1}, \mathbf{d}_{k-1}, k, M] \tag{3.3}$$

The point of rewriting the Likelihood function is that $L(M|\mathbf{c}_N, \mathbf{d}_N)$ does not tend to a finite value as $N \to \infty$, while Q might. The loss function Q defined by (3.3) may obviously be regarded as the sample average of the instantaneous log likelihood function $-\log p[\mathbf{c}(k), \mathbf{d}(k)|\mathbf{c}_{k-1}, \mathbf{d}_{k-1}, k, M]$. If it were instead an ensemble average $E\{-\log L(M|\mathbf{c}_N, \mathbf{d}_N)/N|\mathbf{X}, \mathbf{S}\}$, then the Lemma of the Maximum Likelihood implies that one should minimize $Q(M|\mathbf{c}_N, \mathbf{d}_N)$ to get an estimate of the system $(\mathbf{X}, \mathbf{S})$. Hence one needs a condition that the two averages are equal for long samples:

> ⌐ **Condition: Convergence of the Likelihood** ⌐

The source $(\mathbf{X},\mathbf{S})$ and the widest model structure $\overline{\mathcal{M}}$ must be such that the average of the logarithm of the instantaneous likelihood tends to its mean:

$$Q(M|\mathbf{c}_N,\mathbf{d}_N) \rightarrow E\{Q(M|\mathbf{c}_N,\mathbf{d}_N)|\mathbf{X},\mathbf{S}\} = \overline{Q}_N(M|\mathbf{X},\mathbf{S}),$$

$$\text{all } M \in \overline{\mathcal{M}} \text{ and almost all } (\mathbf{c}_N,\mathbf{d}_N).$$

●●●●●●

Remark: The condition of Convergence of the Likelihood actually contains two parts: 1) that $Q(M|\mathbf{c}_N,\mathbf{d}_N)$ converges, and 2) that it converges to its ensemble average. The first part can obviously be tested from model structure and data, but the second condition depends on the true source and is not testable.

———————— **Example 6.1** ————————

A case where the condition of Likelihood Convergence is not satisfied.

The following model describes the response of a first–order dynamic system, which is measured with a sensor with proportional error:

S: $\mathbf{d}(k) = \Theta^k \mathbf{c}(0) + \lambda\, \Theta^k\, \omega(k), \quad \Theta \in (0,1)$

It follows immediately that

$$\log p[\mathbf{d}(k)|\mathbf{c}_{k-1},\mathbf{d}_{k-1},k,\Theta]$$

$$= -\tfrac{1}{2}\,[\mathbf{d}(k) - \Theta^k\, \mathbf{c}(0)]^2\,(\lambda\,\Theta^k)^{-2} - \log(\lambda\,\Theta^k) - \tfrac{1}{2}\log(2\pi)$$

Obviously, $\dfrac{1}{N}\displaystyle\sum_{k=1}^{N} \log p[\mathbf{d}(k)|\mathbf{c}_{k-1},\mathbf{d}_{k-1},k,\Theta]$ does not converge for

any $\Theta < 1$, since $\dfrac{1}{N}\displaystyle\sum_{k=1}^{N}\log(\lambda\,\Theta^k) = \infty$.

The cause of the failure is that the model depends too strongly on k; in this case the sensor error decreases indefinitely. Notice that if the error variance would have a lower bound (which is actually a more realistic assumption), then the log likelihood would converge. Hence, the case is not modelled adequately.

Remark: Generally, the large–sample theory gets into trouble, when the model is lacking adequate error variables, as a result of overconfidence in the correctness of apriori assumptions. The yields the following *Rule of thumb*: Whenever in doubt concerning the condition of Likelihood Convergence, add noise variables!

———————————— **Example 6.2** ————————————

Illustrating the effect of 'outliers'.

It seems that in order to construct a case where the likelihood does not converge, one will have to conjure up a rather unlikely one:

Assume the data acquisition equipment would normally measure an unknown constant Θ with gaussian error, but occasionally superimpose an 'outlier', which furthermore has linear drift. It may be difficult to conceive a physical fault that would cause this, but the source would be

S: $\mathbf{d}(k) = \Theta + \omega(k) + v(k)\ k$

where $\omega(k)$ is gaussian $(0,1)$, and $v(k)$ is a binary random variable such that $\Pr\{v(k) = 1\} = p_k$.

Assume also that one does not expect any outliers, but uses the model structure

$\mathcal{M}$: $\mathbf{d}(k) = \theta + \omega(k)$, θ real.

The Likelihood loss would be

$$Q(\theta|\mathbf{d}_N) = -\log L(\theta|\mathbf{d}_N)/N = \frac{1}{N}\sum_{k=1}^{N} \frac{1}{2}[d(k) - \Theta]^2 + \frac{1}{2}\log(2\pi).$$

To investigate whether this will converge compute first the function it would possibly converge to: The true source would have the data distribution

$$p[d(k)|\Theta,k] = (1 - p_k)\ \exp\{-\tfrac{1}{2}[d(k) - \Theta]^2\}/\sqrt{2\pi}$$
$$+ p_k\ \exp\{-\tfrac{1}{2}[d(k) - \Theta - k]^2\}/\sqrt{2\pi}$$

This yields

$$E\{Q[\theta|\mathbf{d}_N]\} = \int \frac{1}{N}\sum_{k=1}^{N} \frac{1}{2}(y-\theta)^2 \{(1-p_k) \exp[-\frac{1}{2}(y-\theta)^2]$$

$$+ p_k \exp[-\frac{1}{2}(y-\theta-k)^2]\}/\sqrt{2\pi}\, dy + \frac{1}{2}\log(2\pi)$$

$$= \frac{1}{N}\sum_{k=1}^{N} \frac{1}{2}(1 + k^2\, p_k) + \frac{1}{2}\log(2\pi).$$

This is finite if $p_k \leq 1/k^2$, and the ensemble average exists.

However, since Q is a sum over k, the following recursion holds

$$Q(\theta|\mathbf{d}_k) = Q(\theta|\mathbf{d}_{k-1}) + k^{-1} \{\frac{1}{2}[\mathbf{d}(k)-\theta]^2 + \frac{1}{2}\log(2\pi) - Q(\theta|\mathbf{d}_{k-1})\}$$

and with positive probability p_k the addition to $Q(\theta|\mathbf{d}_{k-1})$ will be

$k^{-1}\{\frac{1}{2}[k-\theta]^2 + \frac{1}{2}\log(2\pi) - Q(\theta|\mathbf{d}_{k-1})\} = O(k)$.

Hence, the sequence $\{Q(\theta|\mathbf{d}_k)|k=1,2,...\}$ does not converge, even if the mean is finite! However large a k, there will always be a big jump in the sequence at some later k–value.

The case illustrates the difference between the two mathematical concepts of 'convergence in probability' and 'convergence with probability one' (Wilks, 1962); the first type of convergence holds, but not the second one. The latter convergence is the stronger, it means that the computed sequence $\{Q(\theta|\mathbf{d}_k)|k=1,2,...\}$ converges as an ordinary sequence for almost all data $\mathbf{d}_N$. And that is what is needed for computing the likelihood.

One more condition is needed

The most one can get out of the limiting function is an average of the probabilistic model $-\log p[\mathbf{c}(k),\mathbf{d}(k)|\mathbf{c}_{k-1},\mathbf{d}_{k-1},k,\mathbf{X},\mathbf{S}]$. If that model may depend too freely on the index k, then the individual terms in the average can never be resolved from the limiting sum. And the latter is what is required for Consistency. Hence, the designer must ensure that the model structure is not too free in that respect.

The condition can be formulated thus:

Condition: Bounded Structure

The widest model structure $\overline{\mathcal{M}}$ must be such that the limiting Likelihood function contains enough information to define a dynamic system. That is, the equation

$$\lim_{N\to\infty} \frac{1}{N}\sum_{k=1}^{N} \log p[c(k),d(k)|c_{k-1},d_{k-1},k,M]$$

$$= \lim_{N\to\infty} \frac{1}{N}\sum_{k=1}^{N} \log p[c(k),d(k)|c_{k-1},d_{k-1},k,\mathbf{M}]$$

$M,\mathbf{M} \in \overline{\mathcal{M}}$, should imply

$$p[c(k),d(k)|c_{k-1},d_{k-1},k,M] = p[c(k),d(k)|c_{k-1},d_{k-1},k,\mathbf{M}],$$

all k, all $(c_k,d_k) \in \mathfrak{R}_k(\mathbf{X})$.

–––––––––– •••••• ––––––––––

Remark: Obviously, it will never be possible to separate, *e.g.* infrequent, unpredictable and not lasting disruptions in the source dynamics from the normal variations in the responses of a stochastic system. Therefore one must either refrain from identifying systems that may exhibit such disruptionss in dynamics, or exclude them from the model structure anyhow, and thus accept that the models obtained will fail for some, infrequent k–values.

The conditions are sufficient for consistency

Now, provided the conditions of Likelihood Convergence and Bounded Structure are satisfied, the Maximum–Likelihood estimate yields a consistent data distribution $\hat{p}[c(k),d(k)|c_{k-1},d_{k-1},k,\mathcal{M},\mathbf{c}_N,\mathbf{d}_N]$. This is a consequence of the proof of the following lemma:

Lemma of the Structured Source: If there is a *Source–Equivalent* model, the condition of *Bounded Structure* is satisfied, and the Likelihood function *Converges*, then the condition of *Consistency* is satisfied.

Proof:
Let n be such that there is a Source–Equivalent model $\mathbf{M} \in \mathcal{M}(n)$,

and let $\hat{M}_N$ be a maximum of the Likelihood function $L(M|\mathbf{c}_N,\mathbf{d}_N)$ within $\mathcal{M}(n)$. Then for the associated loss function, and since there is a Source–Equivalent model $\mathbf{M}$:

$$Q(\hat{M}_N|\mathbf{c}_N,\mathbf{d}_N) \leq Q(\mathbf{M}|\mathbf{c}_N,\mathbf{d}_N) = Q(\mathbf{X},\mathbf{S}|\mathbf{c}_N,\mathbf{d}_N), \text{ all } N.$$

The Lemma of the Maximum Likelihood implies

$$E\{Q(M|\mathbf{c}_N,\mathbf{d}_N)|\mathbf{X},\mathbf{S}\} \geq E\{Q(\mathbf{X},\mathbf{S}|\mathbf{c}_N,\mathbf{d}_N)|\mathbf{X},\mathbf{S}\}, \text{ all } M, \text{ all } N.$$

Convergence of the Likelihood implies that

$$Q(M|\mathbf{c}_\infty,\mathbf{d}_\infty) = E\{Q(M|\mathbf{c}_\infty,\mathbf{d}_\infty)|\mathbf{X},\mathbf{S}\} \triangleq \bar{Q}_\infty(M|\mathbf{X},\mathbf{S})$$

with probability one. Hence,

$$Q(M|\mathbf{c}_\infty,\mathbf{d}_\infty) \geq Q(\mathbf{M}|\mathbf{c}_\infty,\mathbf{d}_\infty) \geq Q(\hat{M}_\infty|\mathbf{c}_\infty,\mathbf{d}_\infty), \text{ all } M, \text{ so that}$$

$$Q(\mathbf{M}|\mathbf{c}_\infty,\mathbf{d}_\infty) = Q(\hat{M}_\infty|\mathbf{c}_\infty,\mathbf{d}_\infty),$$

or

$$\bar{Q}_\infty(\mathbf{M}|\mathbf{X},\mathbf{S}) = \bar{Q}_\infty(\hat{M}_\infty|\mathbf{X},\mathbf{S}).$$

This yields a limiting set of models that cannot be falsified even by an infinite data sample:

$$\mathcal{B}_\infty = \{M|Q(M|\mathbf{c}_\infty,\mathbf{d}_\infty) = Q(\mathbf{M}|\mathbf{c}_\infty,\mathbf{d}_\infty)\}.$$

Write

$$\mathcal{B}_\infty = \{M|D(\mathbf{M},M|\mathbf{c}_\infty,\mathbf{d}_\infty) = 0\},$$

where

$$D(\mathbf{M},M|\mathbf{c}_\infty,\mathbf{d}_\infty) = Q(M|\mathbf{c}_\infty,\mathbf{d}_\infty) - Q(\mathbf{M}|\mathbf{c}_\infty,\mathbf{d}_\infty) \geq 0,$$

and investigate the difference in loss. From the condition of Convergence of the Likelihood

$$D(\mathbf{M},M|\mathbf{c}_\infty,\mathbf{d}_\infty) = E\{Q(M|\mathbf{c}_\infty,\mathbf{d}_\infty) - Q(\mathbf{M}|\mathbf{c}_\infty,\mathbf{d}_\infty)|\mathbf{X},\mathbf{S}\}$$

$$= \lim_{N\to\infty} \int [Q(M|\mathbf{c}_N,\mathbf{d}_N) - Q(\mathbf{M}|\mathbf{c}_N,\mathbf{d}_N)]\, p[\mathbf{c}_N,\mathbf{d}_N|\mathbf{X},\mathbf{S}]\, d(\mathbf{c}_N,\mathbf{d}_N)$$

$$= \lim_{N\to\infty} \int \{-\log p[\mathbf{c}_N,\mathbf{d}_N|M] + \log p[\mathbf{c}_N,\mathbf{d}_N|\mathbf{M}]\}/N$$

$$p[\mathbf{c}_N,\mathbf{d}_N|\mathbf{M}]\, d(\mathbf{c}_N,\mathbf{d}_N).$$

In general, $D(\mathbf{M},M|\mathbf{c}_\infty,\mathbf{d}_\infty)$ may be zero, without $Q(M|\mathbf{c}_\infty,\mathbf{d}_\infty) - Q(\mathbf{M}|\mathbf{c}_\infty,\mathbf{d}_\infty)$ being zero for all $(\mathbf{c}_\infty,\mathbf{d}_\infty)$. However, since $Q(M|\mathbf{c}_\infty,\mathbf{d}_\infty) - Q(\mathbf{M}|\mathbf{c}_\infty,\mathbf{d}_\infty)$ is non–negative, it must be zero for (almost) all $(\mathbf{c}_\infty,\mathbf{d}_\infty)$ that have a positive probability $p[\mathbf{c}_N,\mathbf{d}_N|\mathbf{M}] > 0$ for all N. That is the Range of Model Validity $\mathcal{R}_N(\mathbf{X})$. Hence,

$$\mathcal{B}_\infty = \{M|Q(M|\mathbf{c}_\infty,\mathbf{d}_\infty) = Q(\mathbf{M}|\mathbf{c}_\infty,\mathbf{d}_\infty),$$
$$\text{all } (\mathbf{c}_\infty,\mathbf{d}_\infty) \in \mathcal{R}_\infty(\mathbf{X})\},$$

and

$$Q(\hat{M}_\infty | c_\infty, d_\infty) = Q(M | c_\infty, d_\infty), \text{ all } (c_\infty, d_\infty) \in \mathfrak{R}_\infty(X).$$

Now, the condition of Bounded Structure implies that

$$p[c(k), d(k) | c_{k-1}, d_{k-1}, k, \hat{M}_\infty] = p[c(k), d(k) | c_{k-1}, d_{k-1}, k, M],$$

all k, all $(c_\infty, d_\infty) \in \mathfrak{R}_\infty(X)$,

and hence any limiting ML–estimate $\hat{M}_\infty$ is Source Equivalent. Finally, define the statistic

$$\hat{p}[c(k), d(k) | c_{k-1}, d_{k-1}, k, \mathcal{M}, c_N, d_N]$$
$$= p[c(k), d(k) | c_{k-1}, d_{k-1}, k, \hat{M}_N].$$

Then

$$\hat{p}[c(k), d(k) | c_{k-1}, d_{k-1}, k, \mathcal{M}, c_N, d_N]$$
$$\rightarrow p[c(k), d(k) | c_{k-1}, d_{k-1}, k, M] = p[c(k), d(k) | c_{k-1}, d_{k-1}, k, X, S]$$

with probability one for all k amd all $(c_k, d_k) \in \mathfrak{R}_k(X)$, which is the condition of Consistency. ∎

It is a trivial, but sometimes useful extension of the theory to note that the data (c_N, d_N) may be replaced by a 'parameter–free statistic', which is a function of (c_N, d_N), bijective, and independent of M.

─────────────── **Example 6.3** ───────────────

Illustrating the convenience of using a parameter–free statistic.

Let S: $d(k) = \Theta \, \omega(k) \, d(k-1)$

where $\omega(k)$ is gaussian $(0,1)$ and Θ an unknown parameter.

The Likelihood loss of the model $d(k) = \theta \, \omega(k) \, d(k-1)$ is

$$Q(\theta | d_N) = \frac{1}{N} \sum_{k=1}^{N} \left\{ \frac{1}{2} [d(k)/d(k-1)]^2 \, \theta^{-2} + \frac{1}{2} \log[2\pi \, d(k-1)^2 \, \theta^2] \right\}.$$

It obviously does not converge, since for instance

$$\frac{1}{N} \sum_{k=1}^{N} \log|d(k-1)| = \frac{1}{N} \sum_{k=1}^{N} \sum_{i=1}^{k-1} \log|\Theta \, \omega(i)| \rightarrow \infty.$$

However, defining the sufficient statistic $s(k) = d(k)/d(k-1)$ yields the model $s(k) = \theta \, \omega(k)$ with the Likelihood loss

$$Q(\theta|\mathbf{s}_N) = \frac{1}{N}\sum_{k=1}^{N} \frac{1}{2}\,\mathbf{s}(k)^2\,\theta^{-2} + \frac{1}{2}\log(2\pi\,\theta^2)$$

which converges. Notice that the term $\log \mathbf{d}(k-1)^2$ has disappeared, which is what makes the case converge. However, the term does not affect the optimal θ, even for $Q(\theta|\mathbf{d}_N)$.

Hence, one can still determine an estimate

$$\hat{\theta}_N = \frac{1}{N}\sum_{k=1}^{N} \mathbf{d}(k)/\mathbf{d}(k-1)$$

that is consistent, and the statistic $p[\mathbf{d}(k)|\mathbf{d}_{k-1},\hat{\theta}_N]$ will be consistent, even if the condition of Likelihood Convergence (as defined) is not satisfied.

A potential hazard

If $\mathcal{I} \cap \mathcal{M}(\bar{n}) \subseteq \mathcal{G}$, *i.e.* all Source–Equivalent models are purposive (and if $\mathcal{I} \cap \mathcal{M}(\bar{n})$ is not empty), this ensures that a purposive likelihood maximum is reached within the feasible structure $\mathcal{M}(\bar{n})$. However, the Likelihood function is normally evaluated within a much more restricted structure $\mathcal{M}(n)$, and the hazard is that the global maximum is not in $\mathcal{M}(n)$. In principle one has therefore to expand $\mathcal{M}(n)$ until the minimum of the loss function $Q(M|\mathbf{c}_N,\mathbf{d}_N)$ no longer decreases, which is when $\mathcal{I} \cap \mathcal{M}(n)$ becomes non–empty. That procedure would eventually *falsify* all other models than those in $\mathcal{I} \cap \mathcal{M}(n)$.

A potential problem is that one cannot use the criterion that "the minimum has stopped decreasing" for deciding when to stop the expansion of $\mathcal{M}(n)$, since the last expansion may have been in the wrong direction, and one can never know whether there is a better one (this is a fundamental problem in any search method). However, in this particular search there is another way to decide when to stop: There are statistical test procedures to falsify any model that is not in $\mathcal{B}$ (see section 8.2 on "Unconditional falsification"). Hence one will have to go on expanding the structure $\mathcal{M}(n)$ only as long as all models are still falsified, *i.e.* until $\mathcal{M}(n) \cap \mathcal{B}$ becomes non–empty. The Venn diagram is depicted in *Fig. 31*.

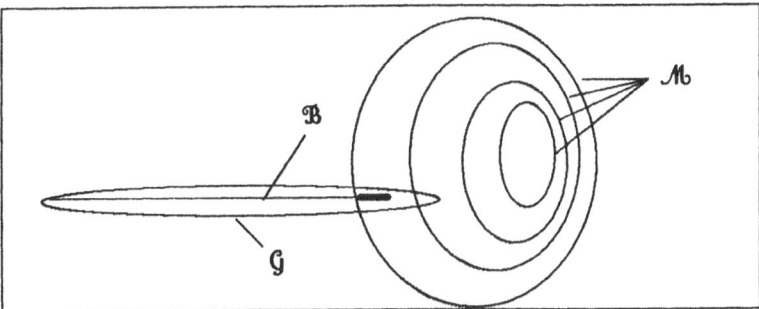

Fig. 31: Venn diagram for proper identification of structured models. Expanding the structure $\mathcal{M}$ stops when $\mathcal{M} \cap \mathcal{B} \neq \emptyset$

However, the falsification tests are valid only if the Condition of Source Equivalence is satisfied, *i.e.* that $\mathcal{I} \cap \mathcal{M}(\bar{n}) \neq \emptyset$. If there is no equivalent (or approximately equivalent) model in $\mathcal{M}(\bar{n})$, then even the widest model structure is not flexible enough to allow the 'structured approach'. Since, the condition involves the Object, and is uncheckable, a 'pitfall' is threatening here.

There are separate losses for object and experimenter

When the experimenter is *Isolated* from the object, the loss function (3.3) separates into two terms

$$Q(M|\mathbf{c}_N,\mathbf{d}_N) = Q^c(M^c|\mathbf{c}_N,\mathbf{d}_N) + Q^d(M^d|\mathbf{c}_N,\mathbf{d}_N) \tag{6.5}$$

where

$$Q^c(M^c|\mathbf{c}_N,\mathbf{d}_N) \triangleq -\frac{1}{N}\sum_{k=1}^{N} \log p[\mathbf{c}(k)|\mathbf{c}_{k-1},\mathbf{d}_k,k,M^c]$$

$$Q^d(M^d|\mathbf{c}_N,\mathbf{d}_N) \triangleq -\frac{1}{N}\sum_{k=1}^{N} \log p[\mathbf{d}(k)|\mathbf{c}_{k-1},\mathbf{d}_{k-1},k,M^d]$$

The unbiassed Likelihood function becomes

$$\log L(M|\mathbf{c}_N,\mathbf{d}_N) = -N\,Q(M|\mathbf{c}_N,\mathbf{d}_N)$$

$$= -N\,Q^c(M^c|\mathbf{c}_N,\mathbf{d}_N) - N\,Q^d(M^d|\mathbf{c}_N,\mathbf{d}_N) \tag{6.6}$$

The two terms in the Likelihood loss function Q are independent, regarded as functions of M^c and M^d, and one has to minimize only the second one Q^d to determine the ML-estimate of $\mathbf{S}$. The first loss term

Q^c does not even have to be computed, which would have been impossible anyhow, if the stimulating process X is not well defined.

Normally, one is only interested in identifying the object S, but if one would want to identify the experimenter too, that can be achieved by minimizing Q^c. An important conclusion, however, it that irrespective of the stimulating process X, the Likelihood function should be evaluated and maximized, *as if* the stimulus were a deterministic sequence and given in advance. The *properties* of the resulting model, however, will depend very much on how the stimulus was generated.

6.2.2 The unstructured approach

The major advantage of using a parametric model is that one can apply the Bayesian approach to compute the model $M = (\mathcal{A}, \nu, \theta)$ from the Likelihood function $L(\nu, \theta | \mathcal{A}, c_N, d_N) = p[c_N, d_N | \mathcal{A}, \nu, \theta] \, p[\nu, \theta | \mathcal{A}]$. In essence, one has only to insert the data values (c_N, d_N) into the given probability density functions. Since the Likelihood function will then be a function only of a finite number of integers ν and reals θ, it is manageable by computers.

When a parametric model is not desirable, or not feasible (because one does not have the necessary apriori structural knowledge), the Bayesian approach cannot be used. The likelihood would then be a function of the model M, which is unstructured, and it is usually impractical to try and evaluate it for all M in $\mathcal{M}$. More fundamentally, the likelihood will yield be a good approximation of the data distribution only as long as there are much fewer degrees of freedom in $\mathcal{M}$ than the given number of relevant data.

The model must not change with time

An alternative that makes it feasible to evaluate the data distribution $\hat{p}[c_K, d_K | \mathcal{M}, c_N, d_N]$ for long samples is to assume that the object S and the experimenter X do not change their properties with time:

Condition: Time Invariance

The widest model structure $\bar{\mathcal{M}}$ must be such that the same description will hold throughout the identification phase. That is,

$$p[c(k),d(k)|c_{k-1},d_{k-1},k,M] = p[c(k),d(k)|c_{k-1},d_{k-1},M],$$

all $M \in \bar{\mathcal{M}}$, all k, all $(c_k,d_k) \in \mathfrak{R}_k(\mathbf{X})$.

●●●●●●

Remark: The basic condition of Reproducibility implies that there must be a certain 'invariance' of the object for identification to be meaningful at all, and the condition of Bounded Structure serves the same purpose. One obviously cannot infer anything about objects that may possibly change their properties too fast and too unpredictably. The condition of Time Invariance, however, is much stronger; the source must not change its properties at all, not even predictably. It means that time (or the sample counter k) must not enter explicitly into the model description.

The model must not be too complex

The idea is in this case to try and estimate the conditional data distribution $p[c(k),d(k)|c_{k-1},d_{k-1},\mathbf{X},\mathbf{S}]$ pointwise by evaluating 'histograms' from the data sample $(\mathbf{c}_N,\mathbf{d}_N)$. By the restriction of Time Invariance one has now a large number of observations $\{c(k),d(k)\}$ with the *same* distribution $p[\bullet,\bullet|\bullet,\bullet,\mathbf{X},\mathbf{S}]$. There is still a number of obstacles to this, which require further conditions to overcome:

Condition: Information State

The widest model structure $\bar{\mathcal{M}}$ must be such that there is an *Information State*. That is,

$$p[c(k),d(k)|c_{k-1},d_{k-1},M] = p[c(k),d(k)|x(k),M]$$
$$x(k+1) = \mathfrak{I}[x(k),c(k),d(k)],$$

all $M \in \bar{\mathcal{M}}$, all k, all $(c_k,d_k) \in \mathfrak{R}_k(\mathbf{X})$.

●●●●●●

Requiring that there exist a finite–dimensional Information State x reduces the number of stochastic variables in the model to a number that does not increase, when the sample length N increases (see section 5.7 on "Finite–memory models").

Remark: The condition of an Information–State may not be satisfied in a real case. However, if it is not, then the possibilities to identify the case from data will depend on how fast complexity grows. Obviously, that will have to be much slower than the length of the data sample grows, since the number of information–carrying parameters will have to be much smaller than the number of data points to allow the averaging out of the effects of randomness.

Remark: The concept of 'complexity' is used in different meanings in the literature (Lofgren, 1987). It may involve concepts of randomness, coding, and semantics. Complexity may be interpreted here as the dimension of the information state. It does however play a fundamental part in some strategies for choosing between models of different structure (*e.g.* Gaines, 1977). Caines (1986) introduces a complexity indicator, defined as the minimum–length description of the model (in bits), into the modelling loss function (besides parameters), and in this way achieves that the loss is minimum for a model of finite complexity. In contrast, the strategy used in this book for choosing between differently complex models hinges on the number n, which is coupled to the number of *parameters* needed to describe the probabilistic model (see chapter 9 on "Structure identification"). The difference reflects the different purposes of modelling. A selection strategy resting on complexity of *description* suits Data Descriptions and 'black–box' models, while the number of parameters becomes important, when the structure is better known apriori ('grey box'), and some or all parameters have a physical meaning. In the 'black–box' case the number of parameters is generally a function of the state dimension, and the selection strategies should, in essence, become equivalent.

Again, there must be a 'law of large numbers'

The last obstacle to a direct pointwise evaluation of the distribution $p[c,d|x,\mathbf{X},\mathbf{S}]$ is that data $\{\mathbf{c}(k),\mathbf{d}(k)\}$ are not independent. This means that the conventional 'law of large numbers' (Wilks, 1962) does not apply, and will have to be replaced by another condition:

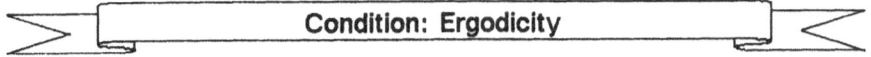

Condition: Ergodicity

The source $(\mathbf{X},\mathbf{S})$ and the widest model structure $\overline{\mathcal{M}}$ must be such that time averages tend to ensemble averages. That is,

$$\frac{1}{N}\sum_{k=1}^{N} f[\mathbf{c}(k),\mathbf{d}(k),\mathbf{x}(k)] \rightarrow E\{f[\mathbf{c}(k),\mathbf{d}(k),\mathbf{x}(k)]|\mathbf{X},\mathbf{S}\},$$

all piece–wise continuous $f[\bullet,\bullet,\bullet]$ and almost all $(\mathbf{c}_N,\mathbf{d}_N)$.

———— •••••• ————

Remark: Ergodicity is the key property that allows one to replace averages over an ensemble of samples by averages over a single long sample. Intuitively, ergodicity means that data points $(\mathbf{c}(k),\mathbf{d}(k))$ and $(\mathbf{c}(k'),\mathbf{d}(k'))$ for indices far from each other should not be too dependent statistically. One can then regard them as almost independent, and hence it would be conceivable to create an 'approximate ensemble' of independent samples, by dividing a very long sample into segments that are so widely spaced that most of the dependence between them will have vanished.

———————— **Example 6.4** ————————

A stochastic process that is stationary but not ergodic:

Let $\mathbf{d}(k) = \omega$, where ω is a gaussian variable with zero mean and unit variance. Then the ensemble average of $\mathbf{d}(k)$ is $E\,\mathbf{d}(k) = E\,\omega = 0$. But the sample average is

$$\frac{1}{N}\sum_{k=1}^{N} \mathbf{d}(k) = \omega,$$

and does not tend to zero when $N \rightarrow \infty$.

Remark: The example shows that 'ergodicity' is a stronger condition than 'stationarity'. It may seem 'academic' and an uninteresting case, but it is not. In practice, objects which contain some very slow dynamics cannot be treated as 'ergodic', even if they would satisfy the mathematical definition. In practice the sample used to evaluate the average has a finite N, even if it may be very large. If the longest sample one can produce by an experiment is still shorter than the slowest phenomenon affecting the object's response, then the evaluation of the average will result in significantly different values, if one repeats the experiment on a later occasion. And none would be equal to an ensemble average. In fact, very slow phenomena (such as instrument 'drift') and very slow changes in object dynamics (for instance due to wear, or from seasonal variations in the environment) constitute a serious problem in practice, which limits the usefulness of the unstructured approach. They originate in the difficulties to ascertain ergodicity.

Remark: In principle, ergodicity *never* holds for the object, and no model remains valid forever (with the possible exception of some fundamental natural laws, like the laws of classical mechanics). However, in practice models do not have to hold forever. This means that one can do with a limited 'ergodicity', holding only for the faster responses of the object, in the hope that the slower phenomena will not change significantly during the application phase. If one can keep that short enough, and instead repeat the identification often enough, there may still be a workable solution to the model design problem. The idea is used for adaptive control (see also section 1.3 on "The purpose").

The Conditions of Time Invariance, Information State, and Ergodicity are enough restriction to make a pointwise evaluation of the data distribution feasible:

Lemma of the Ergodic Source: If there is a *Source–Equivalent* model, and the conditions of *Time Invariance, Information State,* and *Ergodicity* are satisfied, then the condition of *Consistency* is satisfied.

Proof:
Define the cumulative distribution function by
$$P[c,d,x|X,S] = E\{\text{Ind}[c(k) \leq c, \ d(k) \leq d, \ x(k) \leq x]|X,S\}.$$
This suggests the following statistic for the distribution:
$$\hat{P}[c,d,x|\mathcal{M},c_N,d_N] = \frac{1}{N}\sum_{k=1}^{N}\text{Ind}[c(k) \leq c, \ d(k) \leq d, \ x(k) \leq x].$$
Invoking the Ergodicity condition yields
$$\hat{P}[c,d,x|\mathcal{M},c_N,d_N] \rightarrow E\{\text{Ind}[c(k) \leq c, \ d(k) \leq d, \ x(k) \leq x]|X,S\}$$
$= P[c,d,x|X,S]$ with probability one.
Hence, consistency holds for the cumulative distribution for all points (c,d,x).

To prove that consistency holds also for the estimated density $\hat{p}[c,d,x|\mathcal{M},\mathbf{c}_N,\mathbf{d}_N]$, obtained by taking derivatives of the cumulative distribution $\hat{P}[c,d,x|\mathcal{M},\mathbf{c}_N,\mathbf{d}_N]$, requires some regularity assumptions for the density. If one divides the space of (c,d,x) into a finite number of disjoint domains, then the probabilities that $(\mathbf{c}(k),\mathbf{d}(k),\mathbf{x}(k))$ is within each particular domain will converge to the corresponding true probabilities. If the sizes of those domains decrease with increasing n, then one can estimate arbitrarily well the probabilities for arbitrarily small domains, and this is as much as one would usually want, and can ever hope for. Notice however, that mathematically, there may still be densities $p[c,d,x|\mathbf{X},\mathbf{S}]$ such that the corresponding estimate $\hat{p}[c,d,x|\mathcal{M},\mathbf{c}_N,\mathbf{d}_N]$ does not converge pointwise. However, if one excludes those, the lemma is proved. ∎

Remark: The extra regularity assumption on the density $p[c,d,x|\mathbf{X},\mathbf{S}]$ reflects the fact that one can never estimate a nonparametric (completely free) probability density with higher resolution than the amount of data allows. If, for the sake of argument, n indicates the resolution and N the amount of data, then N must obviously go to infinity before n does. In order to prove mathematically that the derivatives of the cumulative distribution converge, one would have to reverse the order of taking limits, and that is not necessarily allowed. Hence, there is a sound practical reason for the particular mathematical difficulty.

--------------- **Example 6.5** ---------------

Illustrating the distinction between Time Invariance and Ergodicity.

Consider again the system in Example 6.3:

S: $\mathbf{d}(k) = \Theta\,\omega(k)\,\mathbf{d}(k-1)$

The probabilistic model is

$$p[d(k)|d_{k-1},k,\mathbf{S}] = \exp\{-\tfrac{1}{2}[d(k)/d(k-1)]^2\,\theta^{-2}\}\,[2\pi\,d(k-1)^2\,\theta^2]^{-1/2}$$

Since it does not depend explicitly on k, the case is Time Invariant.

It is not Ergodic. For instance,

$$E\{\log|\mathbf{d}(k)|\} = E\{\log|\mathbf{d}(k-1)|\} + \log|\Theta| + E\{\log|\omega(k)|\}$$
$$\neq E\{\log|\mathbf{d}(k-1)|\}.$$

However, as in Example 6.3 one can relax the condition of Ergodicity to hold for the sufficient statistic $s(k) = d(k)/d(k-1)$, and still prove Consistency.

───────────────────────── **Example 6.6** ─────────────────────────

Illustrating 'unstructured' identification.

A classical method of identifying an unstructured Ergodic time-series $\{d(k)\}$ is to compute its spectrum by the formulas

$$\hat{r}_N(\tau) = \frac{1}{N}\sum_{k=1}^{N} d(k)\ d(k-\tau)$$

$$\hat{S}_N(f) = \sum_{\tau=0}^{T} \hat{r}_N(\tau)\ [2\cos(2\pi f\tau) - \delta(\tau)], \qquad f \in (-0.5, 0.5)$$

What about the consistency of that estimate?

The condition of Time Invariance is immediately satisfied. To investigate the condition of an Information State one has to find the Information State, and in order to do that one has to restrict the model set $\mathcal{M}$ in some way. The restriction that is implicit in the algorithm above is

$$p[d(k)|d_{k-1},k,M] = p[d(k)|d(k-1),\ldots,d(k-T),M]$$

This means that $d(k)$ is independent of data older than $k-T$, so that $r(\tau) = E\{d(k)\ d(k-\tau)\} = 0$ for $|\tau| > T$. If this restriction is not imposed in some way, the spectrum estimate $\hat{S}_N(\tau)$ will not converge (Blackman and Tukey, 1958).

The Information State follows immediately:

$$x(k) = \{d(k-1),\ldots,d(k-T)\}$$

and has dimension T.

Now, consider the spectral estimate

$$\hat{S}_N(\tau) = \sum_{\tau=0}^{T} [2 \cos(2\pi f\tau) - \delta(\tau)] \frac{1}{N} \sum_{k=1}^{N} \mathbf{d}(k) \, \mathbf{d}(k-\tau)$$

$$= \frac{1}{N} \sum_{k=1}^{N} g[\mathbf{d}(k), x(k)]$$

where $g[d,x] \triangleq d^2 + d \sum_{\tau=1}^{T} 2 \cos(2\pi f\tau) \, x_\tau$.

By the condition of Ergodicity

$$\hat{S}_N(\tau) \rightarrow E\{g[\mathbf{d}(k), \mathbf{x}(k)]\}$$

$$= E\{\mathbf{d}(k)^2 + \sum_{\tau=1}^{T} \mathbf{d}(k) \, \mathbf{d}(k-\tau) \, 2 \cos(2\pi f\tau)\}$$

$$= r(0) + \sum_{\tau=1}^{T} r(\tau) \, 2 \cos(2\pi f\tau) = S(f),$$

and the spectral estimate is consistent if and only if the true spectrum is bandlimited.

Usually one does not assume that $\mathbf{d}(k)$ actually has a limited memory, but forces an equivalent restriction onto the *algorithm*, by multiplying $\hat{r}_N(\tau)$ by a so-called 'lag window', for instance the Hanning window $H(\tau) = \cos^2(0.5 \, \pi \, \tau/T)$ for $|\tau| \leq T$, and 0 for $|\tau| > T$. This makes $\hat{S}_N(\tau)$ converge, but $\hat{S}_N(\tau)$ is now a distorted estimate of the spectrum $S(f)$ for all N, and hence not consistent.

The ARMA model yields another way to identify a spectrum, and introduces other restrictions on $\mathcal{M}$ to achieve Consistency. (See Example 5.5.)

6.3   Identifiability

It is now possible to give some definitions of identifiability. What one should mean by this concept depends obviously on what one has to know about the object to satisfy the purpose of identification, and this creates a number of possible identifiability concepts. Two definitions will be used here, pertinent to two of the five general purposes defined in section 5.8 on "Classification of models by purpose":

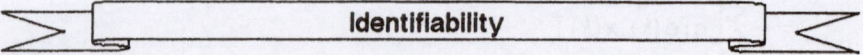

Identifiability

● *Object Identifiability*: The equation

$$\hat{p}[c(k),d(k)|c_{k-1},d_{k-1},k,\mathcal{M},\mathbf{c}_{\infty},\mathbf{d}_{\infty}]$$

$$= p[c(k)|c_{k-1},d_k,k,M^c]\, p[d(k)|c_{k-1},d_{k-1},k,M^d],$$

all k, all $(c_k,d_k) \in \mathfrak{R}_k(\mathfrak{C})$, almost all $(\mathbf{c}_{\infty},\mathbf{d}_{\infty})$,

has the unique solution

$$p[d(k)|c_{k-1},d_{k-1},k,M^d] = p[d(k)|c_{k-1},d_{k-1},k,\mathbf{S}].$$

● *Parameter Identifiability*: The equation

$$\hat{p}[c(k),d(k)|c_{k-1},d_{k-1},k,\mathcal{M},\mathbf{c}_{\infty},\mathbf{d}_{\infty}] = p[c(k),d(k)|c_{k-1},d_{k-1},k,M],$$

all k, all $(c_k,d_k) \in \mathfrak{R}_k(\mathfrak{C})$, almost all $(\mathbf{c}_N,\mathbf{d}_N)$, $M = (\mathcal{A},\nu,\theta)$,

$$\nu \in \mathcal{N}_n,$$

has a unique solution (ν,θ), such that $(\mathcal{A},\nu,\theta) \in \mathcal{S}$.

—————— ●●●●●● ——————

The definitions suit External Object Descriptions and Physical Object Descriptions for the purposes of feedback control and system design. Other model types require other definitions:

● Data Descriptions are always identifiable.

● Source Descriptions would require that the relation

$$\hat{p}[c(k),d(k)|c_{k-1},d_{k-1},k,\mathcal{M},\mathbf{c}_{\infty},\mathbf{d}_{\infty}] = p[c(k),d(k)|c_{k-1},d_{k-1},k,\mathbf{X},\mathbf{S}]$$

hold for all k, all $(c_k,d_k) \in \mathfrak{R}_k(\mathbf{X})$, and almost all $(\mathbf{c}_{\infty},\mathbf{d}_{\infty})$. But that is the condition of Consistency.

● Internal Object Descriptions for the purpose of feedforward control would require that the equation

$$\hat{p}[c(k),d(k)|c_{k-1},d_{k-1},k,\mathcal{M},\mathbf{c}_{\infty},\mathbf{d}_{\infty}]$$

$$= p[c(k)|c_{k-1},d_k,k,M^c]\, p[d(k)|c_{k-1},d_{k-1},k,M^d]$$

yield $p[z(k)|c_{k-1},d_{k-1},k,M^d] = p[z(k)|c_{k-1},d_{k-1},k,\mathbf{S}]$ (see section 5.9.1 on "The requirements of the control purpose").

Remark: A large number of other definitions of identifiability have been suggested in the literature, more or less loosely coupled to particular purposes. Nguyen and Wood (1982) have written a lucid review of concepts of (parameter) identifiability for linear models. They also relate the concepts to each other, and suggest a unification. 'Parameter Identifiability' and 'System Identifiability' (here called 'Object Identifiability' to distinguish this case from that of identifying the experiment system or 'Source') are the most common concepts (Ljung, 1987a). When no attribute is specified, this usually means Parameter Identifiability.

Remark: Kubrusly (1984) defines 'identifiability' as uniqueness of the identification problem based on a given 'equivalence' concept. Since there are several definitions of 'equivalence', this means that there are also several kinds of 'identifiability'. If, in particular, 'equivalence' means that model structure as well as parameter values agree, then this means 'Parameter Identifiability'. The concept of 'Structural Identifiability' (Bellman and Åström, 1970) is essentially the same, since both are mainly a question of choosing a suitable structure and parametrization.

Remark: Vadja (1983) investigates Parametric Identifiability conditions for time–variable linear and time–invariant bilinear deterministic model structures. Vadja and Rabitz (1989) extend this to nonlinear structures. (The style is strict mathematical.)

Remark: Peterka (1981) has shown that the appearance of unidentifiable parameters in the model structure does not change the possibilities of identifying the conditional distribution of the object response, *i.e.* of having Object Identifiability. This is clearly important when the purpose is feedback control.

Remark: Identifiability is related to the more well–known concepts of controllability and observability. For instance, the latter concepts are necessary but not sufficient for parameter identifiability of linear state–vector models (Nguyen and Wood, 1982).

Object Identifiability follows from the general conditions for proper experimentation and model structure design:

Lemma of Object Identifiability: If the experimenter **X** is *Isolated* and *Sufficiently Stimulating*, the object **S** has an *Object–Equivalent* model, and the model structure $\mathcal{M}$ allows a *Consistent* statistic for the distribution of experiment data, then the case is *Object Identifiable*.

Proof:
From the conditions of Consistency, an Isolated Experimenter, and Object Equivalence follows

$\hat{p}[c(k),d(k)|c_{k-1},d_{k-1},k,\mathcal{M},\mathbf{c}_{\infty},\mathbf{d}_{\infty}]$

$= p[c(k),d(k)|c_{k-1},d_{k-1},k,\mathbf{X},\mathbf{S}]$

$= p[c(k)|c_{k-1},d_k,k,\mathbf{X}]\, p[d(k)|c_{k-1},d_{k-1},k,\mathbf{S}],$

$= p[c(k)|c_{k-1},d_k,k,\mathbf{X}]\, p[d(k)|c_{k-1},d_{k-1},k,\mathbf{M}^d]$

all k, all $(c_k,d_k) \in \mathcal{R}_k(\mathbf{X}) \cap \mathcal{R}_k(\mathcal{M})$.

From the condition of Sufficient Stimulation:

$\mathcal{R}_k(\mathbf{X}) \supseteq \mathcal{R}_k(\mathcal{M})$ and $p[c_k,d_k|\mathbf{X},\mathbf{S}] > 0$, $(c_k,d_k) \in \mathcal{R}_k(\mathcal{M})$.

Hence, both factors are positive, and there is a unique solution

$p[d(k)|c_{k-1},d_{k-1},k,\mathbf{M}^d] = p[d(k)|c_{k-1},d_{k-1},k,\mathbf{S}],$

all k, all $(c_k,d_k) \in \mathcal{R}_k(\mathcal{M})$,

which is the definition of Object Identifiability. ∎

Remark: 'Object Identifiability' is a much weaker quality than 'Parameter Identifiability'. In the first case one is only interested in getting a model that behaves *as if* it were the true object. The conditions are therefore mild. Besides the basic requirement that one must be able to compute the data distribution, one needs only a positive denominator to be able to compute the probabilistic model $p[d(k)|c_{k-1},d_{k-1},k,\hat{M}^d]$

$= \hat{p}[c_k,d_k|\mathcal{M},\mathbf{c}_N,\mathbf{d}_N]/\hat{p}[c_k,d_{k-1}|\mathcal{M},\mathbf{c}_N,\mathbf{d}_N].$

Remark: The conditions for Object Identifiability are sufficient. It is possible, of course, to derive them from other sufficient conditions. However, it is not possible to do so without specifying more about the structure, and thus making the conditions more restrictive. Whether or not it is worth while to do that depends on whether these new conditions are easier for the designer to verify, and also on how restrictive they have to be. The text book by Ljung (1987a) presents a number of conditions for consistency and identifiability of linear systems identified by the MPE method.

Remark: Identification in closed loop implies special identifiability conditions (e.g. Gustavsson, Ljung, and Söderström, 1977).

Remark: The asymptotic properties of Minimum Prediction Error identification have been studied by Caines and Ljung (1976), Caines (1978), Ljung (1978, 1987a), and Ljung and Caines

(1979). They treat both the cases when there is an equivalent model in the feasible set, and when there is not.

Remark: Proving consistency of ML models $(\mathcal{A},\nu,\theta)$ usually involves three steps (Ljung, 1987a):
1) Proving that the Likelihood loss function converges to its expected value, *i.e.* $Q(M|\mathbf{c}_N,\mathbf{d}_N) \rightarrow E\{Q(M|\mathbf{c}_N,\mathbf{d}_N)|\mathbf{X},\mathbf{S}\}$.
2) Proving that the expected value converges to a limiting function:
$E\{Q(M|\mathbf{c}_N,\mathbf{d}_N)|\mathbf{X},\mathbf{S}\} \rightarrow \bar{Q}(M|\mathbf{X},\mathbf{S})$.
This yields that the ML model converges into a set minimizing the limiting loss
$$Q(\hat{M}_N|\mathbf{c}_N,\mathbf{d}_N) \rightarrow \min_{M\in\mathcal{M}} \bar{Q}(M|\mathbf{X},\mathbf{S}),$$
but not necessarily that the ML estimate converges.
3) If there is a Source–Equivalent parametric model $\mathbf{M} = (\mathcal{A},\nu,\Theta)$ in $\mathcal{M}$, then by the Lemma of the Maximum Likelihood:
$$Q(\mathcal{A},\nu,\hat{\theta}_N|\mathbf{c}_N,\mathbf{d}_N) \rightarrow \bar{Q}(\mathcal{A},\nu,\Theta|\mathbf{X},\mathbf{S}).$$

If, further, $\bar{Q}$ has a unique minimum, then $\hat{\theta}_N \rightarrow \Theta$, *i.e.* the ML models are consistent.

Generally, identifiability for models higher up in the hierarchy than External Object Descriptions requires that other conditions be satisfied to limit the model structure $\mathcal{M}$. In such cases one wants more information about the object and has also more apriori information in the form of a model structure. In essence, identifiability answers the question: "Has the structure been restricted enough, so that what remains unknown can be inferred from a sufficiently long data sample?"

––––––––––––––––––––––– **Example 6.7** –––––––––––––––––––––––

Illustrating Parameter Identifiability.

Assume the output of a sensor is $d(k) = v_1(k)/v_2(k)$, where $v_i(k)$ are independent gaussian variables with unknown means m_i and standard deviations σ_i. What parameters are possible to estimate from a sufficiently long sample?

The correct model structure is obviously, since $\sigma_1,\sigma_2 > 0$,

S: $d(k) = \sigma_1/\sigma_2\,[m_1/\sigma_1 + \omega_1(k)]/[m_2/\sigma_2 + \omega_2(k)]$

Hence, all models with the same values of σ_1/σ_2, m_1/σ_1, and m_2/σ_2 are Source Equivalent. Hence

M: $d(k) = \theta_3 \left[\theta_1 + \omega_1(k)\right]/\left[\theta_2 + \omega_2(k)\right]$

and the problem is to determine whether this is Parameter Identifiable.

Try approaching the problem naively, and start looking at the case when $m_i \neq 0$ and $\sigma_i \ll |m_i|$, that is the case when $v_i(k)$ are approximately constant. Then

$$d(k) \approx \theta_3 \, \theta_1 \, \theta_2^{-1} \left[1 + \theta_1^{-1} \, \omega_1(k) - \theta_2^{-1} \, \omega_2(k)\right].$$

It is an Internal Object Description. The corresponding External Object Description is

$$d(k) \approx \theta_3 \, \theta_1 \, \theta_2^{-1} \left[1 + (\theta_1^{-2} + \theta_2^{-2})^{1/2} \, \omega(k)\right].$$

Hence, one ought to be able to estimate $\theta_3 \, \theta_1 \, \theta_2^{-1} = m_1/m_2$ from the sample mean and $(\sigma_1^2/m_1^2 + \sigma_2^2/m_2^2)^{1/2}$ from the sample variance of $\mathbf{d}_N$.

In the opposite case that $m_1 = m_2 = 0$

M: $d(k) = \theta_3 \, \omega_1(k)/\omega_2(k)$

It seems possible to determine θ_3 from the sample variance of $\mathbf{d}_N$, since the variance of $\omega_1(k)/\omega_2(k)$ should be given by the properties of the gaussian distribution only.

However, none of the naive methods will work! In neither of the cases will the sample variance converge, and the reason is that the theoretical variances are infinite!

To see this consider the cumulative distribution of $\mathbf{d}(k)$

$$F[d|\theta] = \Pr\{\theta_3 \, (\omega_1 + \theta_1)/(\theta_2 + \omega_2) \leq d\} = \int_D \phi(\omega_1) \, \phi(\omega_2) \, d\omega_1 \, d\omega_2$$

where D is the region depicted in *Fig. 32*.

To evaluate the double integral introduce polar coordinates centered at $(-\theta_1, -\theta_2)$:

$\omega_1 + \theta_1 = r \cos \psi, \quad \omega_2 + \theta_2 = r \sin \psi,$
$\theta_1 = \rho \cos \chi, \quad \theta_2 = \rho \sin \chi, \quad \eta = d/\theta_3,$
$D = \{r,\psi | -\infty < r < \infty, \; \cot \eta \leq \psi \leq \pi\}$

Then

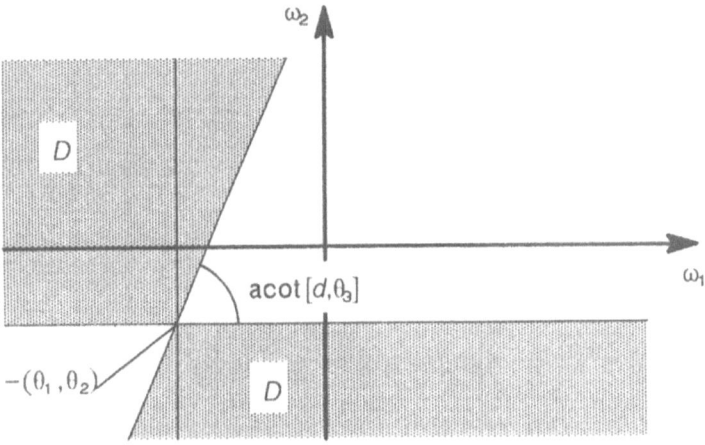

Fig. 32: The region of integration.

$F[d(k)|\theta]$

$$= \int_{acot\,\eta}^{\pi} d\psi \int_{-\infty}^{\infty} \phi[r\cos\psi - \rho\cos\chi]\,\phi[r\sin\psi - \rho\sin\chi]\,|r|\,dr$$

$$= -\frac{1}{2\pi} \int_{acot\,\eta}^{\pi} d\psi \int_{-\infty}^{\infty} \exp\{-\tfrac{1}{2}[r^2 - 2\,r\rho\cos(\psi-\chi) + \rho^2]\}\,|r|\,dr$$

$$= -\frac{1}{2\pi} \int_{acot\,\eta}^{\pi} d\psi \int_{-\infty}^{\infty} \exp[-\tfrac{1}{2}(r-\beta)^2]\,\exp[-\tfrac{1}{2}(\rho^2 - \beta^2)]\,|r|\,dr$$

$$= -\frac{1}{2\pi} \int_{acot\,\eta}^{\pi} d\psi\,\exp[-\tfrac{1}{2}(\rho^2 - \beta^2)]$$
$$\cdot \int_{0}^{\infty} \{(r-\beta)\,\exp[-\tfrac{1}{2}(r-\beta)^2] + (r+\beta)\,\exp[-\tfrac{1}{2}(r+\beta)^2]$$
$$+ \beta\,\exp[-\tfrac{1}{2}(r-\beta)^2] - \beta\,\exp[-\tfrac{1}{2}(r+\beta)^2]\}\,dr$$

$$= \int_{acot\,\eta}^{\pi} g(\psi|\theta_1,\theta_2)\,d\psi$$

where $\beta = \rho\cos(\psi-\chi)$ and

$$g(\psi|\theta_1,\theta_2) = \frac{1}{\pi} \exp(-\tfrac{1}{2}\,\rho^2)$$
$$+ \sqrt{1/2\pi}\ \{2\ \Phi[\rho\cos(\psi-\chi)] - 1\}\ \rho\cos(\psi-\chi)$$
$$\exp[-\tfrac{1}{2}\,\rho^2\sin^2(\psi-\chi)]$$

The probabilistic model for **M** is

$$p[d|\theta] = \theta_3^{-1}\ \frac{d}{d\eta} \int_{\mathrm{acot}\,\eta}^{\pi} g(\psi|\theta_1,\theta_2)\ d\psi$$
$$= \theta_3^{-1}\ g(\mathrm{acot}\,\eta|\theta_1,\theta_2)\ (1+\eta^2)^{-1}$$

In particular, when $m_1 = m_2 = 0$, then $g(\psi,0,0) = 1/\pi$, and

$$p[d|\theta] = \frac{1}{\pi}\ [1 + d^2\,\theta_3^{-2}]^{-1}\ \theta_3^{-1}$$

which means that the ratio of two standard gaussian variables has a Cauchy distribution.

In any case the variance of $d(k)$ is

$$E\{d(k)^2\} = \theta_3 \int_{-\infty}^{\infty} \eta^2\ g(\mathrm{acot}\,\eta|\theta_1,\theta_2)(1+\eta^2)^{-1}\ d\eta = \infty,$$

since $g(\mathrm{acot}\,\eta|\theta_1,\theta_2) \rightarrow constant \neq 0$ when $\eta \rightarrow \infty$.

This means that the sample variance $\frac{1}{N}\sum_{k=1}^{N} d(k)^2$ will not converge. (Try computing it experimentally for increasing N and see what happens!)

All is not lost, however; the identifiability hinges on whether the log likelihood will converge. To investigate this consider the possible limiting value

$$E\{\log p[d(k)|\theta]\}$$
$$= E\{-\log\theta_3 + \log g[\mathrm{acot}(d(k)/\theta_3)|\theta_1,\theta_2] - \log[1 + d(k)^2\,\theta_3^{-2}]\}$$

The first term is obviously finite, and the second term is bounded for all $d(k)$, and hence finite. In order to evaluate the third term let $\kappa = \Theta_3/\theta_3$. Then

$$E\{\log[1 + \kappa^2\,\eta^2]\}$$

$$= \theta_3^{-1} \int_{-\infty}^{\infty} \log[1 + \kappa^2\,\eta^2]\, g\,(\mathrm{acot}\,\eta|\theta_1,\theta_2)(1 + \eta^2)^{-1}\,d\eta$$

$$\leq constant \int_{-\infty}^{\infty} \log[1 + \kappa^2\,\eta^2]\,(1 + \eta^2)^{-1}\,d\eta$$

The integral can be evaluated analytically, for instance by the 'calculus of residues'. The value is $2\pi \log(1 + \kappa)$. Hence, there is a possible finite value to converge to.

Now from the 'strong law of large numbers' (Wilks, 1962) follows that the sample average converges (with probability one) to its expected value, provided all $\log p[\mathbf{d}(k)|\theta]$ are independent and have finite standard deviations $\sigma(k)$ satisfying the condition

$$\sum_{k=1}^{\infty} \sigma(k)^2/k^2 < \infty$$

Since it is possible to show in the same way as before that also $E\{\log p[\mathbf{d}(k)|\theta]^2\}$ is finite, the condition is satisfied, and the Likelihood function converges (pointwise) in this case:

$$\bar{Q}\,(\theta|\Theta) = -\,E\{\log p[\mathbf{d}(k)|\theta]\}$$

$$= -\int_{-\infty}^{\infty} \log\{g[\mathrm{acot}\,(\kappa\eta)|\theta_1,\theta_2](1 + \kappa^2\,\eta^2)^{-1}\,\kappa\,\Theta_3^{-1}\}$$

$$\cdot g[\mathrm{atan}\,\eta|\Theta_1,\Theta_2](1 + \eta^2)^{-1}\,\Theta_3^{-1}\,d\eta$$

It remains to investigate the dependence of the limiting function on the parameters. That is a tedious task and in practice done numerically. In the simple case of $m_1 = m_2 = 0$

$$\bar{Q}\,(\theta|\Theta)$$

$$= -\int_{-\infty}^{\infty} \log\{\pi^{-1}\,(1 + \kappa^2\,\eta^2)^{-1}\,\kappa\,\Theta_3^{-1}\}\,\frac{1}{\pi}\,(1 + \eta^2)^{-1}\,\Theta_3^{-1}\,d\eta$$

$$= -\,\Theta_3^{-1} \log(\pi^{-1}\,\kappa\,\Theta_3^{-1}) + \Theta_3^{-1}\,2\log(1 + \kappa)$$

It has minimum for $-\kappa^{-1} + 2/(1 + \kappa) = 0$, *i.e.* for $\theta_3 = \Theta_3$.

Hence, the case is parameter identifiable, at least for $m_1 = m_2 = 0$.

Generally, it is prohibitively difficult to investigate Parameter identifiability analytically in nonlinear cases. However, since an approximate value of $\bar{Q}\,(\theta|\Theta)$ is computable from the sample average, it is fairly easy to investigate numerically whether its second–order derivative matrix is singular or not (Goodrich and Caines, 1979a).

6.4   Falsification in the limit

The asymptotic analysis in this chapter is relevant for the validation and falsification problems. In particular, it makes sense to ask what will be the set $\mathfrak{B}$ of unfalsified models, when the sample length tends to infinity.

Lemma of Asymptotically Unfalsified Models: If the condition of *Consistency* is satisfied, then the set $\mathfrak{B}$ of unfalsified models tends to the set $\mathfrak{S}$ of Source Equivalents when the sample length $N \to \infty$. If the conditions of *Consistency, Sufficient Stimulation,* and an *Isolated Experimenter* are satisfied, then the set $\mathfrak{B}^d$ of unfalsified object models tends into the set $\mathfrak{S}^d$ of Object Equivalents when the sample length $N \to \infty$.

Proof:
Consistency implies that
$$\hat{p}\,[c(k),d(k)|c_{k-1},d_{k-1},k,\mathcal{M},\mathbf{c}_\infty,\mathbf{d}_\infty]$$
$$= p[c(k),d(k)|c_{k-1},d_{k-1},k,\mathbf{X},\mathbf{S}],$$
and hence
$$\hat{p}\,[c_K,d_K|\mathcal{M},\mathbf{c}_\infty,\mathbf{d}_\infty] = p[c_K,d_K|\mathbf{X},\mathbf{S}]$$
all k, all $(c_K,d_K) \in \mathfrak{R}_K(\mathbf{X})$, almost all $(\mathbf{c}_\infty,\mathbf{d}_\infty)$.
Define
$$\mathfrak{B}_\infty \triangleq \mathfrak{B}(\mathbf{c}_\infty,\mathbf{d}_\infty)$$
$$= \{M|\,\hat{p}\,[c_K,d_K|\mathcal{M},\mathbf{c}_\infty,\mathbf{d}_\infty] = p[c_K,d_K|M],\ \text{all } (c_K,d_K) \in \mathfrak{R}_K(\mathbf{X})\}$$
It follows that *i)* any Source Equivalent model $\mathbf{M}$ (if one exists) is in $\mathfrak{B}_\infty$, *ii)* one can never distinguish between models in $\mathfrak{B}_\infty$, and *iii)* one can falsify in the limit all models that are not in $\mathfrak{B}_\infty$, since $\hat{p}\,[c_K,d_K|\mathcal{M},\mathbf{c}_\infty,\mathbf{d}_\infty]$ is computable from an infinite sample $(\mathbf{c}_\infty,\mathbf{d}_\infty)$, and $p[c_K,d_K|M]$ is computable from the model M to be tested. Hence $\mathfrak{B}_\infty$ is the set of models unfalsifiable in the limit.

But from Consistency follows
$$\hat{p}\,[c_K,d_K|\mathcal{M},\mathbf{c}_\infty,\mathbf{d}_\infty] = p[c_K,d_K|\mathbf{X},\mathbf{S}], \text{ and hence}$$

$\mathcal{B}_{\infty} = \{M | p[c_K, d_K | \mathbf{X}, \mathbf{S}] = p[c_K, d_K | \mathbf{M}], \text{ all } (c_K, d_K) \in \mathcal{R}_K(\mathbf{X})\}$.
This agrees with the set $\mathcal{S}$ (see the definition of Source Equivalence), which proves the first part of the lemma.

When the condition of an Isolated Experimenter is satisfied, define
$\mathcal{B}_{\infty} \triangleq \{M | \hat{p}[c(k), d(k) | c_{k-1}, d_{k-1}, k, \mathcal{M}, c_{\infty}, d_{\infty}]$
$$= p[c(k) | c_{k-1}, d_k, k, M^c] \, p[d(k) | c_{k-1}, d_{k-1}, k, M^d],$$
$$\text{all } k, \text{ all } (c_k, d_k) \in \mathcal{R}_k(\mathbf{X})\}.$$

Consistency implies that
$\mathcal{B}_{\infty} \triangleq \{M | p[c(k) | c_{k-1}, d_k, k, \mathbf{X}] \, p[d(k) | c_{k-1}, d_{k-1}, k, \mathbf{S}]$
$$= p[c(k) | c_{k-1}, d_k, k, M^c] \, p[d(k) | c_{k-1}, d_{k-1}, k, M^d],$$
$$\text{all } k, \text{ all } (c_k, d_k) \in \mathcal{R}_k(\mathbf{X})\}.$$

and Sufficient Stimulation implies that all probabilities are positive.
Hence, $\mathcal{B}_{\infty} = \mathcal{B}^c_{\infty} \times \mathcal{B}^d_{\infty}$
$\mathcal{B}^c_{\infty} = \{M^c | p[c(k) | c_{k-1}, d_k, k, \mathbf{X}]$
$$= p[c(k) | c_{k-1}, d_k, k, M^c], \text{ all } k, \text{ all } (c_k, d_k) \in \mathcal{R}_k(\mathbf{X})\}$$
$\mathcal{B}^d_{\infty} = \{M^d | p[d(k) | c_{k-1}, d_{k-1}, k, \mathbf{S}]$
$$= p[d(k) | c_{k-1}, d_{k-1}, k, M^d], \text{ all } k, \text{ all } (c_k, d_k) \in$$
$\mathcal{R}_k(\mathbf{X})\}.$

From the definition of Object Equivalence follows that
$\mathcal{S}^d = \{M^d | p[d(k) | c_{k-1}, d_{k-1}, k, \mathbf{S}]$
$$= p[d(k) | c_{k-1}, d_{k-1}, k, M^d], \text{ all } k, \text{ all } (c_k, d_k) \in$$
$\mathcal{R}_k(\mathcal{S})\}.$
But $\mathcal{R}_k(\mathcal{S}) \subseteq \mathcal{R}_k(\mathbf{X})$ (Lemma of the Range of Model Validity).
Hence $\mathcal{B}^d_{\infty} \subseteq \mathcal{S}^d$. This proves the second part of the lemma. ■

Remark: The lemma reveals that the set $\mathcal{B}_{\infty}$ has ideal properties in the limit, and this suggests that the corresponding sets $\mathcal{B}(c_N, d_N)$ for finite N should be defined in such a way that they tend to $\mathcal{S}$ when $N \to \infty$ (see chapter 8 on "Falsification techniques").

———————————— **Example 6.8** ————————————

Illustrating the 'pitfall' of having a not Isolated experimenter.

Consider again the system in Example 3.3:

S: $d = s \, c + v_1$
X: $c = x \, v_2 + \alpha \, v_1$

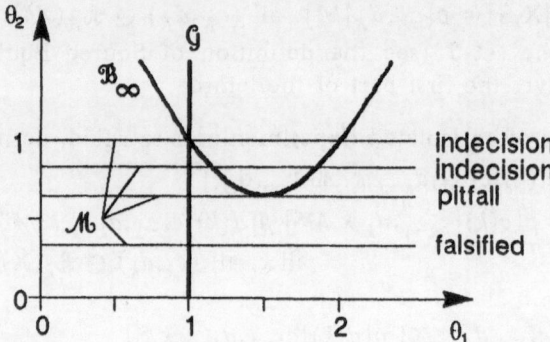

Fig 33: Venn diagram illustrating Example 6.8.

where s is an unknown gain, x is a known amplitude parameter of the stimulus c, and v_1, v_2 are independent gaussian variables with zero means and standard deviations σ_1 and σ_2.

Assume that one knows the correct model structure

$\mathcal{M}$: $d = \theta_1 c + \theta_2 \omega$

Assume also that one knows the experimenter, including the fact that it is not isolated, but one does not know the parameter α or the variances of v_1 or v_2. One wants to estimate s with high accuracy.

In order to see whether that is feasible, assume a long data sample and compute the set $\mathcal{B}_\infty$ of unfalsified models. Since $\mathbf{d}$ and $\mathbf{c}$ are jointly gaussian with zero means, one can compute the three covariances from data (and nothing more). Equating those to the values computed from the model and experimenter yields

$r_{11} = (1+s\alpha)^2 \sigma_1^2 + s^2 x^2 \sigma_2^2$
$r_{12} = \alpha(1+s\alpha) \sigma_1^2 + s x^2 \sigma_2^2$
$r_{22} = \alpha^2 \sigma_1^2 + x^2 \sigma_2^2$.

Eliminating the unknown α and $x \sigma_2$ yields

$.\sigma_1^2 = r_{11} - 2 s r_{12} + s^2 r_{22}$

which defines a hyperbola in the space of (s,σ_1). Since any model in $\mathcal{M}$ is represented by a point in the same space, the hyperbola defines the set of models not possible to falsify (since all yield the same r_{11}, r_{12}, r_{22}). It is depicted in *Fig. 33* together with the set of purposive models $\mathcal{G} = \{\theta_1,\theta_2 | \theta_1 = s\}$ for the case $s = x = \sigma_1 = \sigma_2 = 1$.

Obviously, the case is not parameter identifiable.

A way to improve the situation would be to restrict the structure $\mathcal{M}$ by assuming a value apriori of θ_2. But this would yield no model if $\theta_2 < \sqrt{1/2}$ and two unfalsified models if $\theta_2 > \sqrt{1/2}$. The only case for which there is a unique solution is when θ_2 is minimum. It also corresponds to $\alpha = 0$. And since that coincides with the θ_1–value that makes the variance of the 'model error' $\mathbf{d} - \theta_1\,\mathbf{c}$ minimum (the LS–solution), the assumption is a tempting one.

But it leads into a 'pitfall', since $\mathcal{M} \cap \mathcal{B}_\infty$ is not in $\mathcal{G}$.

6.5 Proper 'black–box' identification

The results so far can be formulated thus:

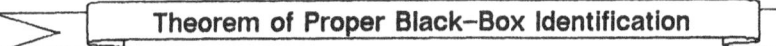

Theorem of Proper Black–Box Identification

If the experiment is *Reproducible,* the experimenter is *Isolated* and *Sufficiently Stimulating,* and there is a *Source–Equivalent* or *Object–Equivalent* model and a *Consistent* statistic for the distribution of experiment data, then the case is *Object Identifiable,* and all not falsified models become *Object Equivalent* and *Applicable* in the limit.

●●●●●●

Proof:
Apply the following lemmas to a dynamic system:
i) Lemma of Object Equivalence:
Sufficient Stimulation, Isolation, Source Equivalence
$\Rightarrow$ *Object Equivalence.*
ii) Lemma of Object Identifiability:
Sufficient Stimulation, Isolation, Object Equivalence, Consistency
$\Rightarrow$ *Object Identifiability.*
iii) Lemma of Asymptotically Unfalsified Models:
Sufficient Stimulation, Consistency
$\Rightarrow \mathcal{B}^d_\infty \subseteq \mathcal{I}^d.$
iv) Lemma of the Closed Loop:
Reproducibility, Sufficient Stimulation, Isolation

⇒ *Representability*

Hence, only models in $\mathcal{B}^d{}_\infty \cap \mathcal{M}^d(n)$ are unfalsified in the limit, and the unfalsified models are also Object Equivalent. Since $p[d(k)|c_{k-1},d_{k-1},k,\hat{M}]$ will be arbitrarily accurate in the limit, the lemma of the Closed Loop implies that it will also be Applicable. These results prove the theorem. ∎

Combining the result in this section with those of the analysis in section 3.6 on "The origin of 'pitfalls'", yields the Venn diagrams in *Fig. 34*. The figure illustrates all possible outcomes of identification in cases when the conditions on Isolation and Sufficient Stimulation are or are not satisfied (remember that the condition for success is $\mathcal{M} \cap \mathcal{B} \subseteq \mathcal{G}$). 'Indecision' occurs when the model is uncertain due to insufficient stimulation. If both conditions are satisfied, the model will be either purposive or falsified. If only the Isolation condition is satisfied, then the model is either falsified or uncertain, but both conditions depend only on data and the assumed model structure. It can possibly be changed to 'decision' by a more stimulating experiment. If the experimenter is neither Isolated nor Sufficiently Stimulating, then the model is either falsified or uncertain, but it will not help to do more experimentation in the same way. If the experimenter is Sufficiently Stimulating but not Isolated, then one has a hazardous case; 'success' or 'pitfall' will depend on whether the model structure $\mathcal{M}$ contains a good model or not (Example 6.8).

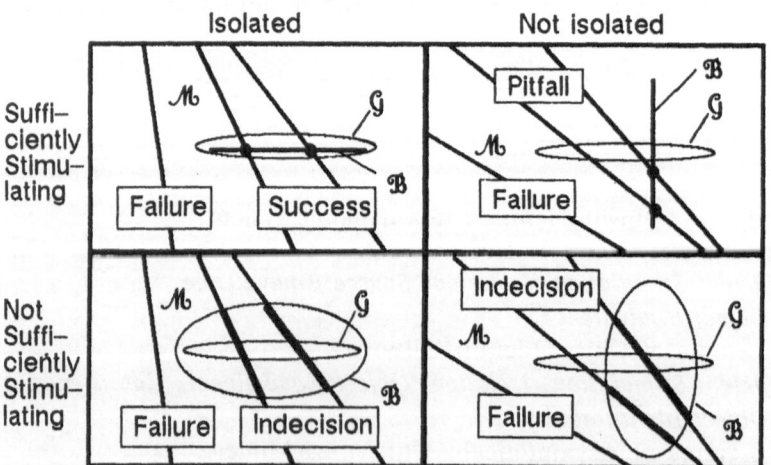

Fig. 34: Illustrating possible outcomes of identification under four experiment conditions. Success, failure, indecision, or pitfall depend on the model structure $\mathcal{M}$. The dots and the heavy bars mark the sets of models $\mathcal{M} \cap \mathcal{B}$

'Indecision' may also result from a proper experiment that is too short. That case is not illustrated in *Fig. 34*. However, it is not a hazardous case, since it will typically be diagnosed by a good identification method, and amended by a longer experiment.

6.6   A concluding example

The following example illustrates different grades of identifiability depending on purpose, experiment conditions, and apriori knowledge, all in the light of the conditions for 'proper identification'.

─────────────── **Example 6.9** ───────────────

Consider the mixing process in *Fig. 35*. Salt is being mixed with water in a tank, and the salt water is pumped out at a given flow rate F. The level in the tank is kept constant by a perfect regulator. The concentration of salt is measured, sampled, and fed into the controller, which manipulates the set point c of a linear valve actuator to control the feed of salt. A reasonable apriori model structure would be:

$$V \dot{z} = -F z + c f(P)$$

where z is the concentration, V is the volume, and $f(P)$ is the influence of feed pressure P on the flow through the valve (f is a

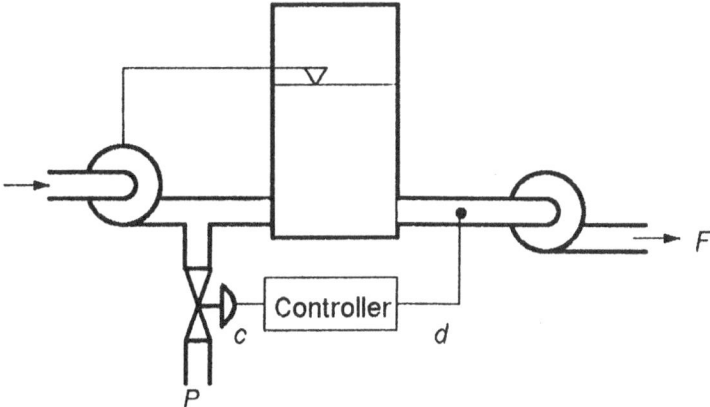

Fig. 35: The problem is to estimate the throuhput F and the feed pressure P under feedback control of the concentration.

known calibration function). Sampling with the interval h and adding a random error ω yields the following object model:

Object: $d(k) = a\,d(k-1) + b\,c(k-1) + \omega(k)$

where $a = \exp(-h\,F)$, $b = (1-a)\,f(P)/F$. It is obviously possible to have $E\{\omega(k)^2\} = 1$ by a suitable scaling of c and d.

One wants to control the salinity to a constant value and has therefore designed the controller

Experimenter/Controller: $c(k) = -\kappa\,d(k)$.

With the particular feedback gain $\kappa = a/b$ one gets a minimum–variance controller adapted to the current throughput F and feed pressure P. However, even if one has designed an optimal controller, one wants to know whether the throughput has changed, in order to adapt the controller to the new throughput. At the same time one wants to check whether the feed pressure has changed, possibly from leakage or overload. A conceivable way would be to identify the object from the operating data (c,d). What kind of model will be feasible under these conditions?

Generally, that depends on apriori knowledge $\mathcal{M}$ and purpose $\mathcal{G}$ (since the experimenter X is given). Consider the following cases:

Case 1: The throughput F is fixed, and the purpose of identification is to check the feed pressure P. Then

$\mathcal{M}(1,\theta)$: $d(k) = a\,d(k-1) + \theta_1\,c(k-1) + \omega(k)$

where $\theta_1 \geq 0$. The set of purposive models is the single point (and in practice a small area around)

$\mathcal{G} = \{\theta | \theta = (b,a)\}$

The logarithm of the probabilistic model is

$\log p[d(k)|c(k-1),d(k-1)]$
$= -\frac{1}{2}\,[d(k) - a\,d(k-1) - \theta_1\,c(k-1)]^2 - \frac{1}{2}\log(2\pi)$

and the log Likelihood function

$\log L(\theta_1|\mathbf{c}_N,\mathbf{d}_N)$

$= -\frac{1}{2}\sum_{k=1}^{N}[\mathbf{d}(k) - a\,\mathbf{d}(k-1) - \theta_1\,\mathbf{c}(k-1)]^2 - \frac{1}{2}N\log(2\pi)$

$= -\frac{1}{2}\sum_{k=1}^{N}[\mathbf{d}(k) - (a - \theta_1\,\kappa)\,\mathbf{d}(k-1)]^2 - \frac{1}{2}N\log(2\pi)$

Maximizing the Likelihood function yields the estimate

$$ML\{\theta_1\} = [a - \sum_{k=1}^{N} \mathbf{d}(k) \, \mathbf{d}(k-1)/\sum_{k=1}^{N} \mathbf{d}(k-1) \, \mathbf{d}(k-1)]/\kappa$$

Since the data come from the object with controller, they satisfy $\mathbf{d}(k) = (a - b \, \kappa) \, \mathbf{d}(k-1) + \omega(k)$, and $ML\{\theta_1\} \rightarrow b$ and is unbiassed. The case is depicted in *Fig. 36a* for the optimal feedback gain $\kappa = a/b$. The model is of all types and satisfies all purposes.

Case 2: The throughput F may have changed, the feed pressure P is still doubted, and the purpose of identification is to estimate both. Then

$$\mathcal{M}(2,\theta): d(k) = \theta_2 \, d(k-1) + \theta_1 \, c(k-1) + \omega(k)$$

where $\theta_1 \geq 0$, $0 \leq \theta_2 \leq 1$. The set of purposive models is

$$\mathcal{G} = \{\theta | \theta = (b,a)\}$$

The logarithm of the Likelihood function is

$$\log L(\theta_1, \theta_2 | \mathbf{c}_N, \mathbf{d}_N)$$

$$= -\frac{1}{2} \sum_{k=1}^{N} [\mathbf{d}(k) - \theta_2 \, \mathbf{d}(k-1) - \theta_1 \, \mathbf{c}(k-1)]^2 - \frac{1}{2} N \log(2\pi)$$

$$= -\frac{1}{2} \sum_{k=1}^{N} [\mathbf{d}(k) - (\theta_2 - \theta_1 \, \kappa) \, \mathbf{d}(k-1)]^2 - \frac{1}{2} N \log(2\pi)$$

Obviously, the case is not Parameter Identifiable: The Likelihood function is all one can ever get out of the experiment and apriori information, and it is a function only of a linear combination of the parameters. The set of all models that are not falsified by data is for an infinite sample

$$\mathcal{B} = \{\theta | \theta_2 - \theta_1 \, \kappa = a - b \, \kappa\}.$$

One can compute only $\mathcal{M}(2) \cap \mathcal{B}$, and hence one cannot both estimate the throughput and check the feed pressure. The Venn diagram in *Fig. 36b* illustrates the correct classification of the model.

Hence, $\mathcal{M}(2) \cap \mathcal{B}$ is not a Physical Object Description. Is it an Internal Object Description? Obviously not; if the structure is expanded more, then the lack of information in the data makes things even worse. Is it an External Object Description? To investigate that, check the conditions for the experiment to be 'proper', *i.e.* Isolated and Sufficiently Stimulating. It is Isolated, since the control signal $c(k)$ is affected by nothing else than the past and present

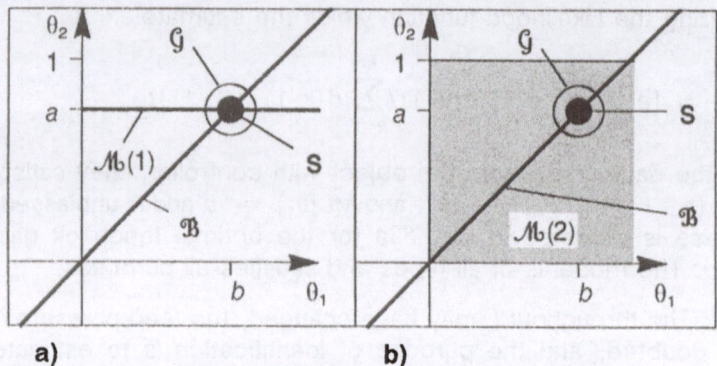

a)   **b)**

Fig. 36: Venn diagrams illustrating cases 1 and 2.
a) $\mathcal{M}(1) \cap \mathcal{B}$ is a good Physical Object Description.
b) $\mathcal{M}(2) \cap \mathcal{B}$ is a Source Description

data $d(k)$. To investigate whether it is also Sufficiently Stimulating, compute the joint probability distribution of data and control signal (section 3.2.1 on "Sufficient stimulation"):

$$\log p[c_N, d_N | X, S]$$

$$= \sum_{k=1}^{N} \{\log p[c(k)|d(k)] + \log p[d(k)|c(k-1),d(k-1)]\}$$

$$= \sum_{k=1}^{N} \log \delta[c(k) + \kappa \, d(k)]$$

$$- \frac{1}{2} \sum_{k=1}^{N} [d(k) - \theta_2 \, d(k-1) - \theta_1 \, c(k-1)]^2 - \frac{1}{2} N \log(2\pi).$$

The distribution is not positive for $c(k) + \kappa \, d(k) \neq 0$, so the experimenter is not Sufficiently Stimulating. Hence, the model is not an External Object Description. It is a Source Description, since the Likelihood function $L(\theta_1, \theta_2 | c_N, d_N)$ is enough to provide an asymptotically unbiassed estimate of the data distribution (see section 6.2.1 on "The structured approach").

The model $\mathcal{M}(1) \cap \mathcal{B}$ is an Internal Object Description, since it is an identifiable Physical Object Description. However, the experimenter is the same, and still not Sufficiently Stimulating. Hence, an insufficiently stimulated experiment may be useful yet, provided the missing information can be supplied apriori. However again, if the information is wrong — if θ_2 would have changed from a — then the case is a 'pitfall'. The identifier based on $\mathcal{M}(1)$ would come out with the value $\theta_1 = (\theta_2 - a)/\kappa + b$ and not with the correct value b.

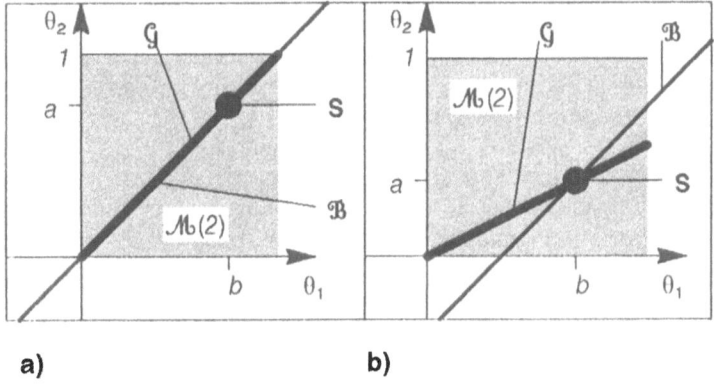

a)   **b)**

Fig. 37: Venn diagrams illustrating case 3.
a) Current control is optimal.
b) Current control is not optimal.

Case 3: If both throughput F and feed pressure P are in doubt, and the purpose is not primarily to estimate their values, but to modify the feedback control accordingly, then

$$\mathcal{M}(2,\theta): \ d(k) = \theta_2 \, d(k-1) + \theta_1 \, c(k-1) + \omega(k)$$

and

$$\mathcal{G} = \{\theta | \theta_2/\theta_1 = b/a\}.$$

The set of models $\mathcal{M}(2) \cap \mathcal{B}$ is the same as in case 2, but $\mathcal{G}$ has changed. The classification of the models depend on whether or not the controller is optimal:

Optimal control: $\kappa = a/b$. *(Fig. 37a)*. $\mathcal{G}$ covers $\mathcal{M}(2) \cap \mathcal{B}$. Hence, all the models in the set are purposive. They are External Object Descriptions (they are not Internal Object Descriptions, nor Physical Object Descriptions). To see how that agrees with the requirement of a Sufficiently Stimulating experimenter, and the fact that the same experimenter in case 2 was labelled 'not Sufficiently Stimulating', notice that the definition of a Sufficiently Stimulating experimenter depends on the set $\mathcal{C}$ of feasible controllers. If the feedback gain κ is not to be changed (the purpose is not adaptive control, but just checking the feedback gain), then $\mathcal{C}$ contains only X, and $p[c_N, d_N | X, S]$ is still positive for $(c_N, d_N) \in \mathcal{R}_N(\mathcal{C}) = \mathcal{R}_N(X) = \{c_N, d_N | c(k) + \kappa \, d(k) = 0\}$. In this particular case the experimenter is theoretically Sufficiently Simulating.

However, that is of little use, since probably $\kappa \neq a/b$. This means that one must consider also the case of

Non–optimal control: $\kappa \neq a/b$ *(Fig. 37b)*. $\mathcal{M}(2) \cap \mathcal{B}$ is not in $\mathcal{G}$, and hence the model is only a Source Description, and useless for the purpose of adapting κ. The experimenter is not Sufficiently Stimulating, since now $\mathcal{R}_N(\mathcal{K}) \supseteq \mathcal{R}_N(\mathbf{X})$.

The fact that the model identified under condition $\mathcal{M}(2)$ cannot be used for designing control, is of course serious. It prevents a naive way of designing adaptive control for this case, namely by repeatedly identifying the object in closed loop, and then updating the controller gain to adapt to the model.

This raises the question of how to make the experimenter Sufficiently Stimulating. Two ways are:

Perturbation signal: An obvious way to render the experimenter Sufficiently Stimulating is to see to it that the controller–output distribution is positive everywhere, for instance by using

Controller: $c(k) = -\kappa\, d(k) + \lambda\, \omega'(k)$.

The added random signal $\lambda\, \omega'(k)$ can be interpreted as having the purpose of increasing the stimulating power of the control signal $c(k)$. The latter then gets a dual purpose of probing and controlling, and the type of control is called 'dual' or 'active'. This yields

$$\log p[c(k)|d(k)] = -\tfrac{1}{2}\,[c(k) + \kappa\, d(k)]^2/\lambda^2 - \tfrac{1}{2}\log(2\pi\lambda^2)\}$$

which is positive for all $c(k), d(k)$, if $\lambda > 0$. The log Likelihood function is

$$\log L(\theta_1,\theta_2|\mathbf{c}_N,\mathbf{d}_N)$$
$$= -\tfrac{1}{2}\sum_{k=1}^{N}[\mathbf{c}(k) + \kappa\, \mathbf{d}(k)]^2/\lambda^2 - \tfrac{1}{2}N\log(2\pi\lambda)$$
$$-\tfrac{1}{2}\sum_{k-1}^{N}[\mathbf{d}(k) - \theta_2\, \mathbf{d}(k-1) - \theta_1\, \mathbf{c}(k-1)]^2 - \tfrac{1}{2}N\log(2\pi)$$

It is a quadratic function of (θ_1,θ_2) and has the unique maximum $\theta_1 = b$, $\theta_2 = a$ for $N = \infty$. The corresponding Venn diagram is shown in *Fig. 38a*.

Changing feedback gain: It is possible to exploit the fact that the set $\mathcal{B} = \{\theta|\theta_2 - \theta_1\,\kappa = a - b\,\kappa\}$ of not falsified models depends on the feedback gain κ. Since (b,a) is the only model that is within $\mathcal{B}$ for all κ, it is possible to falsify all other models by using two experimenters with different values of κ (Gustavsson, Ljung, and Söderström, 1977). The corresponding Venn diagram is shown in *Fig. 38b*.

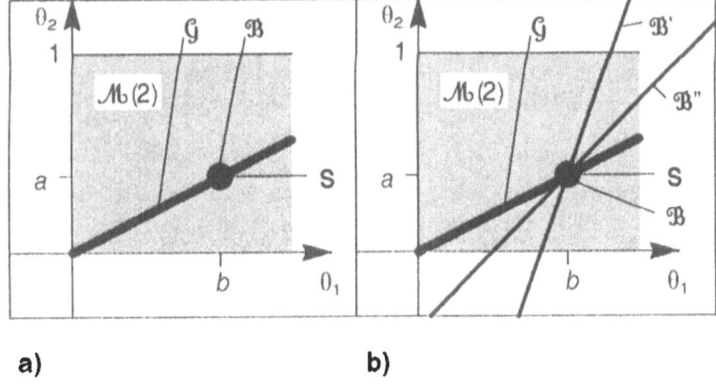

a)   **b)**

Fig. 38: Venn diagrams illustrating two ways to make feedback control a Sufficiently Stimulating experiment:
a) Perturbation signal in the controller.
b) Two feedback gains: $\mathcal{B} = \mathcal{B}' \cap \mathcal{B}''$

7   Validation techniques

In this and the following chapters, the 'modelling' part of the model making is assumed to have been concluded, and interest is focussed on the 'identification' part. This means that $\mathbf{X}$, $\mathcal{M}$, and $\mathcal{G}$ are given (and should satisfy the conditions on proper experimentation and modelling), and the computer takes over from there. The purpose of the following is therefore to review some principles that can be used for designing algorithms for different tasks in identification, in particular validation, falsification, and structure identification.

'Validatability' of an identification problem was defined as the feasibility of determining from data $(\mathbf{c}, \mathbf{d})$ whether a given model M is valid in the sense that $M \in \mathcal{G}$. Only then would one be able to design a rule for stopping the sequential refinement and falsification procedure outlined in section 4.4 on "Basic identification procedures". In the trivial case that $\mathcal{G}$ is not dependent on $(\mathbf{X}, \mathbf{S})$ the problem is of course always validatable, when one has defined $\mathcal{G}$. But $\mathcal{G}$ generally depends on $(\mathbf{X}, \mathbf{S})$, and one cannot compute it, since $\mathbf{S}$ is unknown, and possibly also $\mathbf{X}$. However, in the special but important case that $\mathcal{G}$ is designed to express the region around $(\mathbf{X}, \mathbf{S})$ that the model may belong to and still be considered purposive (and assuming Purpose Equivalence), one can compute the probability $V(M|\mathbf{c}, \mathbf{d}) = \Pr\{M \in \mathcal{G}(\mathbf{X}, \mathbf{S}) | \mathbf{c}, \mathbf{d}\}$ that M is purposive. Some techniques for doing that will be developed in this chapter.

The theory will assume long data samples (but not infinite) and use asymptotic approximations of the Likelihood function, since it is generally prohibitively difficult to compute anything otherwise (with few exceptions), and any effective statistical inference will need long samples anyhow. In addition, parametric models will be used throughout, since in order to compute something one has to have only a finite number of unknowns.

> **Remark:** Other approaches to the problem of validating a model, taking into account its purpose, have been suggested in the literature, even if such contributions are scarce (see section 9.3, "Philosophy revisited"). There are more references treating 'validation' without a specified purpose. In that case it is possible

to confront the model only with experiment data. Methods for this are the topic of Chapter 8 on "Falsification techniques".

Invoking father Bayes again

According to the 'Bayesian approach' the system $(\mathbf{X},\mathbf{S})$ is a stochastic variable, and its probability distribution, given data, is determined by (2.5). That probability distribution is known, and is the only thing that can ever be known about $(\mathbf{X},\mathbf{S})$, given prerequisites. Hence, one should formally be able to compute

$$V(M|\mathbf{c},\mathbf{d}) = \Pr\{M \in \mathcal{G}(\mathbf{X},\mathbf{S})|\mathbf{c},\mathbf{d}\}$$

$$= \int \mathrm{Ind}\{M \in \mathcal{G}(X,S)\}\ dP[X,S|\mathbf{c},\mathbf{d}] \tag{7.1}$$

provided one can define the space to integrate over and the probability measure. However, that creates some problems that have to be overcome first.

The data source is not a feasible model

The system $(\mathbf{X},\mathbf{S})$, the 'true data source', is in 'the set of all models', which at best is a 'space' of very high dimension, or infinite, and certainly not practical. One has to integrate over a space of lower dimension.

> **Remark:** That is possibly an academic problem, since in order to deal with the problem one has to make assumptions that allow the analysis to be carried out *as if* $(\mathbf{X},\mathbf{S})$ were in $\bar{\mathcal{M}}$. The problem is similar to that discussed in chapter 6 on "Large–sample theory", and is solved here by assuming that there a Purpose–Equivalent models $\mathbf{M} \in \bar{\mathcal{M}}$. The distinction between $(\mathbf{X},\mathbf{S})$ and $\mathbf{M}$ is however not uninteresting, even in practice, since it highlights the requirement that $\bar{\mathcal{M}}$ be sufficiently general for the analysis to hold.

It follows from the discussion in section 4.5 on "Conditions for Bayesian validation" that one has to have a model structure $\bar{\mathcal{M}}$ at least so wide that the joint distribution of the 'best' model $\mathbf{M}$ and the 'modelling error' $\mathbf{E}$ conditional on data $(\mathbf{c},\mathbf{d})$ becomes orthogonal, *i.e.* $P[\mathbf{M},\mathbf{E}|\mathbf{c},\mathbf{d}] = P[\mathbf{M}|\mathbf{c},\mathbf{d}]\ P[\mathbf{E}|\mathbf{c},\mathbf{d}],\ \mathbf{M} \in \bar{\mathcal{M}}$. The model to be validated, however, may be simpler, $M \in \mathcal{M} \subset \bar{\mathcal{M}}$.

In all practical cases one must also parametrize.

7.1   Validating parametric models

According to section 5.1 on "Parametrization" a parametric model is a triple $M \triangleq (\mathcal{A}, \nu, \theta)$, where $\mathcal{A}$ is the form of the model, e.g. linear state-vector model, ν is a (possible vector-valued) 'structure index', whose dimension depends on $\mathcal{A}$, and θ is a vector of real variables, whose dimension depends on ν. This allows one to define a sequence of structures of a given degree of complexity n by $\mathcal{M}(n) = \mathcal{M}(n-1) \bigcup_{\nu \in \mathcal{N}_n} \mathcal{M}_\nu$,

where $\mathcal{M}_\nu = (\mathcal{A}, \nu)$, and $\mathcal{N} = \{\mathcal{N}_1, \mathcal{N}_2, \ldots, \mathcal{N}_{\bar{n}}\}$ is a given partial sequencing of the set of feasible structure indices ν, such that the number of parameters required to describe models in $\mathcal{N}_n$ increase with n.

The Purpose–Equivalent model must also be defined

In order to model the 'true source' (for the sake of argument) conceive a set $\mathcal{M}^\infty$ of all possible models, including $(\mathbf{X}, \mathbf{S})$, and represent the models in it by points in an infinite space $(X, S) = (\mathcal{A}^\infty, \infty, \theta)$ (where $\mathcal{A}^\infty$ would include all possible forms). In particular, $(\mathbf{X}, \mathbf{S}) = (\mathcal{A}^\infty, \infty, \Theta)$. The infinite complexity number is a convenient reminder of the fact that $(\mathbf{X}, \mathbf{S})$ is not in any feasible model structure. It will also be conceivable to partition θ into (θ_1, θ_2), where θ_1 has structure index $\bar{\nu} \in \mathcal{N}_{\bar{n}}$, and write any model as a quintuple $(\mathcal{A}^\infty, \bar{\nu}, \infty, \theta_1, \theta_2)$, where θ_2 has infinite dimension.

In particular, the relation between the system $(\mathbf{X}, \mathbf{S})$ and any Purpose–Equivalent model $\mathbf{M}$ is the following: $(\mathbf{X}, \mathbf{S}) = (\mathcal{A}^\infty, \bar{\nu}, \infty, \Theta_1, \Theta_2)$ and $\mathbf{M} = (\mathcal{A}^\infty, \bar{\nu}, \infty, \Theta_1, 0)$, where the joint conditional parameter distribution satisfies the condition $p[\theta_1, \theta_2 | \mathcal{A}^\infty, \bar{\nu}, \infty, c, d] = p[\theta_1 | \mathcal{A}^\infty, \bar{\nu}, \infty, c, d] \, p[\theta_2 | \mathcal{A}^\infty, \bar{\nu}, \infty, c, d]$. A case where the Source has two dimensions and the widest structure one is illustrated in *Fig. 39*.

Since $\mathcal{A}^\infty$ was introduced only for the sake of argument, it has now served its purpose. Furthermore, since $\mathcal{A}$ is the same for all parametric models considered for validation, that too has served its purpose for the argument, and in the interest of simplification it will usually not be written out in the sequel.

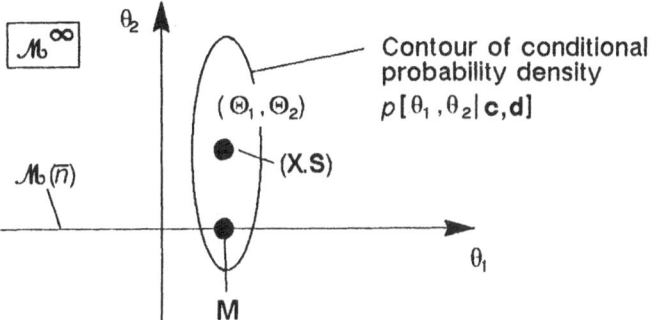

Fig. 39: Illustrating the relation between the source (X,S) and Purpose–Equivalent models **M** in the set $\mathcal{M}_b(\bar{n})$ of feasible models.

A too high complexity causes singularity

If the widest model structure $\overline{\mathcal{M}_b}$ uses more parameters than needed, the evaluation of the validity criterion will run into numerical difficulties due to singularity. This would mean that there are actually models of lower complexity that are Purpose–Equivalent, and obviously, there would be no point in using a Purpose–Equivalent model of higher complexity than needed. Let the minimum complexity be **n**. It may not be known, and the consequences of under– or over–estimating **n** will be analysed in section 7.3.

The 'validity' of a model is now well defined

Assume the condition of Purpose Equivalence holds for $\mathbf{M} = (\nu_{\mathbf{n}}, \Theta)$, and apply (4.7). Then

$$V(M|\mathbf{c},\mathbf{d}) = \Pr\{M \in \mathcal{G}(\nu_{\mathbf{n}},\Theta)|\mathbf{c},\mathbf{d}\}$$

$$= \int \mathrm{Ind}\{M \in \mathcal{G}(\nu,\theta)\}\ dP[\nu,\theta|\mathbf{c},\mathbf{d}] \tag{7.2}$$

and

$$dP[\nu,\theta|\mathbf{c},\mathbf{d}] = p[\mathbf{c},\mathbf{d}]^{-1}\ p[\mathbf{c},\mathbf{d}|\nu,\theta]\ p[\nu]\ p[\theta|\nu]\ d\theta \tag{7.3}$$

since the apriori distribution of Θ has a density.

The factors in (7.3) are known, except $p[\mathbf{c},\mathbf{d}]$. It can formally be computed by summing $p[\mathbf{c},\mathbf{d}|\nu,\theta]$ over ν and integrating over θ (see section 9.5.3 on "The Bayesian approach"). However, this can be expected to mean unsurmountable difficulties in practice, if the structures and the number of parameters are many.

The problem is reduced much if $\mathcal{G}(v_n, \Theta) \subset \mathcal{M}_v$, $v = v_n$. This is a quite reasonable restriction, since it means only that it should be possible to express the purpose using a subset of the same structure as was specified for defining the Purpose–Equivalent model. And this means that v_n is no longer a random variable, but a designer's choice. To indicate this replace v_n by μ, and define

Purpose: $\mathcal{G}(\mu, \Theta) = \{v, \theta | G(\mu, \Theta, v, \theta) \geq 0, v \leq \mu\}$ (7.4)

where $G(\mu, \Theta, v, \theta) \geq 0$ defines the 'tolerance region' in $\mathcal{M}_\mu$ given by the purpose.

> **Remark:** Notice that, formally, one has to assume that only the Purpose–Equivalent model be parametric and with a given structure. In practice, models to be validated must also be parameteric and belong to a substructure, since it will be difficult to specify a reasonable G–function otherwise. However, what is important is that the model to be validated, $M \triangleq (v^M, \theta^M)$, may belong to a *smaller* set than $\mathbf{M}_\mu = (\mu, \Theta)$.

Inserting (7.4) into (7.2) yields

$$V(v^M, \theta^M | \mu, \mathbf{c}, \mathbf{d}) = \int \text{Ind}\{G(\mu, \theta, v^M, \theta^M) \geq 0\}\ dP[\mu, \theta | \mathbf{c}, \mathbf{d}]$$

$$= \int \text{Ind}\{G(\mu, \theta, v^M, \theta^M) \geq 0\}\ p[\theta | \mu, \mathbf{c}, \mathbf{d}]\ d\theta \qquad (7.5)$$

where the new argument μ in V indicates that the structure of the Purpose–Equivalent model is specified by the purpose.

> **Remark:** Notice that (v^M, θ^M) may well depend on data $(\mathbf{c}, \mathbf{d})$, as it usually does when the model has been fitted to data. Hence, there is nothing so far to prohibit both fitting and validating a model using the same data sample. This is in contrast to falsification, where dependence or independence between model and data plays a crucial role.

When data are sequential, the analysis and the result are the same, except that one must hang subscripts N on $\mathbf{c}$ and $\mathbf{d}$, to indicate the length of the sample. A problem is still that even if the purposive domain is well defined by (7.4), the integral in (7.5) is prohibitively difficult to compute in general cases.

7.2 Large–sample techniques

If the aposteriori distribution $p[\theta|\mu, c_N, d_N]$ of the unknown Θ tends to a gaussian distribution as $N \to \infty$, then

$$\log p[\theta|\mu, c_N, d_N]$$
$$\to -\tfrac{1}{2} \log \det(2\pi R_\theta) - \tfrac{1}{2} (\theta - \hat\theta)^T R_\theta^{-1} (\theta - \hat\theta) \qquad (7.6)$$

where $\hat\theta$ is the ML–estimate and R_θ is the covariance matrix of the distribution. Both depend on (μ, c_N, d_N).

> **Remark:** The convergence to a gaussian distribution is not a general property, of course, but holds under mild conditions. See for instance Caines and Ljung (1976), Ljung and Caines (1979), Ljung (1987a).

The probability that (ν^M, θ^M) is purposive is from (7.5) and (7.6)

$$V(\nu^M, \theta^M | \mu, c, d)$$
$$= \int \mathrm{Ind}\{G(\mu, \theta, \nu^M, \theta^M) \geq 0\}$$
$$\exp[-\tfrac{1}{2}(\theta - \hat\theta)^T R_\theta^{-1} (\theta - \hat\theta)]/\mathrm{sqrt}\,\det(2\pi R_\theta)\, d\theta \qquad (7.7)$$

where R_θ is nonsingular.

Now, change coordinates in parameter space: Let $\eta = \Gamma_\theta^T \theta$, where $\Gamma_\theta \Gamma_\theta^T = R_\theta^{-1}$. Then the integral (7.7) will be

$$V(\nu^M, \theta^M | \mu, c_N, d_N)$$
$$= \int \mathrm{Ind}\{G(\mu, \Gamma_\theta^{-T} \eta, \nu^M, \theta^M) \geq 0\}\, \exp[-\tfrac{1}{2}\|\eta - \hat\eta\|^2]\, (2\pi)^{-r/2}\, d\eta \qquad (7.8)$$

where $r = \dim(\theta)$. One has to integrate the gaussian distribution over the domain $G(\mu, \Gamma_\theta^{-T} \eta, \nu^M, \theta^M) \geq 0$. This may still be tedious, if the dimension is large. However, in at least two important special cases the integration can be carried out analytically. And this opens up possibilities for efficient approximations in quite general cases.

7.2.1 Approximations

In the case that $G(\mu,\theta,\nu^M,\theta^M) \geq 0$ is bordered by a single plane

$$G(\mu,\theta,\nu^M,\theta^M) = g(\mu,\nu^M,\theta^M)\ \theta - h(\mu,\nu^M,\theta^M) \geq 0 \qquad (7.9)$$

the integral can be evaluated easily:

Evaluation: Insert (7.9) into (7.8). Then

$$V(\nu^M,\theta^M|\mu,\mathbf{c}_N,\mathbf{d}_N)$$

$$= \int \text{Ind}\{g\ \Gamma_\theta^{-T}\ \eta - h \geq 0\}\ \exp[-\tfrac{1}{2}\|\eta - \hat\eta\|^2]\ (2\pi)^{-\mathbf{r}/2}\ d\eta$$

Do an affine transformation to new coordinates ξ, that first moves the origin to $\hat\eta$, and then turns the coordinate system so that the border plane becomes orthogonal to the ξ_1–axis. Then

$$V(\nu^M,\theta^M|\mu,\mathbf{c}_N,\mathbf{d}_N)$$

$$= \int \text{Ind}\{g\ \hat\theta + \lambda\ \xi_1 - h \geq 0\}\ \exp[-\tfrac{1}{2}\|\xi\|^2]\ (2\pi)^{-\mathbf{r}/2}\ d\xi$$

$$= \int \text{Ind}\{g\ \hat\theta + \lambda\ \xi_1 - h \geq 0\}\ \exp[-\tfrac{1}{2}\xi_1^2]/\sqrt{2\pi}\ d\xi_1$$

$$= \Phi\{[g\ \hat\theta - h]/\lambda\}$$

where Φ is the cumulative normal distribution and

$$\lambda^2 = \|g\ \Gamma_\theta^{-T}\|^2 = g\ \Gamma_\theta^{-T}\ \Gamma_\theta^{-1}\ g^T = g\ R_\theta\ g^T.$$

Now, if $G(\mu,\theta,\nu^M,\theta^M)$ is nonlinear in θ, then it is conceivable to approximate the border by a tangent plane around the point θ^* that has the maximum probability $p[\theta^*|\mu,\mathbf{c}_N,\mathbf{d}_N]$ on the border. The motif for this choice of point to linearize around is that it should yield a small approximation error. The latter is the difference between the probability masses outside $G(\mu,\theta,\nu^M,\theta^M) \geq 0$ and outside the approximating plane respectively. Since one wants to validate with a high confidence, both masses should be small. Since, further, the probability function decreases more than exponentially outside the border and away from the point θ^* where the borders touch, the approximation will even be good. One gets

$$G(\mu,\theta,\nu^M,\theta^M) \approx G_\theta(\mu,\theta^*,\nu^M,\theta^M)\ (\theta - \theta^*)$$

$$= \text{grad}_\theta * G(\mu,\theta^*,\nu^M,\theta^M)(\theta - \theta^*)$$

$$= \text{grad}_\theta * G(\mu,\theta^*,\nu^M,\theta^M)\ \Gamma_\theta^{-T}\ (\eta - \eta^*) \tag{7.10}$$

The point that maximizes the density on the border is

$$\eta^* = \arg\min\{\|\eta - \hat{\eta}\|^2\,|\,G(\mu,\Gamma_\theta^{-T}\eta,\nu^M,\theta^M) = 0\} \tag{7.11}$$

The optimization problem is well known, and can usually be solved by an iterative procedure:

> **Solution:** The vector $\hat{\eta} - \eta^*$ is parallel with the border gradient
> $\Gamma_\theta^{-1}\ G_\theta^T(\mu,\theta^*,\nu^M,\theta^M)$. Hence,
> $\eta^* = \hat{\eta} - \rho\ \Gamma_\theta^{-1}G_\theta^T(\mu,\theta^*,\nu^M,\theta^M)/\|\Gamma_\theta^{-1}\ G_\theta^T(\mu,\theta^*,\nu^M,\theta^M)\|$,
> where ρ is the distance to the border. It should satisfy
> $G(\mu,\Gamma_\theta^{-T}\ \hat{\eta} - \rho\ \Gamma_\theta^{-T}\ \Gamma_\theta^{-1}\ G_\theta^T/\|\Gamma_\theta^{-1}\ G_\theta^T\|,\nu^M,\theta^M) = 0$.
> Solving this by Newton's method yields
> $\rho \leftarrow \rho + G(\mu,\Gamma_\theta^{-T}\ \eta^*,\nu^M,\theta^M)/\|\Gamma_\theta^{-1}\ G_\theta^T(\mu,\Gamma_\theta^{-T}\ \eta^*,\nu^M,\theta^M)\|$
> $\eta^* \leftarrow \hat{\eta} - \rho\ \Gamma_\theta^{-1}G_\theta^T\|\Gamma_\theta^{-1}\ G_\theta^T\|$

When the point η^* has been found, one can use the same procedure as in the linear case and arrive at the approximate probability

$$V(\nu^M,\theta^M|\mu,\mathbf{c}_N,\mathbf{d}_N) \approx \Phi(\|\eta^* - \hat{\eta}\|) = \Phi(\rho) \tag{7.12}$$

Changing back to the original coordinates θ in the iterative procedure yields

$$\rho \leftarrow \rho + G(\mu,\theta^*,\nu^M,\theta^M)/\text{sqrt}(G_\theta\ R_\theta\ G_\theta^T)$$

$$\theta^* \leftarrow \hat{\theta} - \rho\ R_\theta\ G_\theta^T/\text{sqrt}(G_\theta\ R_\theta\ G_\theta^T) \tag{7.13}$$

It remains to compute $\hat{\theta}$ and R_θ. A simple way is to differentiate (7.6). Notice also that from (3.3)

$$\log p(\theta|\mu,\mathbf{c}_N,\mathbf{d}_N) = \textit{constant} - N\ Q(\theta|\mu,\mathbf{c}_N,\mathbf{d}_N) \tag{7.14}$$

where *constant* does not depend on θ. Then

$$N\ \nabla_\theta Q(\theta|\mu,\mathbf{c}_N,\mathbf{d}_N)^T \rightarrow R_\theta^{-1}\ (\theta - \hat{\theta})$$

$$N\ \nabla_{\theta\theta} Q(\theta|\mu,\mathbf{c}_N,\mathbf{d}_N) \rightarrow R_\theta^{-1} \tag{7.15}$$

where ∇_θ denotes the 'gradient', i.e. the row vector of differentiation operators with respect to θ, and $\nabla_{\theta\theta}$ the matrix of second-order differentiation operators.

With $\theta = \theta^*$ inserted

$$R_\theta = Q_{\theta\theta}^{-1}\ N^{-1}$$
$$\hat{0} = \theta^* - Q_{\theta\theta}^{-1}\ Q_\theta^{T} \qquad\qquad (7.16)$$

where Q_θ and $Q_{\theta\theta}$ are the loss function gradients evaluated at the point θ^*.

The approximation of the border by a plane breaks down, if there are several border planes cutting through sets of about the same maximum probability density of the aposteriori density $p[\theta|\mu,c_N,d_N]$. In that case the contribution to the failure probability $\Pr\{M \notin \mathcal{G}(M)|\mu,c,d\}$ from outside the *closest* plane will not dominate over those of the other planes, and the contributions from the latter cannot be neglected. If the other approximating planes were parallel or orthogonal in the η-space, it would be easy to evaluate (7.7) and modify the rule accordingly. However, in other cases of approximating planes the evaluation of the failure probability will be difficult. An approximate rule for such cases will be the following:

When a plane is not a good approximation, use a sphere

If the problem (7.11) of finding the closest point on the tolerance border $G(\mu,\Gamma_\theta^{-T}\eta^*,\nu^M,\theta^M) = 0$ has several local minima with almost the same values of $\|\eta^* - \hat{\eta}\|^2$, then it is a reasonable approximation to substitute a *sphere* (or possibly a spherical cylinder) for the original border.

Replacing $G \geq 0$ in (7.8) with the largest sphere centered at $\hat{\eta}$ yields

$$V(\nu^M,\theta^M|\mu,c_N,d_N)$$
$$\approx \int \mathrm{Ind}\{\rho^2 - \|\eta - \hat{\eta}\|^2 \geq 0\}\ \exp[-\tfrac{1}{2}\|\eta - \hat{\eta}\|^2]\ (2\pi)^{-r/2}\ d\eta \qquad (7.17)$$

where ρ is the shortest distance to the border. But the integral in (7.17) defines the χ^2-distribution with argument ρ^2 and r degrees of freedom.

Collecting the results yields the following alternative rules:

Validating parametric models

Let the domain of purposive models be defined by $\mathcal{G}(M) = \{v, \theta \mid G(\mu, \Theta, v, \theta) \geq 0, v \leq \mu\}$, where $M = (\mu, \Theta)$ is a Purpose–Equivalent model, with r parameters. Let (v^M, θ^M) be the model to be validated.

In order to test the model, compute the point θ^* with maximum aposteriori probability on the border $G(\mu, \theta^*, v^M, \theta^M) = 0$, for instance by the iterative procedure

$$G \leftarrow G(\mu, \theta^*, v^M, \theta^M); \quad G_\theta \leftarrow \mathrm{grad}_{\theta^*} G(\mu, \theta^*, v^M, \theta^M)$$

$$\rho \leftarrow \rho + G/\mathrm{sqrt}(N^{-1} G_\theta Q_{\theta\theta}^{-1} G_\theta^T)]$$

$$\theta^* \leftarrow \theta^* - Q_{\theta\theta}^{-1} [Q_\theta^T + \rho \, G_\theta^T/\mathrm{sqrt}(N \, G_\theta \, Q_{\theta\theta}^{-1} \, G_\theta^T)]$$

where

$Q_\theta = \nabla_\theta Q(\theta^* \mid \mu, c_N, d_N)$, $Q_{\theta\theta} = \nabla_{\theta\theta} Q(\theta^* \mid \mu, c_N, d_N)$.

Then label the model (v^M, θ^M) 'validated' for the purpose $\mathcal{G}$, if $V(v^M, \theta^M \mid \mu, c_N, d_N) > \gamma$.

If there is one solution θ^* with a clearly smallest ρ, then $V(v^M, \theta^M \mid \mu, c_N, d_N) \approx \Phi(\rho)$.

If there are several solutions with about the same minimum ρ, then $V(v^M, \theta^M \mid \mu, c_N, d_N) \approx \mathrm{Chi\_square}[\rho^2, r]$.

Remark: If the region $G(\mu, \theta, v^M, \theta^M) \geq 0$ is convex, the two approximate values of the validation confidence obtained by approximating the border $G(\mu, \theta, v^M, \theta^M) = 0$ with a plane and a sphere respectively, will be upper and lower bounds of the exact value. The 'sphere' approximation is the safe approximation, *i.e.* if a model is computed as 'valid' with a certain probability, it will actually be valid with a higher probability. Unfortunately, it is also a conservative approximation in most cases, since it would be reasonable to presume that $\hat{\eta}$ is *not* close to the center of a spherical domain $G(\mu, \Gamma_\theta^{-T} \eta, v^M, \theta^M) \geq 0$. In that case, on the other hand, the 'plane' approximation should be good, whenever the computed confidence is high, since the probability mass decreases rapidly with increasing distance from $\hat{\eta}$. The two rules are illustrated in *Fig. 40*.

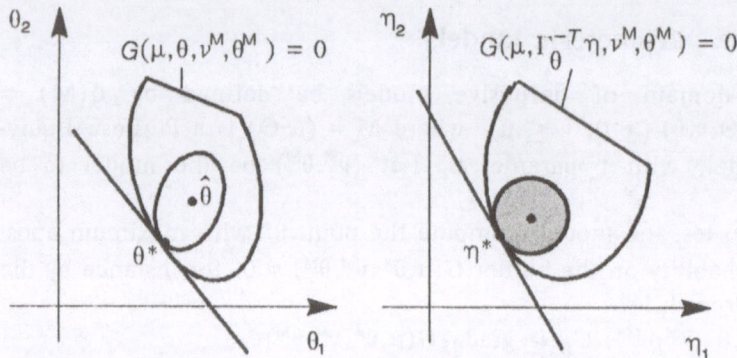

Fig 40: Approximating borders to tolerance regions in parameter space and in the orthogonalized parameter space.
With the spherical approximation the probability mass in the
shaded area is the approximate probability that the model
defined by (ν^M, θ^M) is purposive.

7.2.2 Validating sufficient accuracy

Validation is particularly simple, when the purpose of identification is
to obtain a 'sufficiently accurate' probabilistic model $p[c_K, d_K | \nu, \theta]$ of
the true source $p[c_K, d_K | \mathbf{X}, \mathbf{S}]$. A way to define a suitable set $\mathcal{G}(\mu, \Theta) = \{\nu, \theta | G(\mu, \Theta, \nu, \theta) \geq 0\}$ of purposive models for this case is the following:

Assume for the moment that there is a Source–Equivalent model
$p[c_K, d_K | \nu_{\mathbf{n}}, \Theta]$ of minimal complexity, and also that the model to be
validated may be expanded (by appending zeroes) to the same model
structure. Intuitively, one would want to accept models such that

$$Q(\theta | \nu_{\mathbf{n}}, c_N, d_N) \triangleq -\frac{1}{N} \sum_{k=1}^{N} \log p[c(k), d(k) | c_{k-1}, d_{k-1}, k, \nu_{\mathbf{n}}, \theta]$$

$$\approx -\frac{1}{N} \sum_{k=1}^{N} \log p[c(k), d(k) | c_{k-1}, d_{k-1}, k, \nu_{\mathbf{n}}, \Theta]$$

$$= Q(\Theta | \nu_{\mathbf{n}}, c_N, d_N) \tag{7.18}$$

for large N.

If the condition of Likelihood Convergence is satisfied, then

$$Q(\theta|\nu_{\mathbf{n}},c_N,\mathbf{d}_N) \approx E\{Q(\theta|\nu_{\mathbf{n}},c_N,\mathbf{d}_N)|\nu_{\mathbf{n}},\Theta\}$$

$$= -\frac{1}{N}\int \log p[c_N,d_N|\nu_{\mathbf{n}},\theta]\; p[c_N,d_N|\nu_{\mathbf{n}},\Theta]\; d(c_N,d_N)$$

$$\triangleq \overline{Q}_N(\theta|\nu_{\mathbf{n}},\Theta) \tag{7.19}$$

From the lemma of the Maximum Likelihood:

$$\overline{Q}_N(\theta|\nu_{\mathbf{n}},\Theta) \geq \overline{Q}_N(\Theta|\nu_{\mathbf{n}},\Theta) \tag{7.20}$$

and for large N

$$\overline{Q}_N(\theta|\nu_{\mathbf{n}},\Theta) = [\log \det(2\pi\, R_\theta) + (\theta - \Theta)^T R_\theta^{-1}\,(\theta - \Theta)]/2N \tag{7.21}$$

Hence, it comes natural to accept as 'good' all probabilistic models that are within a given relative margin $\epsilon \geq 0$ from the probability density of a Source–Equivalent model. That is,

$$G(\nu_{\mathbf{n}},\Theta,\nu_{\mathbf{n}},\theta) \triangleq \epsilon + \overline{Q}_N(\Theta|\nu_{\mathbf{n}},\Theta) - \overline{Q}_N(\theta|\nu_{\mathbf{n}},\Theta)$$

$$= \epsilon - (\theta - \Theta)^T R_\theta^{-1}\,(\theta - \Theta)/2N \tag{7.22}$$

> **Remark:** Notice that the definition of G implies a nesting condition; it excludes all models that are not in $\mathcal{M}_b(\nu_{\mathbf{n}})$. In other words, they are apriori not acceptable approximations (for instance, too complex, or of a mathematically unconvenient form for whatever reason). This means that the designer has full freedom to stick to whatever form of model he/she likes, and test whatever approximation that can be modelled as removing structural elements from a widest structure, as long as the latter is wide enough to hold a Source–Equivalent model.

The assumption of Source Equivalence is needed only to define the loss function. If one would (for any other reason) be satisfied with approximating the 'best' model within the class $\mathcal{M}_b(\mu)$, then the derivation serves only to motivate that (7.22) is indeed a reasonable definition of 'good models' even for other structure indices μ. This has nothing to do with the condition for validatability, which is still Purpose Equivalence, and holds for any G.

> **Remark:** Basically, specifications (the G–function) are the responsibility of the designer, which may employ any theory, heuristics, or approximation he/she likes. This is in contrast to the theory, based on the specifications, which is constrained by these specifications.

The probability that an arbitrary parametric model (μ, θ^M) is sufficiently accurate is from (7.5) and (7.6)

$$V(\theta^M | \mu, \mathbf{c}_N, \mathbf{d}_N)$$

$$= \int \text{Ind}\{G(\mu, \theta, \mu, \theta^M) \geq 0\}\ p[\theta | \mu, \mathbf{c}_N, \mathbf{d}_N]\ d\theta$$

$$= \int \text{Ind}\{(\theta - \theta^M)^T R_\theta^{-1}\ (\theta - \theta^M) \leq 2N\epsilon\}$$

$$\exp[-\tfrac{1}{2}\ (\theta - \hat{\theta})^T R_\theta^{-1}\ (\theta - \hat{\theta})]\ [2\pi\ R_\theta]^{-1/2}\ d\theta \qquad (7.23)$$

The expression can be evaluated by a series of variable transformations:

Evaluation: Let $\eta = \Gamma_\theta^T \theta$, where $\Gamma_\theta \Gamma_\theta^T = R_\theta^{-1}$. Then the aposteriori distribution of $\Gamma_\theta^T \Theta$ will be from (7.23)

$$\log p(\eta | \mu, \mathbf{c}_N, \mathbf{d}_N) \rightarrow -\tfrac{1}{2}\, \mathbf{r}\, \log(2\pi)\ -\tfrac{1}{2}\, \|\eta - \hat{\eta}\|^2$$

where $\mathbf{r} = \dim(\theta)$. From (7.23)

$$V(\theta^M | \mu, \mathbf{c}_N, \mathbf{d}_N)$$

$$= \int \text{Ind}\{\|\eta - (\eta^M - \hat{\eta})\|^2 \leq 2N\epsilon\}\ \exp[-\tfrac{1}{2}\|\eta\|^2]\,(2\pi)^{-r/2}\ d\eta$$

Do an orthogonal linear transformation to new coordinates ξ, such that the vector $\eta^M - \hat{\eta}$ becomes parallel with the ξ_1-axis. Let $\rho = \|\eta^M - \hat{\eta}\|$. Then

$$V(\theta^M | \mu, \mathbf{c}_N, \mathbf{d}_N)$$

$$= \int \text{Ind}\{(\xi_1 - \rho)^2 + \xi_2^2 + \ldots + \xi_r^2 \leq 2N\epsilon\}$$

$$\cdot \exp[-\tfrac{1}{2}(\xi_1^2 + \ldots + \xi^2)]\ (2\pi)^{-r/2}\ d\xi_1 \ldots d\xi_r$$

Let $r^2 = \xi_2^2 + \ldots + \xi_r^2$. Then r^2 is χ^2-distributed with $\mathbf{r}-1$ degrees of freedom. Denote its cumulative distribution

$$\Pr\{r^2 \leq \chi^2\} = \text{Chi\_square}(\chi^2, r-1)$$

Then

$$V(\theta^M | \mu, \mathbf{c}_N, \mathbf{d}_N)$$

$$= \int\ \text{Chi\_square}[2N\epsilon - (\xi_1 - \rho)^2, r-1]\ \exp[-\tfrac{1}{2}\xi_1^2]/\sqrt{2\pi}\ d\xi_1$$

The result is an integral over a finite interval of a continuous function, and can be evaluated easily by a numerical rule for quadrature, for instance Simpson's rule.

Denote the resulting function

$$V(\theta^M|\mu,\mathbf{c}_N,\mathbf{d}_N) = \text{Chi\_square}(2N\epsilon,r,\rho^2) \tag{7.24}$$

The right member with triple argument is known as the 'non-central χ^2-distribution function' (Kendall and Stuart, 1976). Various evaluation formulas have been compiled, for instance by Abramowitz and Stegun (1964).

It remains to compute ρ. But (7.15) yields immediately

$$\rho^2 = (\theta^M - \hat{\theta})^T R_\theta^{-1} (\theta^M - \hat{\theta})$$
$$= N \nabla_\theta Q(\theta^M) [\nabla_{\theta\theta} Q(\theta^M)]^{-1} \nabla_\theta Q(\theta^M)^T \tag{7.25}$$

The validation rule will be

Validating sufficient accuracy

Let the domain of purposive models be defined by $\mathcal{G} = \{\theta| \bar{Q}_N(\theta|\mu,\Theta) - \bar{Q}_N(\Theta|\mu,\Theta) \le \epsilon\}$, where (μ,Θ) is a Purpose-Equivalent model, and $\bar{Q}_N(\theta|\mu,\Theta) \triangleq E\{Q(\theta|\mu,\mathbf{c}_N,\mathbf{d}_N)|\mu,\Theta\}$.

Then a given parametric model (μ,θ^M) is validated for the purpose $\mathcal{G}$, if Chi_square$[2N\epsilon,\dim(\theta),\rho^2] > \gamma$,

where $\rho^2 = N Q_\theta(\theta^M) Q_{\theta\theta}(\theta^M)^{-1} Q_\theta(\theta^M)^T$,
$Q_\theta(\theta) = \nabla_\theta Q(\theta|\mu,\mathbf{c}_N,\mathbf{d}_N)$,
$Q_{\theta\theta}(\theta) = \nabla_{\theta\theta} Q(\theta|\mu,\mathbf{c}_N,\mathbf{d}_N)$.

Two special cases are of particular interest:

A correct approximation will eventually be validated, when the data sample increases

When N becomes large, then

$$\rho^2 = N Q_\theta(\theta^M) Q_{\theta\theta}(\theta^M)^{-1} Q_\theta(\theta^M)^T \to N \beta^2(\theta^M|\mu,\Theta) \tag{7.26}$$

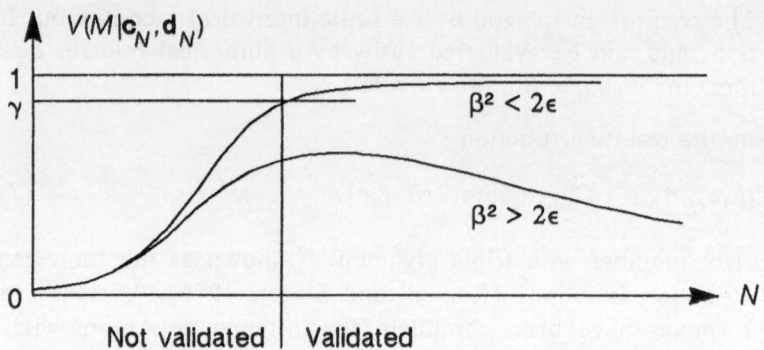

Fig. 41: Validation criterion as a function of sample length N for valid and invalid model.

where β^2 is the limiting function, and

Chi_square$[2N\epsilon, r, N\beta^2]$

$$= \int \text{Chi\_square}\{2N\epsilon - (\xi - \beta\sqrt{N})^2, r-1\} \; \sqrt{1/2\pi} \; \exp(-\tfrac{1}{2}\xi^2) \; d\xi \quad (7.27)$$

Now, if $N \to \infty$

Chi_square$[2N\epsilon - (\xi - \beta\sqrt{N})^2, r-1]$

$\to 1$, if $2\epsilon - \beta^2 > 0$, and $\to 0$, if $2\epsilon - \beta^2 < 0$ $\qquad\qquad (7.28)$

From (7.25) the first condition is equivalent to

$$(\theta^M - \hat{\theta})^T \, Q_{\theta\theta}(\theta^M) \, (\theta^M - \hat{\theta}) < 2\epsilon \qquad\qquad (7.29)$$

It follows that models θ^M within the ellipsoid defined in this way will eventually be validated when N becomes large enough. Those outside will never. This is not surprising, considering the definition (7.22) of the set of good models. Hence, it is possible to sketch the behaviour of Chi_square$[2N\epsilon, r, \rho^2]$ as a function of N, as in *Fig. 41*. Notice that if $\beta^2 > 2\epsilon$, the validation rule will not necessarily stop the inner loop.

There is a structure–independent stopping rule for the experiment

In the second particular case that one wants to validate the ML–estimate $\hat{\theta}$ of in the 'true' structure the offset is $\rho = 0$. The result yields a measure of the goodness of the ML–estimate that *is independent of structure*, and also a simple rule for when to stop collecting data: The accuracy is sufficient with a given risk α, and a given tolerance margin, when

$$N > \chi_r^2(\alpha)/2\epsilon \qquad (7.30)$$

where $\chi_r^2(\alpha)$ is the chi-square variable.

--------------------------- **Example 7.1** ---------------------------

Illustrating the validation of parametric models for a specific purpose.

In Example 4.2 the set of purposive models was computed for the case that the purpose of the identification was to design a stable controller. The closed loop is modelled by

$$d(k) = (\Theta_2 - \Theta_1 \, \theta_2/\theta_1) \, d(k-1) + \omega(k)$$

where $(\Theta_1, \Theta_2) = (1, 0.8)$ are the true parameters, and (θ_1, θ_2) those in the model used for designing the controller. Hence, the loop is stable, if $|\Theta_2 - \Theta_1 \, \theta_2/\theta_1| < 1$, and there are two border lines to the domain $G(\theta, \theta^M) \geq 0$ defining the set of purposive models:

$$G_1(\theta, \theta^M) \triangleq 1 - \theta_2 + \kappa \, \theta_1$$
$$G_2(\theta, \theta^M) \triangleq 1 + \theta_2 - \kappa \, \theta_1$$

where $\kappa = \theta_2^M/\theta_1^M$. Notice that the borders are expressed in terms of the first argument in $G(\theta, \theta^M)$, in contrast to those of the domain of good models $\mathcal{G}(\Theta)$, which is expressed in terms of the second argument. The domain between the borders is therefore different from that shown in *Fig. 15*.

The Likelihood loss is

$$Q(\theta|\mathbf{c}_N \cdot \mathbf{d}_N)$$
$$= \frac{1}{2} \sum_{k=1}^{N} [\mathbf{d}(k) - \theta_2 \, \mathbf{d}(k-1) - \theta_1 \, \mathbf{c}(k-1)]^2 + constant$$

and its second-derivative matrix is

$$Q_{\theta\theta}(\theta|\mathbf{c}_N, \mathbf{d}_N) = \frac{1}{N} \sum_{k=1}^{N} \begin{bmatrix} \mathbf{c}(k-1)^2 & \mathbf{c}(k-1) \, \mathbf{d}(k-1) \\ \mathbf{d}(k-1) \, \mathbf{c}(k-1) & \mathbf{d}(k-1)^2 \end{bmatrix}$$

Assume for simplicity that the stimulating sequence $\{\mathbf{c}(k)\}$ in the identification phase (open loop) is a 'white-noise' sequence with standard deviation σ. Assume further that N is so large that

$Q_{\theta\theta}(\theta|\mathbf{c}_N,\mathbf{d}_N)$ is approximated well by its limiting value (this will happen if **c** is a weak signal). Then

$$\frac{1}{N}\sum_{k=1}^{N} \mathbf{c}(k-1)^2 \approx \sigma^2,$$

$$\frac{1}{N}\sum_{k=1}^{N} \mathbf{c}(k-1)\,\mathbf{d}(k-1) \approx 0,$$

$$\frac{1}{N}\sum_{k=1}^{N} \mathbf{d}(k-1)^2 \approx \sigma^2\,\Theta_1^2/(1-\Theta_2^2) = \sigma^2\,25/9.$$

Now, the simplest model structure in Example 4.2 is $\mathcal{M}_b(0)$ with the single model defined by $\theta^M = (1,1)$. With the true value $\Theta = (1,0.8)$ it is obviously a purposive model, since $|0.8 - 1\cdot 1/1| < 1$. Can it be validated (without knowledge of Θ), and if so, how large a sample is needed?

The case is simple enough to solve analytically. Introduce the transformation $\eta = \Gamma_\theta^T\,\theta$, where

$$\Gamma_\theta\,\Gamma_\theta^T = R_\theta^{-1} = N\,Q_{\theta\theta} = \beta^2\begin{bmatrix} 1 & 0 \\ 0 & 25/9 \end{bmatrix},\text{ where }\beta^2 = N\,\sigma^2,$$

which yields

$$\Gamma_\theta = \beta\begin{bmatrix} 1 & 0 \\ 0 & 5/3 \end{bmatrix},\quad \eta = \beta\begin{bmatrix} \theta_1 \\ \theta_2\,5/3 \end{bmatrix}$$

$$G_1(\Gamma_\theta^{-T}\,\eta,\theta^M) \triangleq 1 - (0.6\,\eta_2 - \eta_1)/\beta$$

$$G_2(\Gamma_\theta^{-T}\,\eta,\theta^M) \triangleq 1 + (0.6\,\eta_2 - \eta_1)/\beta$$

The border lines in η–space (the space of orthogonalized aposteriori distribution of object parameters) are depicted in *Fig. 42*.

The points on the two borders closest to the peak value $\hat\eta$ of the aposteriori distribution of $\Gamma_\theta^T\,\Theta$ are

$$\eta_1^* = \pm 0.735\,\beta + 0.265\,\hat\eta_1 + 0.441\,\hat\eta_2$$

$$\eta_2^* = \mp 0.441\,\beta + 0.441\,\hat\eta_1 + 0.735\,\hat\eta_2$$

The distances ρ to the border are given by

$$\rho^2 = \|\hat\eta - \eta^*\|^2 = 0.735\,(\pm\beta - \hat\eta_1 + 0.6\,\hat\eta_2)^2$$
$$= 0.735\,\beta^2\,(\pm 1 - \hat\theta_1 + \hat\theta_2)^2$$

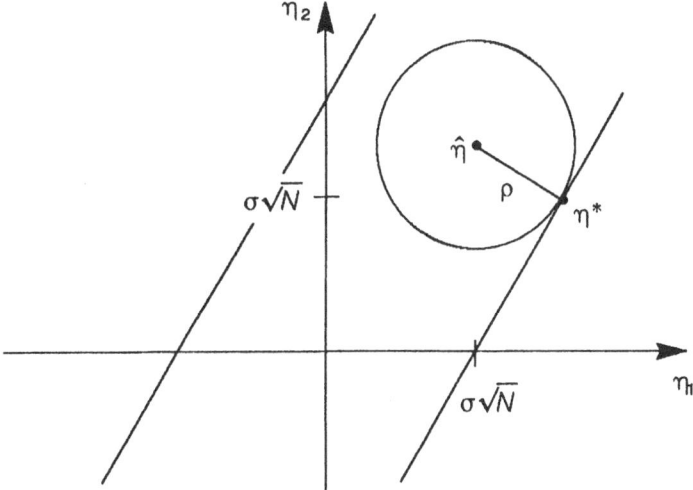

Fig. 42: Borders for good models in the orthogonalized space of the aposteriori distribution.

Hence, the probability that the model θ^M is in $\mathcal{G}(\Theta)$ is from (7.12)

$$V(\theta^M|\mathbf{c}_N,\mathbf{d}_N) = \Phi(0.857 \min|\pm 1 - \hat{\theta}_1 + \hat{\theta}_2|\sigma\sqrt{N})$$

For large N the peak value $\hat{\theta} \to \Theta = (1, 0.8)$ and $V(\theta^M|\mathbf{c}_N,\mathbf{d}_N) \to \Phi(0.686\ \sigma\sqrt{N})$. It obviously increases with N, and $(\mathbf{c}_N,\mathbf{d}_N)$ is sufficient data to validate the model defined by $\theta^M = (1, 1)$ with a given high confidence γ, when $\Phi(0.686\ \sigma\sqrt{N}) > \gamma$.

7.3 Two 'pitfalls'

The validation rule requires that a sufficient structure index $\nu_{\mathbf{n}}$ of a Purpose–Equivalent model be known. The value decides which derivatives of G and Q that are to be evaluated for the test. Obviously, one may have difficulties in determining $\nu_{\mathbf{n}}$, and it is therefore of interest to know what will be the effects of using a value $\mu \neq \nu_{\mathbf{n}}$. There is also a nesting condition: The model to be validated must have a structure index $\nu \leq \nu_{\mathbf{n}}$.

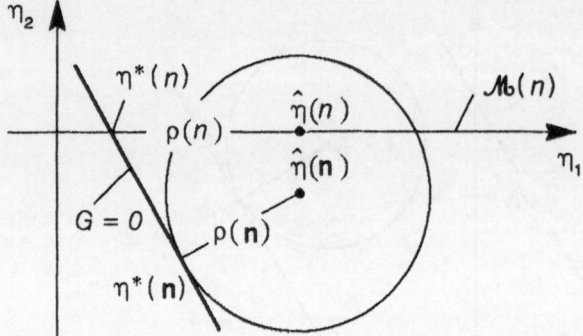

Fig 43: Illustrating the effect of understimating the sufficient complexity for Purpose Equivalence. The distance to the border depends on the value of the "error" parameter $\hat{\eta}(n) - \hat{\eta}(n)$.

A too low order may be serious

When the condition $\mu \geq \nu_n$ is not satisfied, the result of evaluating $V(\theta^M|\mu,c,d)$ will be wrong. The value may be either over- or underestimated; this depends on the G-function. The case of simple nesting is illustrated in *Fig. 43* (see also Example 4.4).

This is the 'pitfall' of violating the validatability condition, and there is little theory can do to help, except suggest that in validation tests one should not be too parsimonious with evaluating derivatives.

A too high value is not serious

When the order μ is *higher* than needed for $\mathcal{M}_{b\mu}$ to contain a Purpose-Equivalent model, the condition is still valid. However, the value may be so high that the hessian Q_{00} becomes singular and cannot be inverted. Theoretically, a too large μ would therefore be detectable from the singularity, and one would not have to consider that case. However, in practice the hessian will gradually become more illconditioned as μ increases, and determining a smallest sufficient μ will be uncertain. The case is illustrated in *Fig. 44*. It is clear that if the aposteriori distribution is illconditioned, then the ellipsoid touching the border $G = 0$ will contain little probability mass (ρ will be small), and the probability that the model is validated will be small. In that case the sequential refinement and falsification procedure will continue with more experimentation,

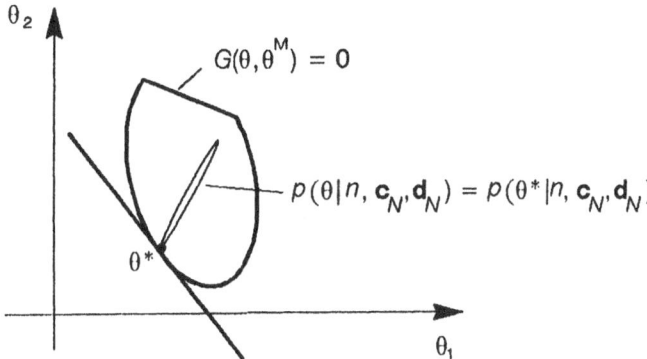

Fig 44: Illustrating the effect of a too high order in validation. The ellipsoid inside $G \geq 0$ will be small, and so will the computed confidence.

which is according to reason; the case is barely identifiable and requires more and better data to be possible to validate.

An exception to this case is when the region $G \geq 0$ is elongated in the same direction as the contour maps of $p[\theta|\mu,c,d]$. That is a case where the experiment may still be satisfactory for the purpose; the experiment is not enough for estimating all the parameters θ, but that will not be needed for the purpose. The model is over–parametrized. The obvious remedy would be to reduce μ. Notice, however, that as long as one can still invert the hessian (as long as the arithmetic precision suffices), one can use the validation rule even with over–parametrization.

One can validate models of simpler structure than the Purpose Equivalent

Notice that the model to be validated (ν^M, θ^M) may actually have a lower complexity than the Purpose–Equivalent model *(Fig. 45)*. This is extremely important, since it provides a theoretical basis for the common-sense belief that one should not use more complicated models than needed for the purpose (irrespective of what might be the 'true model'). It also provides a basis for *model reduction* schemes. When the purpose of the model has been defined, the validation rule decides when the reduction should *stop*.

 Remark: When validating sufficient accuracy, the non–central chi-square distribution appears instead of the ordinary chi–square dis-

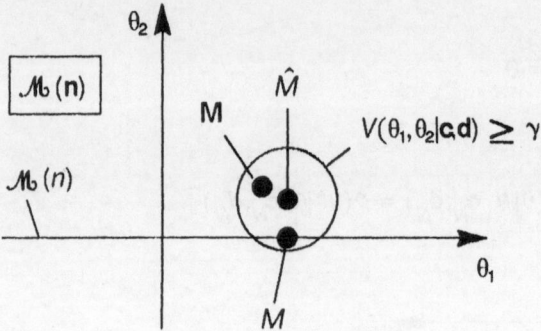

Fig. 45: Illustrating the validation of a model M of lower order than the Purpose–Equivalent **M**

tribution, when the complexity is lower than the Purpose–Equivalent (section 7.2.2).

Beware of another 'pitfall'

It was argued that violation of Purpose Equivalence, for instance by using a too small structure index μ in one of the validation rules derived in section 7.2 on "Large–sample techniques", creates a 'pitfall' — the probability value that is computed will not be correct.

However, apart from that, it might seem that validation answers all the questions concerning a model, whether obtained by identification or otherwise. The rule stops the identification procedure when and only when the model is purposive. Why then bother about the other requirements on a proper identification? If a model M is in $\mathcal{G}$ with high probability (which the test procedure computes), why should it matter how M was obtained, or whether it fits any data?

One answer is that if the experiment is not proper (for instance not Isolated, or not Reproducible), then the definition $\mathcal{G}$ of 'good models' still will not say anything about the actual application of the model. Hence, a second 'pitfall' comes from the difficulty to formulate an adequate criterion for a purposive model.

Proper identification still needs a falsification rule

Now, provided the experiment, the structure, and the purpose are indeed properly designed, so that a validation rule will give a correct result, *then* what is the purpose of the second stopping rule, that of falsification?

It is evident that both loops are necessary; one refines the experimentation, and the other the model structure. But why cannot the same rule stop both loops?

The following example shows that validation will in fact not stop the experimentation, when the model is wrong. That is not surprising either; a wrong model cannot be validated by more experimentation. Conversely, falsification will not stop the refinement of the structure in practice; any model can be falsified, if one tests it enough by experiment. And falsification prompts further refinement.

—————————————— **Example 7.2** ——————————————

This example illustrates why *two* stopping rules are needed.

Let S: $d(k) = \Theta_1\, c_1(k) + \Theta_2\, c_2(k) + \omega(k)$

where $\{\omega(k)\}$ are gaussian $(0,1)$. Let $c_i(k)$ be free to choose for the purpose of simplifying the analysis.

Assume the correct sequence of structures is known

$$\mathcal{M}(n) = \{\theta | d(k) = \sum_{\nu=1}^{n} \Theta_\nu\, c_\nu(k) + \omega(k)\}, \, \mathsf{n} = 2.$$

Let the purpose of identification be to find a model with a relative error $< \epsilon = 0.01$.

Even if one knows the correct complexity n, it may be desirable to see whether a simpler model with only $\theta_1 \neq 0$ would suffice — the influence of c_2 may actually be small.

It is convenient to introduce

$d_N \triangleq \mathrm{col}\{d(1),...,d(N)\}, \, c_N \triangleq \mathrm{col}\{c(1),...,c(N)\}$

where $c(k) = \mathrm{row}\{c_1(k),c_2(k)\}$.

The Likelihood loss is

$$Q(\theta|2,\mathbf{d}_N) = \frac{1}{2N} \sum_{k=1}^{N} [d(k) - \mathbf{c}(k)\, \theta]^2 \; + \textit{constant}$$

$$= \frac{1}{2N}\, \|\mathbf{d}_N - \mathbf{c}_N\, \theta\|^2 \; + \textit{constant}$$

$$= \tfrac{1}{2N} \, [\theta - (\mathbf{c}_N{}^T\mathbf{c}_N)^{-1} \, \mathbf{c}_N{}^T\mathbf{d}_N]^T \, \mathbf{c}_N{}^T\mathbf{c}_N \, [\theta - (\mathbf{c}_N{}^T\mathbf{c}_N)^{-1} \, \mathbf{c}_N{}^T\mathbf{d}_N]$$
$$+ \tfrac{1}{2N} \, \mathbf{d}_N{}^T \, [I - \mathbf{c}_N \, (\mathbf{c}_N{}^T\mathbf{c}_N)^{-1} \, \mathbf{c}_N{}^T] \, \mathbf{d}_N \; + constant$$

This is simplified considerably, if one chooses the stimulus so that $\mathbf{c}_N{}^T\mathbf{c}_N = N \, I$, for instance $c_1(k)$ switches between $+1$ and -1 every k, and $c_2(k)$ every other k, with N a multiple of 2. Then

$$Q(\theta|2,\mathbf{d}_N) = \tfrac{1}{2}\|\theta - \hat{\theta}\|^2 \; + constant$$

where $\hat{\theta} = \mathbf{c}_N{}^T\mathbf{d}_N/N$, i.e. its components are the two cross-correlations. The first two derivatives are

$$Q_\theta(\theta|2,\mathbf{d}_N)^T = - \, (\mathbf{d}_N{}^T - \theta^T \, \mathbf{c}_N{}^T) \, \mathbf{c}_N/N = \theta - \hat{\theta}$$

$$Q_{\theta\theta}(\theta|2,\mathbf{d}_N) = \mathbf{c}_N{}^T \, \mathbf{c}_N/N = I$$

Start the outer loop by hypothesizing that $n = 1$, and the inner loop by making a short experiment $N = 10$ (for instance), and computing the ML-estimate

$$\hat{\theta}_1 = \arg \min\|\theta_1 - \mathbf{c}_{1N}{}^T \, \mathbf{d}_N/N\|^2 = \mathbf{c}_{1N}{}^T \, \mathbf{d}_N/N$$

In order to validate the model use the rule for 'validating sufficient accuracy':

$$\rho^2 = N \, Q_\theta(\hat{\theta}_1) \, Q_{\theta\theta}(\hat{\theta}_1)^{-1} \, Q_{\theta\theta}(\hat{\theta}_1)^T$$
$$= N \, \|\mathbf{c}_{1N}{}^T \, \mathbf{d}_N/N - \hat{\theta}_1\|^2 + N \, \|\mathbf{c}_{2N}{}^T \, \mathbf{d}_N/N - 0\|^2$$
$$= (\mathbf{c}_{2N}{}^T \, \mathbf{d}_N)^2/N = N \, \hat{\theta}_2{}^2$$

Set a threshold $\gamma = 0.99$, and compute the value of the test variable Chi_square$(2N\epsilon, 2, N\hat{\theta}_2{}^2)$. Since $2N\epsilon = 2 \cdot 10 \cdot 0.01 = 0.2$ is small, the value is less than γ, unless $N \, \hat{\theta}_2{}^2$ is very large. The model is not validated.

Now, if there were no falsification rule, the inner loop would go on increasing the sample length to $N = 20, 30, 40,\dots$ (for instance), and the value of the test variable would change, and probably increase (that depends on the ω-sequence and the parameter values Θ_2). As long as the test variable increases with N, there is hope that $\hat{\theta}_1$ might still be validated from a sufficiently long N, and one would be able to keep the simple model. However, it is also possible that the test variable will never surpass the threshold γ for a valid model.

To see whether the procedure will end, investigate the limiting case when $N \rightarrow \infty$. Then

$$\text{Chi\_square}\,(2N\epsilon, 2, N\hat{\theta}_2{}^2) \rightarrow \text{Chi\_square}\,(2N\epsilon, 2, N\Theta_2{}^2).$$

But from (7.29) this tends to one if $\Theta_2{}^2 < 2\epsilon$, which is the (not surprising) limiting condition for being able to accept a model of too low complexity. However, if $\Theta_2{}^2 > 2\epsilon$, the inner loop will not stop, no model $(\hat{\theta}_1, 0)$ will be validated, and it requires a falsification rule to stop the inner loop and force the identification procedure to take θ_2 into account.

---

8   Falsification techniques

Falsifiability of an identification problem was defined as the feasibility of determining whether a given model M is contradicted by data $(\mathbf{c},\mathbf{d})$, that is, the feasibility of computing a set $\mathcal{B}(\mathbf{c},\mathbf{d})$ of 'unfalsified models'. That condition is obviously important in its own right for the feasibility of testing a final model. The falsification rule is then $M \notin \mathcal{B}(\mathbf{c},\mathbf{d})$. But it is also the basis for designing a stopping rule for the refinement of the experiment (the inner loop) in the sequential identification procedure outlined in section 4.4. In that case one wants to test whether *all* models in a class $\mathcal{M}$ are contradicted by data, and the falsification rule becomes $\mathcal{M} \cap \mathcal{B}(\mathbf{c},\mathbf{d}) = \emptyset$. Due to the randomness in the data the set $\mathcal{B}(\mathbf{c},\mathbf{d})$ is not unique and not necessarily the same for model and structure. Falsification techniques deal with the problems of finding suitable sets $\mathcal{B}(\mathbf{c},\mathbf{d})$ for the two cases.

The analysis so far of the set of unfalsified models has been of the *limiting case*, when the length N of the sample tends to infinity, and $\mathcal{B}(\mathbf{c}_{\infty},\mathbf{d}_{\infty})$ becomes a deterministic set (see section 6.4 on "Falsification in the limit"). In that case there is also a natural principle for defining $\mathcal{B}(\mathbf{c}_{\infty},\mathbf{d}_{\infty})$, namely as all models whose output distributions equal that of the sample distribution of experiment data. 'Identifiability' deals with the properties of $\mathcal{B}(\mathbf{c}_{\infty},\mathbf{d}_{\infty})$ under various conditions on experiment and model structure (see section 6.3).

With stochastic models and finite-length samples the set $\mathcal{B}(\mathbf{c}_N,\mathbf{d}_N)$ is stochastic (since it depends on data), and the problem is first to decide how to *define* $\mathcal{B}(\mathbf{c}_N,\mathbf{d}_N)$ in the stochastic case. Clearly, less data cannot falsify more models. Hence, one has the condition that $\mathcal{B}(\mathbf{c}_N,\mathbf{d}_N) \supseteq \mathcal{B}(\mathbf{c}_{\infty},\mathbf{d}_{\infty})$, and one has ways to define the limiting set $\mathcal{B}(\mathbf{c}_{\infty},\mathbf{d}_{\infty})$. But one has also to find a way to determine *how wide* the set $\mathcal{B}(\mathbf{c}_N,\mathbf{d}_N)$ has to be for a given sample length N. In addition, $\mathcal{B}(\mathbf{c}_N,\mathbf{d}_N)$ can no longer be regarded as 'the set of all models that accord with data', but becomes 'the set of all models not (yet) falsified by data'. The test becomes one-sided: $M \notin \mathcal{B}(\mathbf{c}_N,\mathbf{d}_N)$ falsifies, but $M \in \mathcal{B}(\mathbf{c}_N,\mathbf{d}_N)$ does not verify, not even in theory.

The theory of 'statistical inference' offers a number of principles for defining what would be suitable sets $\mathfrak{B}(c_N, d_N)$ for the falsification of models.

> **Remark:** There are several other approaches to the falsification problem in the literature. In addition to various statistical tests applicable to linear systems Kashyap and Ramachandra Rao (1976) suggest that one compare both various characteristics of model and data, computed in different ways, and models fitted to different parts of the data sample. See also section 9.3, "Philosophy revisited".

In the face of uncertainty one must take a calculated risk

A general way to deal with the falsification problem for stochastic models and finite data samples is to assign *probabilities* α, preferably small, to a 'reject' decision being wrong (Bohlin, 1987c). This can be expressed as $\Pr\{M \notin \mathfrak{B}(c, d) | M\} = \alpha$. The probability is called *'risk of the first kind'* — the risk that one will reject a correct model M, because $\mathfrak{B}(c, d)$ is a random set. The complementary probability that one will not reject a correct model, $\gamma = 1 - \alpha$, is called *'confidence'* in the reject decision. Thus, the set $\mathfrak{B}$ of not (yet) falsified models depends on γ, and the decision 'reject if $M \notin \mathfrak{B}(\gamma | c, d)$' has confidence γ.

A consequence of the definition is that even M may not be in $\mathfrak{B}(\gamma | c, d)$ for certain; there is a small probability α that it be outside. However, if one would prescribe no risk, $\alpha = 0$, then $\mathfrak{B}(\gamma | c, d)$ would be too wide to be of any use for a falsification rule. It would not falsify a correct model, but it would not falsify any incorrect models either. Since models inside $\mathfrak{B}(\gamma | c, d)$ cannot be rejected with any confidence, they must be accepted (so far), and $\mathfrak{B}(\gamma | c, d)$ will therefore also be referred to as the 'acceptance domain'.

> **Remark:** If there would be a Source–Equivalent model M in $\mathcal{M}_b(\bar{n})$, one would be able to use the set $\mathfrak{B}(\gamma | c, d)$ to make statements of the following type: "The probability is γ that $\mathfrak{B}(\gamma | c, d)$ contains a Source–Equivalent model". In this interpretation the set $\mathfrak{B}(\gamma | c, d)$ is called 'confidence region', since if γ is high, one may be confident in stating that M is actually somewhere in $\mathfrak{B}(\gamma | c, d)$. Hence, if a given model M turns out not to be in $\mathfrak{B}(\gamma | c, d)$, there is only a small risk $1-\gamma$ that it is not false.

> **Remark:** Apparently $\mathfrak{B}(\gamma | c, d)$ widens when the confidence γ increases, since fewer models are possible to reject, when the

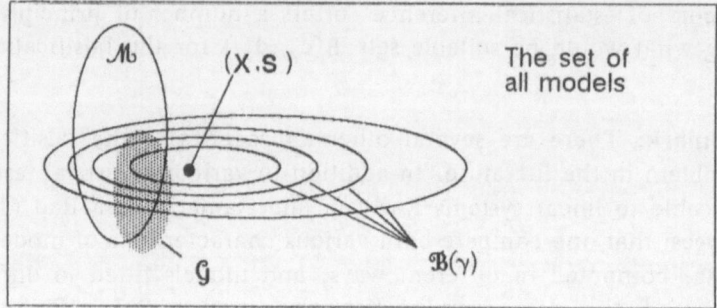

Fig. 46: Illustrating sets 𝕭 of not falsified models with in-creasing confidence γ in the reject decision. The sets in-crease with γ.

demand on the correctness of that decision is higher. An illustra-tion of $\mathfrak{B}(\gamma|\mathbf{c},\mathbf{d})$ is a contour map with γ as parameter, as in *Fig. 46*. The map depicts a pit with 1−γ as the depth coordinate. Since γ is a probability, the maximum depth is one.

Falsifying with or without apriori knowledge

In the sequel two general approaches will be used to determining the set $\mathfrak{B}(\gamma|\mathbf{c},\mathbf{d})$ of models likely not to disagree with data $(\mathbf{c},\mathbf{d})$, *viz.* one 'con-ditional', which exploits what one knows about the structure apriori, and one 'unconditional', which does not. Thus the attribute 'condi-tional' means 'conditional on the structure information'. The 'condi-tional' approach uses Bayes' idea and the properties of the Likelihood function (which carries the structure information), while the 'uncondi-tional' approach is based on classical parameter–free techniques for testing statistical hypotheses.

With or without independent data

Both techniques use probabilistic concepts to derive the tests for fal-sification, and this means that they assume *statistical independence* be-tween the models to be tested and the data used for the test. This means that it is not immaterial *how* the model was obtained. If it was calcu-lated, wholly or partially from data, then one must in principle use other and *independent* data for its falsification. This is called *'cross–fal-sification'*. If one does not have an independent data sequence for fal-sification, it may still be feasible to carry out an *'auto–falsification'*, *i.e.* to falsify a model using the same data as the model was based on. But that changes the 'acceptance domains'. And it requires that the *struc-*

ture of the model that was fitted to data be independent of the data (see section 8.4 on "Conditional falsification of structures"). In that case it is possible to test also the structure. In any case, one should ask oneself, before applying a method for cross–falsification: "Where did the model and the assumed structure come from, and could they have been influenced by the data to be used for their falsification?"

> **Remark:** Leontaritis and Billings (1987) review general hypothesis testing of dynamical models. Söderström (1977) has analysed and compared a number of conventional tests of model structure, *viz.* Akaike's criteria, Parzen's criteria, the F–test, methods based on singularity of the information matrix, methods based on pole cancellation, tests of residuals, and plots of signals. The analysis is carried out for linear, multivariable, transfer–function models.

8.1   Statistical tests

To understand the 'unconditional' solution to the falsification problem it is necessary first to understand the basic ideas of statistical tests. The fundamental procedure of statistical hypothesis testing runs like this (Lehman, 1959):

> **Procedure:**
> 1) Formulate a *null hypothesis H_0*
> 2) Find a *domain $\mathfrak{D}(\gamma|H_0)$* in data space, such that the probability that data will be in $\mathfrak{D}(\gamma|H_0)$ has a given high value γ, when H_0 is true. Hence, $\Pr\{(c,d) \in \mathfrak{D}(\gamma|H_0)|H_0\} = \gamma$. The domain outside $\mathfrak{D}(\gamma|H_0)$ is called 'the critical domain'.
> 3) *Reject* the hypothesis H_0 if $(\mathbf{c},\mathbf{d}) \notin \mathfrak{D}(\gamma|H_0)$

One may define different domains $\mathfrak{D}(\gamma|H_0)$, all which have the same risk $1 - \gamma$ of rejecting H_0 when it is actually true. However, all $\mathfrak{D}(\gamma|H_0)$ do not have the same probability of rejecting H_0 when it is not true. Thus, all $\mathfrak{D}(\gamma|H_0)$ are not equally efficient.

The concept of *power* was introduced in statistical theory to measure the discriminating efficiency of a test. In order to do that it is necessary first to specify an alternative hypothesis H_1. The 'power' is then defined as the probability $\Pr\{(c,d) \notin \mathfrak{D}(\gamma|H_0)|H_1\}$ of rejecting H_0 when H_1 is true. When H_1 has been specified, it is meaningful to define an *optimum* or *maximum–power* test as one which has the highest reject probability (within a given class of tests and for a given alternative hypothesis).

When Lehman's procedure is applied to the model falsification problem, it may be formulated thus:

Testing statistical hypotheses

1) Formulate a *null hypothesis* H_0: "The model M is Source Equivalent; it could have produced the data $(\mathbf{c}, \mathbf{d})$ actually coming from $(\mathbf{X}, \mathbf{S})$".
2) Find a *domain* $\mathfrak{D}(\gamma|M)$ in data space, such that the probability that data will be in $\mathfrak{D}(\gamma|M)$ has a given high value γ, when data has been generated by M.
3) *Reject* the hypothesis H_0 if $(\mathbf{c}, \mathbf{d}) \notin \mathfrak{D}(\gamma|M)$.

●●●●●●

The problem is now to find $\mathfrak{D}(\gamma|M)$. That domain is neither uniquely determined by γ, nor quite free, since it must satisfy the equation $\gamma = \Pr\{(c, d) \in \mathfrak{D}(\gamma|M)|M\}$. Except for this, one can define $\mathfrak{D}(\gamma|M)$ freely, but that is not only an advantage, since the possibilities are overwhelmingly many. The following way of defining $\mathfrak{D}(\gamma|M)$ is often practical and will be used in the sequel:

Look for a parameter–free statistic

Let $p[c, d|M]$ be the probability density of the data (c, d) produced by the model M. If it is possible to find a 'statistic' $\xi(c, d, M)$, *i.e.* a function of data and model, such that *its probability distribution is known*, when H_0 is true, then the statistic is called *'parameter–free'*, and one can use it to define $\mathfrak{D}(\gamma|M)$ indirectly: Define a parametric set of domains $\{\Xi(\rho)|\rho \in (0,1)\}$ in the space of ξ, such that $\Xi(\rho) \subset \Xi(\rho')$ when $\rho < \rho'$, and determine $\Xi(\gamma)$ from $\gamma = \Pr\{\xi \in \Xi(\gamma)\}$. That is normally much easier to do than finding the corresponding domain $\mathfrak{D}(\gamma|M)$ in data space, since the idea of a good statistic ξ is that it should compress data, while retaining the information in the data (actually it holds the information in *both* model and data), and its distribution is known and should be simple. The set $\mathfrak{D}(\gamma|M)$ is defined indirectly by the relation

$$\mathfrak{D}(\gamma|M) = \{c, d|\xi(c, d, M) \in \Xi(\gamma)\} \tag{8.1}$$

If a continuous distribution is known, it can be transformed into a standard gaussian distribution by change of coordinates, and hence one can assume without loss of generality that $\xi(c, d, M)$ is a gaussian vector of dimension $r = \dim(\xi)$, and with zero mean and unit covariance matrix.

Then a natural family of domains $\Xi(\gamma)$ are those bounded by the spheres of equal probability density:

$$\Xi(\gamma) = \{\xi | \xi^T \xi < \chi^2_r(1-\gamma)\} \tag{8.2}$$

where $\chi^2_r(1-\gamma)$ is the 'chi–square' value with r degrees of freedom and risk $1-\gamma$. Notice that the domains $\mathfrak{D}$, Ξ, and $\mathfrak{B}$ all define the same event, although they are defined in different spaces: $\mathfrak{B}(\gamma) = \{M | \xi(c,d,M) \in \Xi(\gamma)\}$.

Now the problem is instead to find a 'parameter–free statistic', but there are several good ideas on how to do that.

--------------------------- **Example 8.1** ---------------------------

Illustrating the basic test procedure.

Consider again the generic linear model in Example 2.2:

M: $d = c\, \theta + \omega\, \lambda$,
c is a deterministic stimulus (matrix)
d is a stochastic response (vector)
ω is an orthonormal gaussian vector
θ is a vector parameter
λ is a scalar parameter.

1) The null hypothesis is H_0: $\theta_0 = \theta$, $\lambda_0 = \lambda$

2) It is seen immediately that under H_0 the statistic $\xi = \lambda^{-1}(d - \theta\, c)$ is parameter–free, since if $\theta_0 = \theta$, $\lambda_0 = \lambda$, then $\xi = \omega$, which is gaussian. Hence $\xi^T\xi$ is chi–square distributed with r degrees of freedom, where r is the length of d. The domain is $\mathfrak{D}(\gamma) = \{c,d | \xi^T\xi \leq \chi^2_r(1-\gamma)\}$, where $\chi^2_r(1-\gamma)$ is the value of a chi–square variable with r degrees of freedom for which the cumulative distribution function has the value γ.

3) Reject (θ,λ) with risk $1-\gamma$ if $\xi^T\xi > \chi^2_r(1-\gamma)$.

However, the test does not have maximum efficiency; in fact, it may be quite poor. If $\lambda \neq \lambda_0$, the test variable is $\xi^T\xi = \omega^T\omega\, \lambda_0^2/\lambda^2$. Hence, if $\lambda > \lambda_0$, then M is likely to be rejected, but not if $\lambda < \lambda_0$. Therefore, the χ^2–test usually requires accurate estimation of λ_0.

8.2   Unconditional falsification

External algorithmic dynamic models of the form

$$d(k) \leftarrow A^d[c_{k-1}, d_{k-1}, \omega^d(k), k | M^d]$$

$$c(k) \leftarrow A^c[c_{k-1}, d_k, \omega^c(k), k | M^c] \tag{8.3}$$

$\{\omega^c(k), \omega^d(k)\}$ gaussian, independent, and orthonormal, with their inverse forms

$$\omega^d(k) \leftarrow W^d[c_{k-1}, d_k, k | M^d]$$

$$\omega^c(k) \leftarrow W^c[c_k, d_k, k | M^c] \tag{8.4}$$

can be falsified conveniently by statistical hypothesis testing. The reason is that their innovations sequences

$$\mathbf{w}^d(k | M^d) \leftarrow W^d[\mathbf{c}_{k-1}, \mathbf{d}_k, k | M^d]$$

$$\mathbf{w}^c(k | M^c) \leftarrow W^c[\mathbf{c}_k, \mathbf{d}_k, k | M^c] \tag{8.5}$$

offer a basis for defining 'parameter–free statistics'.

The hypothesis to be falsified is the model M, where the primitive random vector sequences $\{\omega^d(k), \omega^c(k)\}$ are independent, white, and gaussian with zero means and unit covariance matrices. Hence, the innovations sequences $\{\mathbf{w}(k | M)\} = \{\mathbf{w}^d(k | M^d), \mathbf{w}^c(k | M^c)\}$ would be gaussian and orthonormal (zero mean, unit covariance, and white), if M were a Source–Equivalent model, but not otherwise. One gets the criterion that the model M is Source–Equivalent, *if and only if innovations* $\{\mathbf{w}(k | M)\}$ *are gaussian and orthonormal*.

Proof:
To see this, recall the definition of Source Equivalence:
$\mathcal{S} = \{X, S | p[c_K, d_K | X, S] = p[c_K, d_K | \mathbf{X}, \mathbf{S}], \text{ all } (c_K, d_K) \in \mathcal{R}_K(\mathbf{X})\}$.
Since there is a bijective correspondence between data $(\mathbf{c}_K, \mathbf{d}_K)$ and innovations $\mathbf{w}_K = (\mathbf{w}^c{}_K, \mathbf{w}^d{}_K)$ the condition for Source Equivalence may also be expressed in the distributions of the innovations:
$\mathcal{S} = \{X, S | p[w_K | X, S] = p[w_K | \mathbf{X}, \mathbf{S}], \text{ all } (c_K, d_K) \in \mathcal{R}_K(\mathbf{X})\}$.

But $w(k|X,S) = W[c_k, d_k|X,S] = \omega(k)$, and hence the joint distribution $p[w_K|X,S] = \phi[w_K]$, where ϕ is the gaussian distribution. ∎

Remark: Object Equivalence is *not* determined by the properties of $\{w^d(k|M^d)\}$; it requires orthonormality of the *joint* sequences $\{w^c(k|M^c), w^d(k|M^d)\}$ *and* an experimenter that is Isolated and Sufficiently Stimulating (see section 6.1 on "Equivalent dynamic models").

────────────────── **Example 8.2** ──────────────────

Illustrating that testing the innovations of the object model is not enough.

Let (X,S): $c(k) = \omega_1(k)$, $d(k) = c(k) + \omega_2(k)$,
and consider the model
(M^c, M^d): $c(k) = \omega_1(k)$, $d(k) = \sqrt{2}\, \omega_2(k)$.

This yields $w^c(k|M^c) = c(k)$, $w^d(k|M^d) = d(k)/\sqrt{2}$. Hence, $\{w^d(k|M^d)\}$ is gaussian with zero mean and unit covariance. In addition, the experiment is Isolated and Sufficiently stimulating.

But $\{w^c(k|M^c), w^d(k|M^d)\} = \{c(k), d(k)/\sqrt{2}\}$ is not orthonormal, since $E\{c(k)\, d(k)/\sqrt{2}\} = E\{\omega_1(k)[\omega_1(k) + \omega_2(k)]/\sqrt{2}\} = 1/\sqrt{2} \neq 0$. Hence, M^d and S are not Object Equivalent.

──

The falsification problem boils down to testing whether all the components in the given stochastic vector sequence $\{w(k|M)|k=1,...,N\}$ are independent and gaussian with zero means and unit variances. This does however meet with serious fundamental problems (Kendall and Stuart, 1967, p. 473). One has to be content with testing a number of *characteristics* of gaussian sequences, such as moments and correlations.

Remark: The condition that innovations be gaussian and orthonormal apparently covers everything, since it is independent of whether the model itself is based on little or much of hypothesizing, parametric or non–parametric, or time–variable or not. It is therefore no surprise that the condition cannot be tested conclusively. In fact, parameter–free statistics require estimates of a number of characteristics, and hence the basic problem is the same as that of finding consistent estimates (see section 6.2 on "Consistency"). In both cases one has to limit the number of unknowns to something much smaller than the number of data points (by structural restrictions).

8.2.1 Testing the parameter–free statistics

If the vector components $w_i(k|M)$ are independent and N is large, there are elementary statistical tests (χ^2–tests) that falsify a non–standard amplitude distribution.

Testing normality (gaussianness) is conventionally done by plotting histograms in a special diagram scaled nonlinearly to yield a linear curve for the gaussian distribution and a nonlinear curve for other distributions. Since graphs generally reveal more to the eye than numbers, they are useful tools for the model designer.

Usually, it is more important to test 'orthogonality', *i.e.* whether the components $w_i(k|M)$ and $w_j(k'|M)$ are uncorrelated for $k \neq k'$ or $i \neq j$.

For that purpose compute a number of correlations

$$\xi_{ij}(l) = \frac{1}{N}\sum_{k=1}^{N} w_i(k-l)\, w_j(k) \tag{8.6}$$

Since $\xi_{ij}(l) = \xi_{ji}(l)$ and $\xi_{ij}(l) \approx \xi_{ij}(-l)$, there is no point in investigating other cases than $i \geq j$ and $l \geq 0$.

All ξ defined by (8.6) consititute 'parameter–free statistics', since their distributions are known, when the hypothesis is true. For large N they are approximately gaussian distributed. Their means and covariances are

$$E\{\xi_{ij}(l)\} = \frac{1}{N}\sum_{k=1}^{N} E\{w_i(k-l)\, w_j(k)\} = \delta(i-j)\,\delta(l) \tag{8.7}$$

$$E\{\xi_{ij}(l)\, \xi_{i'j'}(l')\}$$

$$= \frac{1}{N}\frac{1}{N}\sum_{k=1}^{N}\sum_{k'=1}^{N} E\{w_i(k-l)\, w_j(k)\, w_{i'}(k'-l')\, w_{j'}(k')\}$$

$$= \frac{1}{N}\frac{1}{N}\sum_{k=1}^{N}\sum_{k'=1}^{N}\{E[w_i(k-l)\ w_j(k)]\ E[w_{i'}(k'-l')\ w_{j'}(k')]$$

$$+ E[w_i(k-l)\ w_{i'}(k'-l')]\ E[w_j(k)\ w_{j'}(k')]$$

$$+ E[w_i(k-l)\ w_{j'}(k')]\ E[w_{i'}(k'-l')\ w_j(k)]\}$$

$$= \frac{1}{N}\frac{1}{N}\sum_{k=1}^{N}\sum_{k'=1}^{N}[\delta(i-j)\ \delta(l)\ \delta(i'-j')\ \delta(l')$$

$$+ \delta(i-i')\ \delta(k-k'-l+l')\ \delta(j-j')\ \delta(k-k')$$

$$+ \delta(i-j')\ \delta(k-k'-l)\ \delta(j-i')\ \delta(k-k'+l)]$$

$$= \delta(i-j)\ \delta(l)\ \delta(i'-j')\ \delta(l')$$

$$+ \frac{1}{N}[\delta(i-i')\ \delta(j-j')\ \delta(l-l') + \delta(i-j')\ \delta(j-i')\ \delta(l+l')]$$

$$\mathrm{Cov}\{\xi_{ij}(l)\ \xi_{i'j'}(l')\} = \frac{1}{N}[\delta(i-i')\ \delta(j-j')\ \delta(l-l')$$

$$+ \delta(i-j')\ \delta(j-i')\ \delta(l+l')]$$

$$= \begin{cases} 1/N & \text{if } l = l' > 0 \text{ and } (i,j) = (i',j') \\ 1/N & \text{if } l = l' = 0 \text{ and } (i,j) = (i',j'),\ i > j \\ 2/N & \text{if } l = l' = 0,\ i = j = i' = j' \\ 0 & \text{otherwise} \end{cases} \tag{8.8}$$

It follows that all $\xi_{ij}(l)$, $i \geq j$, $l \geq 0$, are uncorrelated with means and variances

$$m_\xi(i,i,0) = 1,\ \sigma_\xi(i,i,0)^2 = 2/N$$

$$m_\xi(i,j,l) = 0,\ \sigma_\xi(i,j,l)^2 = 1/N,\ (i > j \text{ or } l > 0) \tag{8.9}$$

This yields two cases: In essence, $\xi_{ii}(0)$ test the unit variance of w_i, and $\xi_{ij}(l)$ for other (i,j,l) test orthogonality.

Remark: Correlation tests suit the purpose of forecasting and control. The 'residuals' $e(k) = [\nabla_d w^d(k)]^{-1}\ w^d(k)$ can be interpreted as prediction errors one step (see section 5.5.2 on "Innovations"). Hence, they should be uncorrelated with previous data. If they were not, the dependence on previous data could have been used to estimate the innovation from previous data, and hence to predict better. Hence, checking uncorrelatedness ensures that the model is a good predictor.

Remark: Checking for the presence of feedback can be done by testing the hypothesis that the experiment is open loop. This means that $w^d(k|M^d)$ should be uncorrelated with future input c.

Remark: The text book of Kashyap and Ramachandra Rao (1976) contains a number of tests based on innovations.

Remark: Time invariance is often tested better by the eye than by computing. Generally, plotted innovations should look 'white' and have an even amplitude distribution. Any 'transients' and 'coincidences' between different sequences often reveal themselves to the eye, while they are lost in the averaging process of computing covariances or other estimates. Whenever there seems to be a 'pattern' in the innovations sequence, this indicates that the model may not accord with data (it may still be valid for the purpose).

It simplifies the analysis to collect all test variables $\xi_{ij}(l)$ into a vector ξ, where the order of the components are arbitrary, and write

$$\xi = \frac{1}{N}\sum_{k=1}^{N} \mathbf{Z}(k)^T \mathbf{w}(k) \tag{8.10}$$

The matrix $\mathbf{Z}(k)$ has columns $\delta_i\, w_j(k-l)$, where δ_i is the vector with a unit in the i-position and zeroes elsewhere.

However, it is sometimes convenient to use other functions of data for the testing: $\mathbf{Z}(k) = Z(c_k, d_k|M)$. This is the case, in particular, when one has no model M^c of the process generating the stimulus. In that case the input innovations $w^c(k|M^c)$ cannot be computed, so that $\mathbf{Z}^c(k) = 0$, and $\mathbf{Z}^d(k) = Z^d(c_k, d_k|M^d)$ is some arbitrary function of data. In essence, all components in $\mathbf{Z}(k)$ that depend only on *previous* values should be uncorrelated with $w^d(k|M^d)$.

Some special choices of $\mathbf{Z}^d$ are customary:

● *Auto-correlation*: $\mathbf{Z}^d_{ijl}(k) = \delta_i\, w^d_j(k-l)$ $(l \geq 1)$
● *Cross-correlation*: $\mathbf{Z}^d_{ijl}(k) = \delta_i\, c_i(k-l)$ $(l \geq 0)$

This tests whether the output innovations are uncorrelated with previous innovations and input, which is usually enough for identification of linear objects in open loop.

A disadvantage with these more general tests is that the components of ξ become correlated. This holds also for the simple cross–correlation test (unless c is white), and generally complicates a correct falsification (see Rule 3 below).

If one excludes from $\mathbf{Z}(k)$ those columns that are correlated with the innovations $\mathbf{w}(k)$ (under the null hypothesis), then

$$m_\xi = E\{\xi|M\} = \frac{1}{N}\sum_{k=1}^{N} E\{\mathbf{Z}(k)^T \mathbf{w}(k)|M\} = 0$$

$$R_\xi = E\{\xi\,\xi^T|M\}$$

$$= E\{\frac{1}{N}\frac{1}{N}\sum_{k=1}^{N}\sum_{k'=1}^{N} \mathbf{Z}(k)^T \mathbf{w}(k)\,\mathbf{w}(k')^T \mathbf{Z}(k')|M\}$$

$$= E\{\frac{1}{N}\frac{1}{N}\sum_{k=1}^{N}\sum_{k'=1}^{N} \mathbf{Z}(k)^T E[\mathbf{w}(k)\,\mathbf{w}(k')^T|M]\,\mathbf{Z}(k')|M\}$$

$$= E\{\frac{1}{N}\frac{1}{N}\sum_{k=1}^{N} \mathbf{Z}(k)^T \mathbf{Z}(k-|M\}$$

$$\rightarrow \frac{1}{N}\frac{1}{N}\sum_{k=1}^{N} \mathbf{Z}(k)^T \mathbf{Z}(k) \tag{8.11}$$

Notice again that if all columns in $\mathbf{Z}(k)$ are independent of $\mathbf{w}(k)$, the test must be supplemented by one testing the unit variance of $\mathbf{w}$. Otherwise the test will not be efficient. For instance, erroneous models yielding small values of $\mathbf{w}$ will be accepted (see Example 8.1).

However, computing m_ξ and R_ξ in this case will be cumbersome, unless the dependent columns in $\mathbf{Z}(k)$ are designed to simplify the testing of variance, *e.g.* $\mathbf{Z}_{ii}(k) = \delta_i\,\mathbf{w}_i(k)$, as defined in (8.6). In that case the formulas (8.9) apply.

Multiple tests increase the risk

The correlations in ξ still make a great many statistics to test, even if much fewer than the innovations $\{\mathbf{w}(k)\}$. In principle one can test each one of them at a time, or combine them into a single test statistic. Notice however, that if r statistics will be tested, each of them has the same risk $\alpha = 1 - \gamma$ that even a correct model will produce a statistic that will reject it. Hence, one should preferable increase the confidence level, if several statistics should be tested for the same model. How much one should increase the confidence in each test depends on how dependent the tested statistics are. If they are *independent*, like those based on auto–correlation, then it is easy to calculate the change in confidence: If the confidence in one test is denoted γ_1, then the confidence of r repeated tests is γ_1^r. Equating this with a specified confidence γ yields $\gamma_1 = \gamma^{1/r}$. This yields falsification rule 1 (see below) for r *scalar* statistics.

In case the r statistics in ξ are independent (and gaussian) it is possible to form the combined statistic: Let $\xi(l)$, $m_\xi(l)$, and $\sigma_\xi(l)$ be the scalar components in ξ, their means, and their standard deviations. Then

$$\chi^2 = \sum_{l=1}^{r} [\xi(l) - m_\xi(l)]^2/\sigma_\xi(l)^2 \qquad (8.12)$$

is chi–square distributed with r degrees of freedom. This yields rule 2.

> **Remark:** The difference between the two rules is the forms of the acceptance domains $\Xi(\gamma)$ in the space of parameter–free statistics. In the first case it is a super–cube, and in the second case a super–sphere.

If the components in ξ are dependent, then the statistic

$$\chi^2 = [\xi - m_\xi]^T R_\xi^{-1} [\xi - m_\xi] \qquad (8.13)$$

is chi–square distributed with $\dim(\xi)$ degrees of freedom. This yields rule 3. In summary, the three rules are:

Unconditional falsification rules

Rule 1: If r scalar statistics $\xi(l)$ are uncorrelated, then reject model M with confidence γ, if any of
$$|\xi(l) - m_\xi(l)| > \lambda(\gamma^{1/r}) \, \sigma_\xi(l)$$
where $\lambda(\gamma)$ is defined by $\Phi(\lambda) = \gamma$.

Rule 2: If r scalar statistics $\xi(l)$ are uncorrelated, then reject model M with confidence γ, if

$$\sum_{l=1}^{r} [\xi(l) - m_\xi(l)]^2/\sigma_\xi(l)^2 > \chi^2_r(1-\gamma).$$

Rule 3: Reject model M with confidence γ, if
$$[\xi - m_\xi]^T \, R_\xi^{-1} \, [\xi - m_\xi] > \chi^2_r(1-\gamma)$$
where $r = \dim(\xi)$.

Remark: If r is not extremely large (less than a hundred, say), since the tests are not very sensitive to the risk level. For example, changing the risk $1 - \gamma$ in Rule 1 from 0.01 to 0.0001 increases the threshold λ by only 60%. Hence, there is not much motif in practice to refrain from testing several aspects of one model.

Example 8.3

Illustrating the effect of using the same data for estimation and for falsification of the resulting model.

Apply cross–correlation tests to falsify the model
$$d(k) = \hat{\theta}_1 \, c(k-1) + \hat{\theta}_2 \, c(k-2) + \omega(k)$$
where $\{c(k)\}$ is a 'white' sequence with zero mean and unit variance, and $(\hat{\theta}_1, \hat{\theta}_2)$ is estimated from data $\mathbf{d}_N$.

The innovations are
$$\mathbf{w}(k) = \mathbf{d}(k) - \hat{\theta}_1 \, \mathbf{c}(k-1) + \hat{\theta}_2 \, \mathbf{c}(k-2)$$
and the cross–correlations are
$$\xi(l) = \frac{1}{N} \sum_{k=1}^{N} \mathbf{c}(k-l) \, \mathbf{w}(k)$$
$$= \frac{1}{N} \sum_{k=1}^{N} \mathbf{c}(k-l) \, [\mathbf{d}(k) - \hat{\theta}_1 \, \mathbf{c}(k-1) + \hat{\theta}_2 \, \mathbf{c}(k-2)].$$

Their covariances are
$$\sigma_\xi(l,l') = \frac{1}{N}\frac{1}{N} \sum_{k=1}^{N} \mathbf{c}(k-l) \, \mathbf{c}(k-l') \rightarrow \frac{1}{N} \, \delta(l-l')$$

Case 1: Independent data is used for the test.

Then
$$\xi(l) = \frac{1}{N} \sum_{k=1}^{N} \mathbf{c}(k-l) \, [(\Theta_1 - \hat{\theta}_1) \, \mathbf{c}(k-1) + (\Theta_2 - \hat{\theta}_2) \, \mathbf{c}(k-2) + \omega(k)]$$
$$\rightarrow \Theta_l - \hat{\theta}_l + \eta(l), \text{ where } \eta(l) = \frac{1}{N} \sum_{k=1}^{N} \mathbf{c}(k-l) \, \omega(k).$$

By the 'central limit theorem' $\eta(l)$ tends to a gaussian variable with zero mean and variance N^{-1}. Furthermore, $\eta(l)$ and $\eta(l')$ are uncorrelated for $l' \neq l$. From Rule 2 the model is rejected if

$$N \sum_{l=1}^{p} \xi(l)^2 = N\ [(\Theta_1 - \hat{\theta}_1)^2 + (\Theta_2 - \hat{\theta}_2)^2] + \chi^2_p > \chi^2_p(1-\gamma)$$

where $\chi^2_p = N \sum_{l=1}^{p} \eta(l)^2$ is asymptotically chi–square distributed with p degrees of freedom.

Hence, if $\hat{\theta}_i = \Theta_i$ it is rejected with probability $1-\gamma$; if $\hat{\theta}_i \neq \Theta_i$ it is rejected with higher probability. The latter tends to one when $N \to \infty$.

Case 2: The test is based on the same data.

Then

$$\hat{\theta}_l = \sum_{k=1}^{N} [\ \mathbf{c}(k-l)^2]^{-1} \sum_{k=1}^{N} \mathbf{d}(k)\ \mathbf{c}(k-l)$$

and

$$\xi(l) = \frac{1}{N} \sum_{k=1}^{N} \mathbf{c}(k-l)\ [\mathbf{d}(k) - \hat{\theta}_1\ \mathbf{c}(k-1) + \hat{\theta}_2\ \mathbf{c}(k-2)]$$
$$= \eta(l) \text{ if } l \geq 3, \text{ and zero otherwise.}$$

The difference is that fitting $\hat{\theta}_l$ to the same data makes the first two $\xi(l)$ values zero. Hence, it remains $p-2$ stochastic variables $\eta(l)$. Falsification Rule 2 will be the same, except that the chi–square variable now has $p-2$ degrees of freedom:

$$\text{Reject if } N \sum_{l=1}^{p} \xi(l)^2 > \chi^2_{p-2}(1-\gamma).$$

Since the test variable is thus independent of the model parameters, the conclusion is that cross–correlation test cannot be used to falsify the Maximum–Likelihood estimates $\hat{\theta}_l$. However, it can be used to test whether there are any higher–order terms Θ_l, $l \geq 3$ in the object. In that case the test will tend to

$$N \sum_{l=3}^{p} \Theta_l^2 + \chi^2_{p-2} > \chi^2_{p-2}(1-\gamma).$$

Hence, cross–correlation tests the order value (2) of the model, and in general all parameter values (zero), except those that were fitted to data.

8.3 Conditional falsification of models

The unconditional falsification rules do not have maximum power in the sense defined in section 8.1 on "Statistical tests". Hence, they do not need an alternative hypothesis, and that is what makes them 'unconditional'. With an alternative hypothesis the test can be made 'efficient', but it also becomes 'conditional', in the sense that its result hinges on the alternative hypothesis. The latter is therefore an 'assumption', *i.e.* untested, and if the alternative hypothesis is wrong, there will be a 'pitfall'. The consequences of this will be elaborated later (section 8.6 on "Efficiency vs safety").

Now, the assumption of a sequence $\mathcal{M}(1),\dots,\mathcal{M}(\bar{n})$ of expanding structures offers a number of possible alternative hypotheses: H_1: $\mathbf{M} \in \mathcal{M}(n)$. This means that the test becomes dependent on $\mathcal{M}$ and n. With both structure and maximum degree of complexity given by the alternative hypothesis, it will be possible to use Bayes' principle to determine the family of sets of not rejected models. Denote it $\mathcal{B}(\gamma|n,\mathbf{c},\mathbf{d})$ to emphasize the dependence on the degree of complexity.

> **Remark:** Since the calculation of the confidence is based on the presumption that $\mathcal{M}$ is sufficiently general, all $\mathcal{B}(\gamma|n,\mathbf{c},\mathbf{d})$ will be entirely in $\mathcal{M}(n)$, and some model in $\mathcal{M}(n)$ will always remain unfalsified. Hence, $\mathcal{B}(\gamma|n,\mathbf{c},\mathbf{d}) \cap \mathcal{M}(n)$ is never empty, so that the basic criterion for falsifying $\mathcal{M}(n)$ yields nothing. Hence, a conditional falsification scheme can falsify models M within the structure $\mathcal{M}(n)$, but not the structure itself. On the other hand, it can falsify more efficiently (within the structure) than an unconditional scheme.

> **Remark:** If one expands the admissible set $\mathcal{M}(n)$, the conditional falsification will be less efficient, and less 'conditional', *i.e.* safer, since it will use less apriori information. And if one would expand $\mathcal{M}(n)$ so much that it becomes the 'feasible' set $\bar{\mathcal{M}}$, then falsification would be 'unconditional' on everything but $\bar{\mathcal{M}}$.

The rationale for using the 'Bayesian approach' (see section 2.1) is that, given the alternative hypothesis $\mathbf{M} \in \mathcal{M}(n)$, the aposteriori distribution $P[M|\mathcal{M},n,\mathbf{c},\mathbf{d}]$ is known, and is the only thing that can ever be known about **M**, given prerequisites. Hence one can compute the probability that **M** be in any given set. If this set would be a confidence region $\mathcal{B}(\gamma|n,\mathbf{c},\mathbf{d})$, then one would have a solution to the falsification problem: Models outside $\mathcal{B}(\gamma|n,\mathbf{c},\mathbf{d})$ would have the (small) probability $1 - \gamma$ of being true, and could therefore be rejected with the risk $1 - \gamma$. The confidence region would satisfy

$$\gamma = \text{Pr}\{\mathbf{M} \in \mathcal{B}(\gamma|n,\mathbf{c},\mathbf{d})|n,\mathbf{c},\mathbf{d}\}$$

$$= \int \text{Ind}\{M \in \mathcal{B}(\gamma|n,\mathbf{c},\mathbf{d})\} \; dP[M|n,\mathbf{c},\mathbf{d}]$$

$$= \int \text{Ind}\{B(M|n,\mathbf{c},\mathbf{d}) \leq \gamma\} \; dP[M|n,\mathbf{c},\mathbf{d}] \qquad (8.14)$$

where the last equality follows from (4.1). The reject rule would be $B(M|n,\mathbf{c},\mathbf{d}) > \gamma$.

However, two difficulties have to be overcome first:

A parametric structure is needed

The first difficulty is that one has to provide an apriori distribution of the models in $\mathcal{M}(n)$

$$dP[M|\mathcal{M},n] = \sum\nolimits_{k \leq n} \text{Ind}\{v \in \mathcal{N}_k\} \; p[v|k] \; p[\theta|\mathcal{A},v] \; d\theta,$$

where $M = (\mathcal{A},v,\theta)$, and θ is any real vector that characterizes the model with a given structure index v. Generally, the parametric representations introduced in section 7.1 on "Validating parametric models" are suitable to use for distinguishing between the various models and sets of models that play a role in falsification.

Hence, define an expanding sequence of parametric model structures:

$$(\mathcal{A},v,\theta) \in \mathcal{M}(n) = \mathcal{M}(n-1) \bigcup_{v \in \mathcal{N}_n} \mathcal{M}_v, \text{ where } \mathcal{M}_v = (\mathcal{A},v), \text{ and } \mathcal{N} =$$

$\{\mathcal{N}_1,\mathcal{N}_2,\dots,\mathcal{N}_{\bar{n}}\}$ is a given partitioning and sequencing of the set of structure indices v.

The set of data descriptions has to be defined

The second difficulty is that there are many sets $\mathcal{B}(\gamma|n,\mathbf{c},\mathbf{d})$ that yield the same probability γ. One has to find a way of selecting one of them for the falsification rule.

> **Remark:** Notice first that it would not make sense to try and compute the probability for an individual M; the probability that some

model **M** be precisely *M* depends on how many alternatives there are. If the alternatives are infinitely many, like in the case of parametric models $M = (v, \theta)$, where θ is a real vector, the probability for *each particular M* is zero. The Likelihood function yields a 'likelihood', which is a probability *density*, and a density must be integrated to yield a probability. Hence, one has to group all the models *M* in a domain together to get a positive probability for the group.

A free choice is also a design possibility

The fact that one can choose $\mathfrak{B}(\gamma|n,\mathbf{c},\mathbf{d})$ in several ways gives a certain freedom of design, and also raises the question of whether there is an 'optimal' domain in some sense (or an unsolved problem, if one prefers the pessimistic viewpoint). Obviously, one obtains a high confidence by having a large set, but obviously one also wants $\mathfrak{B}(\gamma|n,\mathbf{c},\mathbf{d})$ small, so that one will be able to reject a large set of false models. This suggests the following optimization problem, as a possible principle for choosing $\mathfrak{B}(\gamma|n,\mathbf{c},\mathbf{d})$: "For a given confidence γ choose the domain $\mathfrak{B}(\gamma|n,\mathbf{c},\mathbf{d})$ that contains the smallest number of models".

But that is stating the problem too loosely, if one has a parametric model with continuous parameters, since all sets with positive probability contain an uncountably infinite number of models. Hence, $\mathfrak{B}(\gamma|n,\mathbf{c},\mathbf{d})$ must be a domain. One might reformulate the optimization problem to that of minimizing *e.g.* the *area* of the domain. But the 'area' depends on what coordinates one chooses to define the domain. Hence, the falsification will not be invariant under coordinate transformations. Further, the *dimension* of the area, the number of parameters, depends on the degree of complexity *n*, which is unknown. However, if *the parameters have physical meanings*, and *the complexity can be determined in another way*, the principle is still a reasonable one. The designer can use the 'free parameter set' (section 5.2) to determine the coordinates in the area to be minimized.

Optimal design is feasible, if a sufficiently wide structure is known

The solution of the problem of optimizing $\mathfrak{B}$ is now quite simple, at least in principle: Denote the 'acceptance domain' by $\mathfrak{B}(\gamma|n,\mathbf{c},\mathbf{d})$ to indicate that a sufficient degree of complexity *n* is known (if nothing else, one may choose $n = \bar{n}$).

A domain $\mathfrak{B}(\gamma|n,c,d)$ with a given confidence obviously covers the smallest area, if it includes only the models with the highest probability densities. But from Bayes' rule $dP[v,\theta|n,c,d] = p[c,d|n]^{-1} p[c,d|v,\theta] p[v|n] p[\theta|v] d\theta$. In addition, it follows from the nesting condition (section 5.1 on 'Parametrization') that the maximum probabilities are in $(\mathcal{A},\mathcal{N}_n)$, so that it will not pay to search over structures with lower complexity than n. Hence, the optimal domain will be

$\mathfrak{B}(\gamma|n,c,d)$
$$= \{v,\theta | L(v,\theta|n,c,d) \geq L(\hat{v},\hat{\theta}|n,c,d) \exp[-\tfrac{1}{2}\rho^2(\gamma|n,c,d)]\} \qquad (8.15)$$

where $(\hat{v},\hat{\theta})$ is the peak value of $L[v,\theta|n,c,d] = p[c,d|v,\theta] p[v|n] p[\theta|v]$, and $\rho^2(\gamma|n,c,d)$ is an increasing nonnegative real function of γ, defined by the condition

$$\gamma = \int \mathrm{Ind}\{(v,\theta) \in \mathfrak{B}(\gamma|n,c,d)\} \, dP[v,\theta|n,c,d]$$
$$= \int \mathrm{Ind}\{L(v,\theta|n,c,d) \geq L(\hat{v},\hat{\theta}|n,c,d) \exp[-\tfrac{1}{2}\rho^2(\gamma|n,c,d)]\}$$
$$dP[v,\theta|n,c,d] \qquad (8.16)$$

The inverse function is

$\Gamma(\rho^2|n,c,d)$
$$\triangleq \int \mathrm{Ind}\{L(v,\theta|n,c,d) \geq L(\hat{v},\hat{\theta}|n,c,d) \exp[-\tfrac{1}{2}\rho^2]\} \, dP[v,\theta|n,c,d]$$
$$= \int \mathrm{Ind}\{2 \log[L(\hat{v},\hat{\theta}|n,c,d)/L(v,\theta|n,c,d)] \leq \rho^2\} \, dP[v,\theta|n,c,d] \qquad (8.17)$$

Using Bayes' rule for $dP[v,\theta|n,c,d]$ yields

$$\Gamma(\rho^2|n,c,d) = p[c,d]^{-1} \sum_{v \in \mathcal{N}_n} p[v|n]$$
$$\int \mathrm{Ind}\{2 \log[L(\hat{v},\hat{\theta}|n,c,d)/L(v,\theta|n,c,d)] \leq \rho^2\} \, p[c,d|v,\theta] p[\theta|v] \, d\theta \qquad (8.18)$$

where

$$p[c,d] = \sum_{v \in \mathcal{N}_n} p[v|n] \int p[c,d|v,\theta] p[\theta|v] \, d\theta \qquad (8.19)$$

Since the condition in (8.16) is equivalent to $B(v,\theta|n,c,d) \leq \gamma$, this determines $\mathfrak{B}(\gamma|n,c,d)$ by the distance function between model and data (the confidence function):

$$B(v,\theta|n,\mathbf{c},\mathbf{d}) = \Gamma\{2 \log[L(\hat{v},\hat{\theta}|n,\mathbf{c},\mathbf{d})/L(v,\theta|n,\mathbf{c},\mathbf{d})]|n,\mathbf{c},\mathbf{d}\} \qquad (8.20)$$

Remark: The relations between the various functions and constants are illustrated in *Fig. 47*.

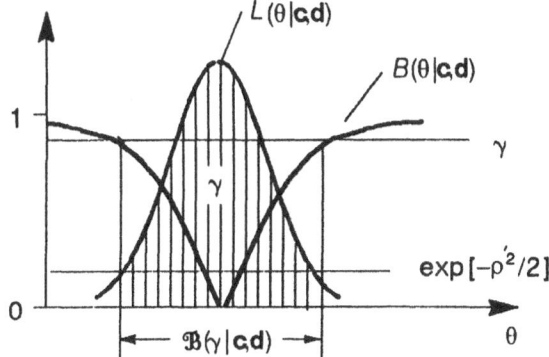

Fig. 47: Illustrating the relations between Likelihood function $L(\theta|\mathbf{c},\mathbf{d})$, confidence $B(\theta|\mathbf{c},\mathbf{d})$, and the acceptance domain $\mathcal{B}(\gamma|\mathbf{c},\mathbf{d})$.

The $\mathcal{B}(\gamma|n,\mathbf{c},\mathbf{d})$–domain yields a way to falsify any parametric model $M = (v^M,\theta^M)$, provided a structure $\mathcal{M}(n)$ is known. The rule will be "Reject M with confidence γ, if $B(v^M,\theta^M|n,\mathbf{c},\mathbf{d}) > \gamma$".

When data are sequential, the analysis and the result are the same, except that one must hang subscripts N on $\mathbf{c}$ and $\mathbf{d}$ to indicate the length of the sample.

Composite structures are a problem in practice

A problem is still that even if the optimal $\mathcal{B}(\gamma|n,\mathbf{c},\mathbf{d})$ is well defined by (8.18–20), the function Γ is prohibitively difficult to compute, except for gaussian Likelihood functions. The obstacles are the same as prevented validation in general cases (chapter 7), namely *i) composite structures* and *ii) evaluating the integral*. It is therefore necessary to resort to the analysis of simple (not composite) structures and long samples.

For this purpose define the *nested subsets* of $\mathcal{M}(n)$, viz. $\mathcal{M}(v) \triangleq \bigcup_{v'\leq v} \mathcal{M}_{v'}$, where $\mathcal{M}_v = (\mathcal{A},v)$, $v \in \mathcal{N}_n$, is non-composite. The relevant elements are:

● The alternative structure $\mathcal{M}(v)$.

- The model to be falsified: $M \triangleq (\nu^M, \theta^M) \in \mathcal{M}(\nu)$. It must have a structure index $\nu^M \leq \nu$. This is a nesting condition.
- The structure to be falsified: $\mathcal{M}_{\nu'}$. The structure index must satisfy $\nu' < \nu$.
- The 'acceptance domain' for *independent* models ('cross–falsification'): $\mathcal{B}(\gamma|\nu, \mathbf{c}, \mathbf{d}) \subset \mathcal{M}(\nu)$.
- The 'acceptance domain' for *ML–models* $(\nu^M, \hat{\theta}^M)$, ('auto–falsification'): $\mathcal{B}(\gamma|\nu^M, \nu, \mathbf{c}, \mathbf{d}) \subset \mathcal{M}(\nu)$. The structure index for the ML–models must satisfy $\nu^M < \nu$.

Assuming a non–composite structure $\mathcal{M}_\nu$ changes only the alternative hypothesis to become $H_1 \colon M \in \mathcal{M}(\nu)$. To emphasize this denote the acceptance domain and the confidence function by $\mathcal{B}(\gamma|\nu, \mathbf{c}, \mathbf{d})$ and $B(M|\nu, \mathbf{c}, \mathbf{d})$ respectively. Hence, it has the advantage of allowing several tests, one for each $\mathcal{M}_\nu$ in $\mathcal{M}(n)$.

This means that it is possible to write

$$\mathcal{B}(\gamma|n, \mathbf{c}, \mathbf{d}) = \bigcap_{\nu \in \mathcal{N}_n} \mathcal{B}(\gamma|\nu, \mathbf{c}, \mathbf{d}) \tag{8.21}$$

which is equivalent to

$$B(\nu^M, \theta^M|n, \mathbf{c}, \mathbf{d}) = \max\{B(\nu^M, \theta^M|\nu, \mathbf{c}, \mathbf{d})|\nu \in \mathcal{N}_n, \nu \geq \nu^M\} \tag{8.22}$$

Notice that if $\nu^M < \nu$, then it it always possible to expand (ν^M, θ^M) to have the structure index ν, normally by augmenting zero parameters. To indicate this write $B(\theta|\nu, \mathbf{c}, \mathbf{d}) \triangleq B(\nu, \theta|\nu, \mathbf{c}, \mathbf{d})$. Hence, evaluating several $B(\nu^M, \theta^M|\nu, \mathbf{c}, \mathbf{d})$ yields a test for composite alternative structures, although not the optimal one given by the more cumbersome formulas (8.16–20).

The non–composite confidence function becomes

$$B(\theta|\nu, \mathbf{c}, \mathbf{d}) = \Gamma\{2 \log[L(\hat{\theta}|\nu, \mathbf{c}, \mathbf{d})/L(\theta|\nu, \mathbf{c}, \mathbf{d})]|\nu, \mathbf{c}, \mathbf{d}\} \tag{8.23}$$

where

$$\Gamma(\rho^2|n, \mathbf{c}, \mathbf{d})$$
$$= \int \mathrm{Ind}\{2 \log[p(\hat{\theta}|\nu, \mathbf{c}, \mathbf{d})/p(\theta|\nu, \mathbf{c}, \mathbf{d})] \leq \rho^2\} \, p[\theta|\nu, \mathbf{c}, \mathbf{d}] \, d\theta \tag{8.24}$$

Remark: Since the tests thus hinges on the alternative structure $\mathcal{M}_\nu$, it is of interest to see what happens, if the model structure with the chosen index v would not contain a better model than that with some other index μ. But in this case the maximum value of the likelihood $L(\hat{\theta}|v,\mathbf{c},\mathbf{d}) < L(\hat{\theta}|\mu,\mathbf{c},\mathbf{d})$, and hence $B(\theta|v,\mathbf{c},\mathbf{d}) < B(\theta|\mu,\mathbf{c},\mathbf{d})$. This means that including a particular $\mathcal{M}_v$ in the sequence of tests will not change the net outcome of the whole test procedure (as long as the better $\mathcal{M}_\mu$ is used too). Hence, so far there is no 'pitfall' here.

A still more important consequence of using non–composite structures for alternatives is that for long samples it opens up a way to evade a usually prohibitive numerical evaluation of the integrals.

8.3.1 Asymptotic confidence regions

If the aposteriori distribution of the unknown Θ in $\mathbf{M} = (v,\Theta)$ tends to a gaussian distribution as $N \to \infty$ (Caines and Ljung, 1976), then

$$\log p[\theta|v,\mathbf{c}_N,\mathbf{d}_N]$$
$$\to -\tfrac{1}{2} \log \det(2\pi R_\theta) - \tfrac{1}{2} (\theta - \hat{\theta})^T R_\theta^{-1} (\theta - \hat{\theta}) \tag{8.25}$$

where $\hat{\theta}$ is the ML–estimate and R_θ is the covariance matrix of the distribution. Inserting this into (8.24) yields

$$\Gamma(\rho^2|n,\mathbf{c},\mathbf{d}) = \int \text{Ind}\{2 \log[p(\hat{\theta}|v,\mathbf{c},\mathbf{d})/p(\theta|v,\mathbf{c},\mathbf{d})] \le \rho^2\} \, p[\theta|v,\mathbf{c},\mathbf{d}] \, d\theta$$

$$= \int \text{Ind}[(\theta - \hat{\theta})^T R_\theta^{-1} (\theta - \hat{\theta}) \le \rho^2]$$
$$\cdot \exp[-\tfrac{1}{2}(\theta - \hat{\theta})^T R_\theta^{-1} (\theta - \hat{\theta})]/\text{sqrt} \det(2\pi R_\theta) \, d\theta$$

$$= \int \text{Ind}[\eta^T \eta \le \rho^2] \exp[-\tfrac{1}{2} \eta^T \eta]/(2\pi)^{-r/2} \, d\eta \tag{8.26}$$

where $r = |v| = \dim(\theta)$. This is the definition of the χ^2–distribution for the argument ρ^2, and with r degrees of freedom. Hence, $\rho^2(\gamma) = \chi_r^2(1-\gamma)$. The asymptotically optimal domain will be defined by

$$B(\theta|v,\mathbf{c}_N,\mathbf{d}_N)$$
$$= \text{Chi\_square}\{2 \log[L(\hat{\theta}|v,\mathbf{c}_N,\mathbf{d}_N)/L(\theta|v,\mathbf{c}_N,\mathbf{d}_N)],r\} \tag{8.27}$$

or, equivalently, by

$$\mathcal{B}(\gamma|v,\mathbf{c}_N,\mathbf{d}_N) = \{\theta | (\theta - \hat{\theta})^T R_\theta^{-1} (\theta - \hat{\theta}) \leq \chi_r^2(1-\gamma)\} \qquad (8.28)$$

Using (8.27) or (8.28) for a stopping rule requires than one first compute either the optimum $\hat{\theta}$, or at least the maximum likelihood value $L(\hat{\theta}|v,\mathbf{c}_N,\mathbf{d}_N)$. From (7.15)

$$N \nabla_{\theta\theta} Q(\theta|v,\mathbf{c}_N,\mathbf{d}_N) \rightarrow R_\theta^{-1}$$

$$N \nabla_\theta Q(\theta|v,\mathbf{c}_N,\mathbf{d}_N)^T \rightarrow R_\theta^{-1} (\theta - \hat{\theta}) \qquad (8.29)$$

Inserting this into (8.28) yields the following rule:

Conditional falsification rule for models

Let the parametric model $M = (\mathcal{A},v,\theta^M)$ be given and independent of data $(\mathbf{c}_N,\mathbf{d}_N)$. Then M is falsified with confidence γ, if

$$\text{Chi\_square}[N \, Q_\theta(\theta^M) \, Q_{\theta\theta}(\theta^M)^{-1} \, Q_\theta(\theta^M)^T, \dim(\theta)] > \gamma$$

where $Q_\theta(\theta) = \nabla_\theta Q(\theta)$, $Q_{\theta\theta}(\theta) = \nabla_{\theta\theta} Q(\theta)$,
$Q(\theta) = - \log L(\theta|\mathcal{A},v,\mathbf{c}_N,\mathbf{d}_N)/N$

This yields a computable way of falsifying any model (v,θ) defined within the alternative structure, provided one has independent and long data sequences, can differentiate the log Likelihood function twice, and evaluate derivatives at the point defined by the model one wants to test. Notice that, if the model is defined by (v',θ'), where $v' < v$, and there is a nesting condition, then one can augment zero parameters to get a model $(v',v,\theta',0) \in \mathcal{M}_v$.

> **Remark:** The criterion is easy to interpret: The model is rejected, if the Likelihood function derivatives in the 'free space' are too far from zero. If they are, then there is a much higher likelihood for a model somewhere up the slope. This is of course a criterion one could have guessed without all this analysis, but the point of the analysis is that it provides a limit to *how* far from zero, in *what* direction, for *how* many parameters, and with *what* credibility.

> **Remark:** The test is called *'locally most powerful'* (LMP) in the statistical literature (Lehman, 1959), on the ground that it has

maximum power (uniformly) for all alternatives in a small neighbourhood around the model to be tested (Bohlin, 1978). For larger deviations between model and alternatives the reject probability will of course be larger, but not maximum. However, it is in the case of small deviations that one needs an efficient test most. This means invoking the 'minimax' principle.

Remark: When the data are *gaussian*, and the models are linear and possible time–variable with gaussian input noise, then it is possible to evaluate the distributions of quadratic test statistics exactly, both under the null–hypothesis and alternative hypotheses. Baram (1980a) derives such distributions for linear time–variable state–vector models, and computes the power of tests for the mean, covariance, and correlation of the residual sequence, *i.e.* the difference between data and model prediction.

One can use a pseudo–inverse instead

When the structure index is larger than needed, the Hessian $Q_{\theta\theta}$ may become singular and cannot be inverted. Factorize $Q_{\theta\theta} = U \wedge U^T$, where $\wedge$ is the diagonal matrix of positive eigenvalues. The matrix U of the corresponding eigenvectors has fewer columns than rows, but is orthogonal, *i.e.* $U^T U = I$. Then at the point θ, the likelihood is a function only of η, where $\theta = U \eta$. Hence

$$\mathfrak{B}(\gamma|\nu,c_N,d_N) \rightarrow \{\theta|\theta = U \eta, Q_\eta Q_{\eta\eta}^{-1} Q_\eta^T \leq \chi_r^2(1-\gamma)/N\} \quad (8.30)$$

where $Q_\eta = \nabla_\eta Q(U \eta|n,c_N,d_N) = Q_\theta U$, $Q_{\eta\eta} = U^T Q_{\theta\theta} U = \wedge$, and r = dim(η). The pseudo–inverse is $Q_{\theta\theta}^\dagger = U \wedge^{-1} U^T$. It follows that $Q_\eta Q_{\eta\eta}^{-1} Q_\eta^T = Q_\theta U \wedge^{-1} U^T Q_\theta^T = Q_\theta Q_{\theta\theta}^\dagger Q_\theta^T$, so that if one substitutes $Q_{\theta\theta}^\dagger$ for $Q_{\theta\theta}^{-1}$ in the conditional falsification rule, then the latter can be used also when the complexity is unnecessarily high.

8.4   Conditional falsification of structures

It was stated in the previous section that "conditional falsification depends on the structure, and therefore cannot falsify the structure itself". However, it can falsify narrower structures $\mathcal{M}(\nu')$ within $\mathcal{M}(\nu)$, $\nu' < \nu$.

In order to compute the acceptance domain $\mathcal{B}(\gamma|\nu',\nu,\mathbf{c},\mathbf{d})$ for model structures $\mathcal{M}(\nu')$ let the structures have a partial nesting, so that $(\nu',\nu,\theta_1,\theta_2)$ is a model in the more complex structure, and (ν',θ_1) a model in the less complex structure to be falsified. Thus, the parameter vector θ_1 specifies a model within the simpler model structure, and the vector θ_2 provides the added detail needed to specify a model within the more complex structure.

Since a model structure is to be tested, the acceptance domain $\mathcal{B}$ must be independent of θ_1, and it is reasonable to let $\mathcal{B}(\gamma|\nu',\nu,\mathbf{c},\mathbf{d}) = \{\theta_1,\theta_2|B(\theta_2|\nu',\nu,\mathbf{c},\mathbf{d}) \leq \gamma\}$.

Following the derivation in section 8.3 on "Conditional falsification of models", one observes that $\mathcal{B}(\gamma|\nu',\nu,\mathbf{c},\mathbf{d})$ yields the highest confidence, if it includes only the models with the highest probability density.

$\mathcal{B}(\gamma|\nu',\nu,\mathbf{c},\mathbf{d})$

$$= \{\theta_1,\theta_2|p[\theta_2|\nu',\nu,\mathbf{c},\mathbf{d}] \geq p[\hat{\theta}_2|\nu',\nu,\mathbf{c},\mathbf{d}] \exp[-\tfrac{1}{2}\rho^2(\gamma|\nu',\nu,\mathbf{c},\mathbf{d})]\} \quad (8.31)$$

which yields the confidence function

$$B(\theta_2|\nu',\nu,\mathbf{c},\mathbf{d}) = \Gamma\{2 \log[p[\hat{\theta}_2|\nu',\nu,\mathbf{c},\mathbf{d}]/p[\theta_2|\nu',\nu,\mathbf{c},\mathbf{d}]]|\nu',\nu,\mathbf{c},\mathbf{d}\} \quad (8.32)$$

The optimal acceptance domain is for long samples

$\mathcal{B}(\gamma|\nu',\nu,\mathbf{c}_N,\mathbf{d}_N)$

$\to \{\theta_1,\theta_2|N\ Q_{\theta_2}(\theta_2)\ Q_{\theta_2\theta_2}(\theta_2)^{-1}\ Q_{\theta_2}(\theta_2)^T < \chi_r^2(1-\gamma)\},$

$r = \dim(\theta_2)$

$Q_{\theta_2}(\theta_2) = \nabla_{\theta_2}Q(\theta_2|\nu',\nu,\mathbf{c}_N,\mathbf{d}_N)$

$$Q_{\theta_2\theta_2}(\theta_2) = \nabla_{\theta_2\theta_2}Q(\theta_2|\nu',\nu,\mathbf{c}_N,\mathbf{d}_N) \quad (8.33)$$

The structure $\mathcal{M}(\nu')$ is falsified if $\theta_2 = 0$ is falsified, *i.e.* if

$$N\ Q_{\theta_2}(0)\ Q_{\theta_2\theta_2}(0)^{-1}\ Q_{\theta_2}(0)^T > \chi_r^2(1-\gamma) \quad (8.34)$$

Remark: The acceptance domains for a model M and a model structure $\mathcal{M}_{\nu'}$, both in $\mathcal{M}_\nu$, are illustrated in *Fig. 48* for the case of $\nu' = 1$, $\nu = 2$.

Remark: The number of degrees of freedom is an important parameter in falsification, and has also an intuitive meaning. A

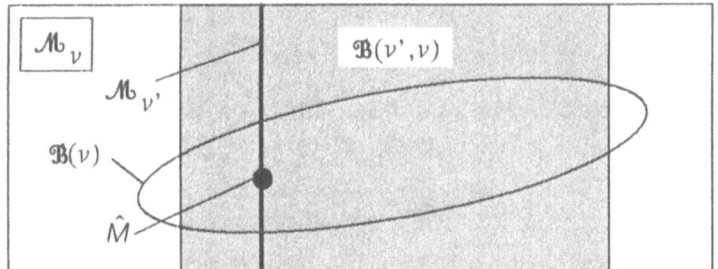

Fig. 48: Optimal acceptance domains for models M and for model structures $\mathcal{M}_{v'}$ (shaded area). $\hat{M}$ is the ML–estimate in the structure $\mathcal{M}_{v'}$,

model M in $\mathcal{M}_v$ is represented by (v,θ), or, equivalently, by (v',v,θ_1,θ_2), *i.e.* by $|v| = \dim(\theta_1) + \dim(\theta_2)$ real parameters, when v and v' are given. Hence there is a real space of dimension $\dim(\theta_1) + \dim(\theta_2)$ of possible models to falsify, and this determines the acceptance domain $\mathcal{B}(\gamma|v,\mathbf{c}_N,\mathbf{d}_N)$ for models in $\mathcal{M}_v$. In comparison, a structure $\mathcal{M}_{v'}$ in $\mathcal{M}(v)$ is represented by $(v',v,\bullet,\theta_2)$, which has $\dim(\theta_2)$ degrees of freedom. Hence there is a space of dimension $\dim(\theta_2)$ of possible structures to falsify, and this determines the acceptance domain $\mathcal{B}(\gamma|v',v,\mathbf{c}_N,\mathbf{d}_N)$ for structures $\mathcal{M}_{v'}$, in $\mathcal{M}(v)$.

The factors in the test variable $N\, Q_\theta(\theta_2)\, Q_{\theta\theta}(\theta_2)^{-1}\, Q_\theta(\theta_2)^T$ defining the acceptance domain $\mathcal{B}(\gamma|v',v,\mathbf{c}_N,\mathbf{d}_N)$ for large N are defined by means of the Likelihood functions for θ_2 only, *viz.* $L(\theta_2|v',v,\mathbf{c}_N,\mathbf{d}_N)$, which is, in essence, a *marginal* distribution. Normally, one has the *joint* distribution of (θ_1,θ_2) to start with, *viz.* $L(\theta_1,\theta_2|v',v,\mathbf{c}_N,\mathbf{d}_N)$, so there is still some derivation to do in order to bring the test into an easily computable form. One has to integrate out θ_1. But that is easy to do when the limiting distribution is gaussian:

Notice that if $\hat{\theta}_1$ is the ML estimate of θ_1 conditional on $\theta_2 = 0$, then $Q_{\theta_1}(\hat{\theta}_1,0) = 0$ and

$$Q_{\theta_2}(0)\, Q_{\theta_2\theta_2}(0)^{-1}\, Q_{\theta_2}(0)^T = Q_\theta(\hat{\theta}_1,0)\, Q_{\theta\theta}(\hat{\theta}_1,0)^{-1}\, Q_\theta(\hat{\theta}_1,0)^T.$$

In conclusion, one can falsify a given structure $\mathcal{M}(v')$ within a wider structure $\mathcal{M}(v)$ in the following way:

Conditional falsification rule for structures

Let the parametric model structure $\mathcal{M}(\nu')$ be given, independent of data $(\mathbf{c}_N, \mathbf{d}_N)$, and a strict subset of some larger structure $\mathcal{M}(\nu) \triangleq \bigcup_{\mu < \nu} (\mathcal{A}, \mu)$.

Compute the ML–estimate $\hat{\theta}_1 = \arg\min Q(\theta_1 | \mathcal{A}, \nu', \mathbf{c}_N, \mathbf{d}_N)$. Then $\mathcal{M}(\nu')$ is falsified with confidence γ, if

$$\text{Chi\_square}[N \, Q_\theta(\hat{\theta}_1, 0) \, Q_{\theta\theta}(\hat{\theta}_1, 0)^{-1} \, Q_\theta(\hat{\theta}_1, 0)^T, r] > \gamma$$

where $Q_\theta(\theta) = \nabla_\theta Q(\theta)$, $Q_{\theta\theta}(\theta) = \nabla_{\theta\theta} Q(\theta)$,
$Q(\theta) = -\log L(\theta | \mathcal{A}, \nu, \mathbf{c}_N, \mathbf{d}_N)/N$,
$\theta = (\theta_1, \theta_2)$, $r = \dim(\theta) - \dim(\theta_1)$.

Remark: Notice that the derivatives need to be evaluated only at the point $(\hat{\theta}_1, 0)$. Notice also that the first $\dim(\theta_1)$ derivatives need not be computed, since the ML estimate satisfies $\nabla_{\theta_1} Q(\hat{\theta}_1, 0) = 0$. However, all second–order derivatives must be computed, if $\nabla_{\theta_1 \theta_2} Q(\hat{\theta}_1, 0) \neq 0$.

Notice the similarity with the test criterion for models; the test statistic for the structure $\mathcal{M}(\nu')$ is of the same form, only evaluated for the ML model in $\mathcal{M}(\nu')$ instead of for a given data–independent model. However, the degrees of freedom are different, yielding different thresholds for the statistics.

The conclusion is that *falsifying an ML model optimally falsifies also the structure*. If the ML model is unlikely, then any other model with the same structure is even more unlikely.

Remark: This yields another way to interpret the concept of 'degrees of freedom' and to understand why the value will be lower for the ML–estimate: What basically determines the probabilistic derivation of acceptance domains, and in particular the number of degrees of freedom in the χ^2–variable is degree of freedom *from data*. It is assumed that a model M and a structure $\mathcal{M}(\nu')$ to be tested in the two cases are both statistically *independent of data*. Since falsifying $\mathcal{M}(\nu')$ is equivalent to falsifying the ML–model $\hat{M}$, one might think that one could test $\hat{M}$ by using the acceptance domain $\mathcal{B}(\gamma|\nu,\mathbf{c},\mathbf{d})$ for *models*. But since $\hat{\theta}_1 = ML\{\theta_1|\nu',\mathbf{c},\mathbf{d}\}$, it is no longer independent of data, and hence $\mathcal{B}(\gamma|\nu,\mathbf{c},\mathbf{d})$ is not valid. Unlike an independent θ_1 the estimate, $\hat{\theta}_1$ has no longer any degrees of freedom from data; all freedom was lost, when $\hat{\theta}_1$ was computed from data. The possible models in $\mathcal{M}b(\nu)$ are specified by $(\hat{\theta}_1, \theta_2)$, which has $\dim(\theta_2)$ degrees of freedom (from data). That determines the size of acceptance domain $\mathcal{B}$ for *ML–models*, and for *structures*.

Remark: Notice again that if *independent* data are used for falsification of the structure (cross–falsification) then the falsification rule for *models* should be used on the ML model. The practical difference is the number of degrees of freedom in the test. Independent data are of course preferable, and tests for this case can generally be made more powerful (Goodrich and Caines, 1979b).

Correct but not conclusive

Notice again that the power of the test depends on the alternative hypothesis, the structure $\mathcal{M}b(\nu)$. A false model or structure will not be falsified, if $\mathcal{M}b(\nu)$ does not contain a significantly better model or structure. Example 8.4 will illustrate this. The test will be valid for all ν, but it will not be conclusive, until $\nu = \nu_n$, where $\mathcal{M}b(\nu_n)$ contains a Source–Equivalent model. This means that it is of interest to investigate what will happen if ν_n is either underestimated or overestimated.

Overestimating the sufficient complexity is not a 'pitfall'

If the sufficient structure index is overestimated, $\nu > \nu_n$, then the derivation of $\mathcal{B}(\gamma|\nu',\nu,\mathbf{c}_N,\mathbf{d}_N)$ is not valid immediately. The Hessian $Q_{\theta\theta}$ becomes singular and cannot be inverted to provide the test quantity $N\, Q_\theta\, Q_{\theta\theta}^{-1}\, Q_\theta^T$.

With the same arguments as used for the validation rule (section 7.3 on "Two 'pitfalls'"), a too large ν would theoretically be detectable from the singularity, and one would not have to consider the case $\nu > \nu_n$. However, in practice there is probably no exactly Source–Equivalent model $\mathbf{M} = (\nu_n, \Theta)$, such that $Q_{\theta\theta}$ becomes singular when $\nu > \nu_n$. The Hessian will gradually become more ill–conditioned as ν increases, and determining the smallest sufficient v will be uncertain. Hence, it is important instead that the conditional falsification rule be robust with respect to the structure index v of the alternative structure. But that is not difficult to show:

Proof:
Do a linear orthogonal transformation of θ to the new coordinates η such that the Hessian $Q_{\eta\eta}(\eta)$ becomes diagonal. Then

$$Q_\theta(\theta)\ Q_{\theta\theta}(\theta)^{-1}\ Q_\theta(\theta)^T = Q_\eta(\eta)\ Q_{\eta\eta}(\eta)^{-1}\ Q_\eta(\eta)^T$$

$$= Q_{\eta_1}\ Q_{\eta_1\eta_1}^{-1}\ Q_{\eta_1}{}^T + Q_{\eta_2}\ Q_{\eta_2\eta_2}^{-1}\ Q_{\eta_2}{}^T$$

where $\eta = \mathrm{col}(\eta_1,\eta_2)$, $\dim(\eta_1)$ is the number of parameters in $(v_{\mathbf{n}},\Theta)$, and η_2 are redundant parameters introduced by choosing $v > v_{\mathbf{n}}$. For large N the Likelihood loss function is from (8.25) and (8.29)

$$Q(\theta|v,\mathbf{c}_N,\mathbf{d}_N)$$

$$\rightarrow \tfrac{1}{2}\ \log \det(2\pi\ N^{-1}\ Q_{\eta_1\eta_1}^{-1}) + \tfrac{1}{2}\ \log \det(2\pi\ N^{-1}\ Q_{\eta_2\eta_2}^{-1})$$

$$+ \tfrac{1}{2}(\eta_1 - \hat{\eta}_1)^T\ Q_{\eta_1\eta_1}\ (\eta_1 - \hat{\eta}_1) + \tfrac{1}{2}\ (\eta_2 - \hat{\eta}_2)^T\ Q_{\eta_2\eta_2}\ (\eta_2 - \hat{\eta}_2).$$

Hence, $Q_{\eta_2}(\eta) = Q_{\eta_2\eta_2}\ (\eta_2 - \hat{\eta}_2)$ and

$$Q_{\eta_2}\ Q_{\eta_2\eta_2}^{-1}\ Q_{\eta_2}{}^T = (\eta_2 - \hat{\eta}_2)^T\ Q_{\eta_2\eta_2}\ (\eta_2 - \hat{\eta}_2).$$

This value is small when $v > v_{\mathbf{n}}$, even if $v_{\mathbf{n}}$ is not well defined, so that the contribution from overparametrization to the test variable $N\ Q_\theta(\theta)\ Q_{\theta\theta}(\theta)^{-1}\ Q_\theta(\theta)^T$ is small. ∎

Hence, in practice one can always invert $Q_{\theta\theta}$ and use the falsification rule (even if it may require high−precision arithmetics to avoid rounding errors). The robustness property implies that the test result will be valid even in ill−conditioned cases. The acceptance domain $\mathcal{B}(\gamma|v',v,\mathbf{c}_N,\mathbf{d}_N)$ will be very wide in some direction. In the (theoretical) singular case it will be infinite in the directions along the surface defined by $\mathcal{S} \cap \mathcal{M}(v)$. But this means only that the domain will still be useful. As in the falsification rule for models one can replace the inverse $Q_{\theta\theta}^{-1}$ by the pseudo−inverse $Q_{\theta\theta}{}^\dagger$ to cover this case.

Underestimating sufficient complexity may or may not be a 'pitfall'

Underestimation ($v < v_{\mathbf{n}}$) is not detectable in the same easy way, and may therefore be a 'pitfall', *if no validation is possible*. The test will not necessarily falsify a wrong structure, regardless of the extension of the

experiment, if ν is not sufficiently large to provide a sufficiently general
alternative structure $\mathcal{M}_b(\nu)$ to hold a Source–Equivalent model.

If the case is a 'pitfall' or not depends on what one will do, when a
structure is not falsified. According to the 'Scientist's rule', suggested in
section 4.4 on "Basic identification procedures", one should expand the
experiment. However, if one would run out of experiments, or has only
one data sample to use for falsification, then there is an obvious risk
that one would consider the case closed and the model 'verified'. *That
is a 'pitfall'.*

─────────────────────── **Example 8.4** ───────────────────────

Illustrates the 'pitfall' of not having a sufficient alternative structure.

Let the limiting Likelihood loss be
$Q(\theta_1,\theta_2) = [\theta_1{}^2 + (\theta_2 - 1)^2]/2$
and one wants to test the structure
$\mathcal{M}_b(0,0) = \{\theta = 0\}$.

Assume that one specifies (incorrectly) the 'sufficient' alternative
$\nu = (1,0)$. Then the optimal test in the previous section will give
$Q_\theta(\hat\theta)\, Q_{\theta\theta}(\hat\theta)^{-1}\, Q_\theta(\hat\theta)^T = \hat\theta_1{}^2 = 0 < \chi_1{}^2(1-\gamma)/N$
and the obviously incorrect structure $\mathcal{M}_b(0,0)$ is *accepted* by the
test. The error is a consequence of the fact that the alternative
structure $\mathcal{M}_b(1,0) = \{\theta_1,0\}$ does not contain a better model, even
though $\mathcal{M}_b(1,1) = \{\theta_1,\theta_2\}$ does.

The sufficient alternative $\nu = (1,1)$ will yield the value
$Q_\theta(\hat\theta)\, Q_{\theta\theta}(\hat\theta)^{-1}\, Q_\theta(\hat\theta)^T = \hat\theta_1{}^2 + (\hat\theta_2 -1)^2 = 1 > \chi_2{}^2(1-\gamma)/N,$
if N is large enough, and $\mathcal{M}_b(0,0)$ will be rejected correctly.

Testing the structure $\mathcal{M}_b(1,0)$ against $\mathcal{M}_b(1,1)$ yields the same
result, except that there is now only one degree of freedom.

───

8.5   The Likelihood–Ratio test

The LMP–test applied to the falsification of structures (section 8.4) was
optimized to the case when there is only a small deviation between the

structure to be tested and that which would be required for describing the data with sufficient confidence. (Bohlin, 1978). An alternative for testing structures is the '*Likelihood–Ratio test*' (Wilks, 1962). It tests the ML–model $\hat{M}'$ in the tentative structure $\mathcal{M}b'$ (hence it tests the structure), and is optimized to the alternative hypothesis that the data is described by the ML–model $\hat{M}$ in the larger structure $\mathcal{M}b \supset \mathcal{M}b'$.

Hence, according to the procedure for testing of statistical hypotheses (section 8.1) one should optimize the domain $\mathfrak{D}(\gamma|\hat{M}')$ to maximize the reject probability $\Pr\{(\mathbf{c},\mathbf{d}) \notin \mathfrak{D}(\gamma|\hat{M}')|\hat{M}\}$ subject to the condition $\Pr\{(\mathbf{c},\mathbf{d}) \in \mathfrak{D}(\gamma|\hat{M}')|\hat{M}'\} = \gamma$. This means

$$\min_{\mathfrak{D}} \; [\Pr\{(\mathbf{c},\mathbf{d}) \in \mathfrak{D}|\hat{M}\} - \lambda \Pr\{(\mathbf{c},\mathbf{d}) \in \mathfrak{D}|\hat{M}'\}]$$

$$= \min_{\mathfrak{D}} \int \mathrm{Ind}\{(c,d) \in \mathfrak{D}\} \; \{p[c,d|\hat{M}] - \lambda \, p[c,d|\hat{M}']\} \; d(c,d) \qquad (8.35)$$

where λ is a Lagrange multiplier. But this is obviously minimized if $\mathfrak{D}$ contains only points (c,d) for which the integrand is negative

$$\mathfrak{D}(\lambda) = \{c,d|p[c,d|\hat{M}] < \lambda \, p[c,d|\hat{M}']\} \qquad (8.36)$$

Hence $\hat{M}'$ is preferred before $\hat{M}$ if

$$p[\mathbf{c},\mathbf{d}|\hat{M}]/p[\mathbf{c},\mathbf{d}|\hat{M}'] < \lambda \qquad (8.37)$$

Since the (unbiassed) likelihood is the data density with data inserted, the left member is the ratio of likelihoods, which gives the test its name.

The value of λ is determined by (8.36) and the equation

$$\int \mathrm{Ind}\{(c,d) \in \mathfrak{D}(\lambda)\} \; p[c,d|\hat{M}']\} \; d(c,d) = \gamma \qquad (8.38)$$

However, the equation is difficult to solve in other than simple cases. For *long samples* and *parametric models* the following asymptotic value was given by Leontaritis and Billings (1987): let $M' = (\nu',\theta')$ and $M = (\nu,\theta)$, $\nu' < \nu$. Then $\lambda \to \exp[\chi_r^2(1-\gamma)/2]$, where $r = \dim(\theta) - \dim(\theta')$. The condition $\nu' < \nu$ is satisfied if the feasible structures have a nesting.

Significant loss reduction

The Likelihood–Ratio test for long samples can also be interpreted in terms of the Likelihood loss function introduced in section 3.4.1:

$$Q(M|\mathbf{c}_N,\mathbf{d}_N) \triangleq -\log L(M|\mathbf{c}_N,\mathbf{d}_N)/N \tag{8.39}$$

Then (8.37) becomes

$$-\log L(\hat{M}'|\mathbf{c}_N,\mathbf{d}_N)/L(\hat{M}|\mathbf{c}_N,\mathbf{d}_N)$$

$$= N\,[Q(\hat{M}'|\mathbf{c}_N,\mathbf{d}_N) - Q(\hat{M}|\mathbf{c}_N,\mathbf{d}_N)] < \chi_r^2(1-\gamma)/2 \tag{8.40}$$

This says that the simpler model is preferred when the reduction in loss achieved by the more complex model $\hat{M} = (\hat{v},\hat{\theta})$ is lower than the 'significant' value $\chi_r^2(1-\gamma)/2$. Notice that even if $\hat{M}'$ would be as good a model as $\hat{M}$, the latter would have a smaller loss, simply because $Q(M|\mathbf{c}_N,\mathbf{d}_N)$ has more free parameters that take part in the minimization than $Q(M'|\mathbf{c}_N,\mathbf{d}_N)$ has. The threshold $\chi_r^2(1-\gamma)/2$ is the loss reduction that would be expected in that case, when there are r more parameters. Hence, only loss reductions larger than this indicate that $\hat{M}$ is a better model with probability γ.

In summary,

Likelihood–Ratio falsification rule

Reject the ML model $\hat{M}' = (\mathcal{A},\hat{v}',\hat{\theta}')$ in favour of $\hat{M} = (\mathcal{A},\hat{v},\hat{\theta})$, $\hat{v}' < \hat{v}$, with risk α, if

$$N\,[Q(\hat{v}',\hat{\theta}'|\mathcal{A},\mathbf{c}_N,\mathbf{d}_N) - Q(\hat{v},\hat{\theta}|\mathcal{A},\mathbf{c}_N,\mathbf{d}_N)] > \chi_r^2(\alpha)/2$$

where N is large, and $\chi_r^2(\alpha)$ has $r = \dim(\hat{\theta}) - \dim(\hat{\theta}')$ degrees of freedom.

The LR test requires overfitting

A comparison with the LMP–test yields that the LR–test has a higher reject probability when the model $\hat{M}$ actually describes data, and if $\hat{M}$ is close to $\hat{M}'$ the LR–test is as good as LMP. The price one has to pay for this, however, is the necessity to compute a model $(\hat{\nu}, \hat{\theta})$ that is more complex than the one to be tested. If the simpler model $(\hat{\nu}', \hat{\theta}')$ is actually adequate, one would have to compute at least one model more than will be necessary with the LMP–test. This 'over–fitting' problem is aggravated by the fact that computing models of higher complexity than necessary is usually an illconditioned problem.

> **Remark:** Overfitting is avoided by the LMP test. How this is achieved, and the relation between LMP and a special LR–test, the F–test, is illustrated by Söderström (1981).

8.6   Efficiency *vs* safety

Optimal tests all assume something apriori, in order to gain efficiency. The more that is assumed, the more is gained, in essence because the test can exclude more alternatives apriori. The latter is the cause of the decreasing safety that accompanies increasing efficiency. If the apriori assumptions exclude too much, the test may accept a wrong model, and there may be a 'pitfall', as discussed previously. *Optimal falsification can never falsify the assumptions the optimality is based on.*

> **Remark:** The (optimal) conditional tests assume Source Equivalence, since they are all conditional at least on the widest structure $\bar{\mathcal{M}}$. This does not mean, however, that the tests will always give the wrong answers when there is no Source Equivalent model in $\bar{\mathcal{M}}$. Both the conditional and the unconditional tests do in fact test the null hypothesis that M is Source Equivalent, and hence, and *a fortiori*, that there is a Source–Equivalent model in $\bar{\mathcal{M}}$. This means that a *reject* decision will always have the prespecified risk (of rejecting a Source–Equivalent model). However, an *accept* decision will depend on the alternative hypothesis, and a correct risk value (of accepting a not Source–Equivalent model) requires

the assumption of Source Equivalence. Notice that the conditional tests do not compute an 'accept' risk.

Remark: The inherent conflict between efficiency and safety is by no means unique in engineering design. In connection with adaptive control Niedzwiecki (1990) observes that "there exists a relationship between the amount of information about time–varying system parameters, which is available apriori, and the robustness of the identification algorithm based on such prior knowledge. The more specialized the estimation algorithm is, the less reliable it might be under nonstandard conditions". A perhaps more well-known case with much the same propery is function optimization:

──────────────── **Example 8.5** ────────────────

Illustrating the conflict between efficiency and safety.

Minimize the function $f(x) = (1 + x^2)^{-1} x^2$ using two conventional search methods.

1) *Gradient method*:

$$x_{k+1} = x_k - \alpha f'(x_k) = x_k - \alpha 2 x_k (1 + x_k^2)^{-2}$$
$$= x_k [(1 + x_k^2)^2 - 2 \alpha] (1 + x_k^2)^{-2}.$$

Since the factor $|[(1 + x_k^2)^2 - 2 \alpha] (1 + x_k^2)^{-2}| < 1$ for $0 < \alpha < 1$, the mapping $x_k \mapsto x_{k+1}$ is contracting, and the algorithm converges to zero for all start values.

In each step the error is reduced by a factor
$$x_{k+1}/x_k = [(1 + x_k^2)^2 - 2 \alpha] (1 + x_k^2)^{-2} \to 1 - 2 \alpha.$$

2) *Newton–Raphson's method*:

$$x_{k+1} = x_k - f'(x_k)/f''(x_k) = x_k - x_k (1 + x_k^2)(1 - 3 x_k^2)^{-1}.$$

In each stepo the error is reduced by a factor
$$x_{k+1}/x_k = -4 x_k^2 (1 - 3 x_k^2)^{-1}.$$

A comparison reveals immediately that the NR method converges much faster when x_k approaches the minimum, unless $\alpha = 1/f''(0) = 0.5$. However, since the minimum is unknown in the first place, one is usually not free to adapt α to the optimal value, and generally $\alpha \neq 0.5$.

The faster convergence is paid for by a smaller convergence domain; If $|4 x_k^2 (1 - 3 x_k^2)^{-1}| > 1$, the routine will not converge. That is, when $x_0^2 \geq 1/7$.

What is best depends on the case

Thus, efficiency is gained at a cost. How much is gained, and at what cost depends of course on the particular case, and cannot be quantified generally. The following examples may serve to illustrate the performances of conditional and unconditional tests, and the fundamental conflict between efficiency and safety in falsification:

--- **Example 8.6** ---

This example is rather long, but the phenomenon it illustrates is basic, and one may draw rather far–reaching conclusions from it. To draw the same conclusions from a general analysis would of course strengthen the result, but it would probably be much more difficult for the reader to absorb. (Besides, I do not yet know how to analyse a general case.)

Consider the system in *Fig. 50.* It consists of a tank, the flow into which is controlled by a valve. The volume $V(t)$ of liquid in the tank is sensed by a level gauge with uncorrelated gaussian errors: $y(t) = V(t) + \omega_1(t)$.

The valve is assumed to be linear, and the flow $f(t)$ is also affected by uncorrelated gaussian errors (probably from fluctuations in the

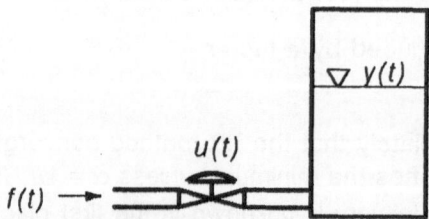

Fig. 50: The problem is to test the valve gain from noisy measurements of tank level and under random fluctua–tion in feed pressure.

feed pressure):
$$f(t) = \mathbf{b} \, u(t) + \omega_2(t).$$

For simplicity, let the variables be deviations from nominal values, so that they may take on both positive and negative values. Also for simplicity, let time be discrete, all measurements be in the data sample, and the valve actuator controlled at all times, so that
$$d(k) = \Delta y(t)$$
$$c(k) = u(t)$$
and describe the tank dynamics by
$$\Delta V(t) = f(t-1)$$
where Δ is the backwards–difference operator. Hence, the model structure is
$$\mathcal{M}_b(1,b): \Delta y(t) = b \, u(t-1) + \omega_2(t-1) + \Delta \omega_1(t).$$

Suppose that specifications claim that the valve is linear and has been calibrated to gain $\mathbf{b} = 1$, and that one wants to check the calibration. For this purpose one perturbs the valve setting by an uncorrelated sequence $\{c(k)\}$ and observes the resulting changes $\{d(k) = \Delta y(t)\}$ in the level readings. What is the optimal test of the hypothesis that $\mathbf{b} = 1$, assuming the model structure?

In order to compute the Likelihood function as the basis for design-ing an optimal test it is convenient first to derive an 'external model' containing a single primitive random variable ω (see section 4.5 on "internal and external models"). The following model has the same output distribution:
$$d(k) = b \, c(k-1) + \lambda \, [\omega(k) - \kappa \, \omega(k-1)]$$

To see this compute and equate the autocorrelation functions of the two models. That will also yield equations for determining λ and κ, and reveal that $\kappa \in [0,1]$.

The Likelihood function would now be possible to compute numeri-cally in a way described in section 5.5.2 on "Innovations". However, in this simple case an analytic evaluation for long data samples is also feasible, and it suits better the purpose of compar-ing efficiencies: The distribution of $\omega(k)$ is gaussian, and hence the conditional output distribution is obtained by a linear transformation $\omega(k) \mapsto d(k)$ (the model)
$$p[d(k)|c_{k-1}, d_{k-1}] = \exp[-\tfrac{1}{2}\omega(k)^2]/(\sqrt{2\pi}\,\lambda)$$
where $\omega(k)$ depends on $d(k)$, c_{k-1}, and d_{k-1} in a way determined by the model. Hence, the Likelihood loss is
$$Q(b|c_N, d_N) = constant + \frac{1}{2N} \sum_{k=1}^{N} w(k|b)^2/\lambda^2$$

where the 'innovations' are computed from
$$w(k|b) = \kappa\, w(k-1|b) + [\mathbf{d}(k) - b\, \mathbf{c}(k-1)]/\lambda.$$

The derivatives with respect to b are
$$\nabla_b Q(b) = \frac{1}{N}\sum_{k=1}^{N} w(k|b)\, \nabla_b w(k|b)/\lambda^2$$
$$\nabla_b^2 Q(b) = \frac{1}{N}\sum_{k=1}^{N} [\nabla_b w(k|b)]^2/\lambda^2$$

The second-order derivative $\nabla_b^2 w(k|b) = 0$, since $w(k|b)$ is linear in b. Differentiating the recursive equation defining the innovations yields
$$\nabla_b w(k|b) = \kappa\, \nabla_b w(k-1|b) - \mathbf{c}(k-1)/\lambda,$$

which has the solution
$$\nabla_b w(N|b) = -\sum_{k=1}^{N} \kappa^{k-1}\, \mathbf{c}(N-k+1)/\lambda.$$

According to the 'conditional falsification rule for models' (see section 8.3) the test to falsify the hypothesis that $b = 1$ is now:
$$N\, Q_b(1)\, Q_{bb}^{-1}(1)\, Q_b(1) > \chi^2_1(1-\gamma).$$

To see the power of the test compute the range of gains $\mathbf{b} \neq 1$ that the test will accept. This gain parameter is no longer an arbitrary value in the Likelihood function, but the unknown gain in the 'true' model that generates the data
$$\mathbf{d}(k) = \mathbf{b}\, \mathbf{c}(k-1) + \lambda\, [\omega(k) - \kappa\, \omega(k-1)].$$

Hence the innovations satisfy
$$w(k|b) = \kappa\, w(k-1|b) + [\mathbf{d}(k) - b\, \mathbf{c}(k-1)]/\lambda.$$

Since $w(k|b)$ is linear in both b and $\mathbf{b}$, and equal to $\omega(k)$ if $b = \mathbf{b}$, it follows that
$$w(k|b) = \omega(k) + (b - \mathbf{b})\, \nabla_b w(k|b).$$

It is now easy to derive asymptotic values of the factors in the test criterion:
$$\nabla_b^2 Q(b) = \frac{1}{N}\sum_{k=1}^{N} [\nabla_b w(k|b)]^2/\lambda^2$$
$$\rightarrow E[\nabla_b w(N|b)]^2/\lambda^2 = \sum_{k=1}^{N} \kappa^{2k-2}\, E\, \mathbf{c}(N-k+1)^2/\lambda^2$$
$$= N\, \sigma_c^2\, (1 - \kappa^{2N})/\lambda^2\, (1 - \kappa^2)$$

$$\nabla_b Q(b) = \frac{1}{N}\sum_{k=1}^{N} [\omega(k) + (b - \mathbf{b}) \nabla_b \mathbf{w}(k|b)] \nabla_b \mathbf{w}(k|b)/\lambda^2$$

$$\rightarrow (b - \mathbf{b}) \sigma_c^2 (1 - \kappa^{2N})/\lambda^2 (1 - \kappa^2) + \frac{1}{N}\sum_{k=1}^{N} \omega(k) \nabla_b \mathbf{w}(k|b)/\lambda^2$$

The last term decreases as $1/\sqrt{N}$, and may be neglected, since $\omega(k)$ and $\nabla_b \mathbf{w}(k|b)$ are independent. Hence, the test quantity tends to

$$N [\nabla_b Q(b)]^2/\nabla_b^2 Q(b) \rightarrow (b - \mathbf{b})^2 N \sigma_c^2/\lambda^2 (1 - \kappa^2)$$

The range is obviously determined by the equation

$$N Q_b(b) Q_{bb}(b)^{-1} Q_b(b) = \chi^2_1(1-\gamma)$$

which yields

$$(b - \mathbf{b})^2 = \chi^2_1(1-\gamma) \lambda^2 (1 - \kappa^2)/\sigma_c^2 N.$$

The narrower range of b around b, the more powerful test. The power apparently increases with the signal–to–noise ratio σ_c/λ and the sample length N, which is to be expected.

In order to compare performances, compute also the power of the unconditional test of cross–correlations: The test statistics are from section 8.2 on "Unconditional falsification"

$$\xi(l|b) = \frac{1}{N}\sum_{k=1}^{N} \mathbf{c}(k-l) \, \mathbf{w}(k|b).$$

The variances are

$$\sigma_\xi(l|b)^2 = \frac{1}{N}\frac{1}{N}\sum_{k=1}^{N} \mathbf{c}(k|b)^2$$

The test to falsify the hypothesis is from Rule 2, section 8.2:

$$\sum_{l=1}^{r} \xi(l|b)^2/\sigma_\xi(l|b)^2 > \chi^2_r(1-\gamma)$$

To evaluate the power, insert

$$\mathbf{w}(k|b) = \omega(k) + (b - \mathbf{b}) \nabla_b \mathbf{w}(k|b).$$

Then

$$\xi(l) = (b - \mathbf{b}) \frac{1}{N}\sum_{k=1}^{N} \mathbf{c}(k-l) \nabla_b \mathbf{w}(k|b)$$

$$\rightarrow (b - \mathbf{b}) E\{\mathbf{c}(N-l) \nabla_b \mathbf{w}(N|b)\}$$

$$\rightarrow (b - \mathbf{b}) \sum_{k=1}^{N} \kappa^{k-1} E\{\mathbf{c}(N-l) \mathbf{c}(N-k+1)\}/\lambda$$

$$= (b - \mathbf{b})\, \sigma_c{}^2\, \kappa^l / \lambda$$

$$\sigma_\xi (l|b)^2 = \frac{1}{N}\frac{1}{N}\sum_{k=1}^{N} c(k|b)^2 \rightarrow \sigma_c{}^2/N$$

The range is determined by the equation

$$\chi_r{}^2(1-\gamma) = \sum_{l=1}^{r} \xi(l|b)^2/\sigma_\xi(l|b)^2 = \sum_{l=1}^{r} (b - \mathbf{b})^2\, \sigma_c{}^2\, \kappa^{2l}\, N/\lambda^2$$

$$= (b - \mathbf{b})^2\, \sigma_c{}^2\, N\, (1 - \kappa^{2r})/\lambda^2\, (1 - \kappa^2)$$

which yields

$$(b - \mathbf{b})^2 = \chi_r{}^2(1-\gamma)\, \lambda^2\, (1 - \kappa^2)/\sigma_c{}^2\, N\, (1 - \kappa^{2r})$$

It is now possible to compare the efficiencies of the two tests, for instance by comparing the data lengths N it takes to achieve a certain acceptance range of $b - \mathbf{b}$. The ratio will be

$$N(\text{unconditional})/N(\text{conditional}) = \chi^2_r(1-\gamma)/\chi^2_1(1-\gamma)(1 - \kappa^{2r})$$

It depends on the number r of cross–correlations one decides to test. But the efficiency has a maximum for a finite r, since the two r–dependent factors are increasing and decreasing respectively. This gives a way to compute the optimal r and compare the efficiencies of the conditional test and the best cross–correlation test.

The efficiency depends critically on κ, which is a property of the object. If $\kappa = 0$, the optimum is $r = 1$, and the tests are equally efficient. However, if κ is close to one, the conditional test is superior. Too see the difference consider the original problem and assume some figures:

$\lambda = 1$, $\kappa = 0.99$, $\sigma_c = 0.1$. This corresponds to a large error ratio $\text{rms}(\omega_1)/\text{rms}(\omega_2) = 100$, and a signal–to–noise ratio $\text{rms}(u)/\text{rms}(\omega_1) = 0.1$. The numbers illustrate a case where the tank volume is large compared to the input flow per unit time. Since it is reasonable to assume that the relative errors are of comparable magnitude, the *absolute* errors in measuring tank volume level would be much greater than the random variation in input flow. The signal–to–noise ratio is kept small, in order not to perturb the normal operation of the object.

With these figures the optimal number of cross–correlations is $r = 27$, and the ratio of sample lengths is for the risk level $1-\gamma = 0.01$:

$$N(\text{unconditional})/N(\text{conditional})$$
$$= \chi^2_{27}(0.01)/\chi^2_1(0.01)(1 - 0.99^{54}) \approx 17.$$

It is clear that the efficiency of unconditional falsification may be superior in quite realistic cases; in this case as soon as the tank volume is large. Technically, the gain comes from the factor κ. That factor plays an important role in the equation computing the innovations from data

$w(k|b) = \kappa\, w(k{-}1|b) + [\mathbf{d}(k) - b\, \mathbf{c}(k{-}1)]/\lambda$

and hence also in the equation predicting the output $d(k)$ from data $\mathbf{c}_{k-1}, \mathbf{d}_{k-1}$, as well as computing the Likelihood function. If κ is close to one, the predictor will have a large time constant, and the innovations and predicted output will change slowly with k. The conditional test compensates for the slow dynamics of the predictor. The unconditional test cannot do that, since it was designed on the basis that there is no reliable a priori information on the value of κ.

In summary, it is reasonable to infer that for objects such that the associated output predictors contain some slow dynamic modes, a conditional test which utilizes that information will be much superior to a unconditional one.

Try the same tests on another example:

──────────────── **Example 8.7** ────────────────

Consider the two-input non-dynamic object

$d(k) = \theta_1\, c_1(k{-}1) + \theta_2\, c_2(k{-}1) + \omega(k)$

where $c_1(k)$, $c_2(k)$, $\omega(k)$ are all uncorrelated gaussian variables with zero means and unit variances, and independent. Let the true parameter value be $\Theta = (0,1)$, and assume that one wants to falsify the model with value $\theta = (0,0)$. Assume also that one believes (falsely) that if there is any influence of input $\mathbf{c}$ on the response $\mathbf{d}$ at all (which one doubts), that influence would possibly come from c_1.

Hence, the largest model structure is

$\mathcal{M}_b(1)\colon d(k) = \theta_1\, c_1(k{-}1) + \omega(k)$.

In order to apply the optimal conditional falsification routine with these prerequisites, first derive the logarithm of the Likelihood loss function and its derivatives in parameter space:

$$Q(\theta|\mathbf{c}_N,\mathbf{d}_N) = constant + \frac{1}{2N}\sum_{k=1}^{N}[\mathbf{d}(k) - \theta_1\,\mathbf{c}_1(k-1)]^2$$

$$\nabla_\theta Q(\theta) = -\frac{1}{N}\sum_{k=1}^{N}[\mathbf{d}(k) - \theta_1\,\mathbf{c}_1(k-1)]\,\mathbf{c}_1(k-1)$$

$$= -\frac{1}{N}\sum_{k=1}^{N}[-\theta_1\,\mathbf{c}_1(k-1) + \mathbf{c}_2(k-1) + \omega(k)]\,\mathbf{c}_1(k-1) \rightarrow \theta_1$$

$$\nabla_\theta^2 Q(0) = \frac{1}{N}\sum_{k=1}^{N}\mathbf{c}_1(k-1)^2 \rightarrow 1$$

The reject criterion is

$$\chi^2_1(1-\gamma) < N\,[\nabla_\theta Q(0)]^2/\nabla_\theta^2 Q(0) = N\,\theta_1^2 = 0$$

Hence, the incorrect model is *accepted*, regardless of the sample length, and the conditional test has obviously failed.

For an unconditional falsification compute the two cross–correlations

$$\xi(l)^T = \frac{1}{N}\sum_{k=1}^{N}[\mathbf{c}_1(k-l)\,\mathbf{c}_2(k-l)]\,\mathbf{w}(k)$$

$$= \frac{1}{N}\sum_{k=1}^{N}[\mathbf{c}_1(k-l)\,\mathbf{c}_2(k-l)]\,[(\Theta_1 - \theta_1)\,\mathbf{c}_1(k-1) + \Theta_2\,\mathbf{c}_2(k-1)$$

$$+\,\omega(k)]$$

$$\rightarrow [-\theta_1\;\;1]\,\delta(1-l) = [0\;\;1]\,\delta(1-l)$$

$$R_\xi(l,l') \rightarrow \frac{1}{N}\,\delta(l-l')$$

The reject criterion is

$$\chi^2_{2r}(1-\gamma) < \sum_{l=1}^{r}\xi(l)^T R_\xi(l)^{-1}\,\xi(l) \rightarrow N$$

and the model is *rejected* correctly for large N.

Hence, in this case the unconditional test is the better. The conditional test failed because the 'condition', *i.e.* the alternative hypothesis $M \in \mathcal{M}(1)$ was incorrect. $\mathcal{M}(1)$ contains all models $(\theta_1, 0)$, while the correct model is in $\mathcal{M}(2) = \{\theta_1, \theta_2\}$.

Testing auto- and crosscorrelation is not enough

The conventional unconditional tests of autocorrelation of innovations $w(k)$ and its crosscorrelation with the stimulus sequence $c(k)$ are not sufficient for detecting nonlinearities, not even for infinitely long samples (Leontaritis and Billings, 1987). It may not falsify a linear model when the object is nonlinear, in contrast to a conditional maximum-power test. Example 8.8 (Billings and Voon, 1983) will demonstrate that.

However, the conditional test still needs an assumption about the type of nonlinearity to be efficient, and one may therefore want an unconditional test that will reveal nonlinearities, even if not maximally efficient. Billings and Voon (1983) have developed such a test. In essence, one will have to test also the third moments

$$r_{eec}(l) = \frac{1}{N} \sum_{k=1}^{N} e(k) \, e(k-l) \, c(k-l),$$

for all τ, in addition to the conventional $r_{ee}(l)$ and $r_{ec}(l)$.

─────────────────── **Example 8.8** ───────────────────

Illustrating the shortcomings of the auto- and crosscorrelation tests.

Let $(\mathbf{X},\mathbf{S})$: $d(k) = \Theta_1 \, c(k) + \Theta_2 \, c(k-1) \, \omega(k-1) + \omega(k)$,
(c,ω) gaussian $(0,1)$ and uncorrelated,
but M: $d(k) = \theta_1 \, c(k) + \lambda \, \omega(k)$.
Assume the best linear model $\theta_1 = \Theta_1$.

The residual is
$e(k|M) = \mathbf{d}(k) - \Theta_1 \, \mathbf{c}(k) = \Theta_2 \, \mathbf{c}(k-1) \, \omega(k-1) + \omega(k)$.

Auto- and crosscorrelation:

The autocovariance function is
$r_{ee}(l) = E\{e(k|M) \, e(k-l|M)\}$
$= E\{[\Theta_2 \, \mathbf{c}(k-1) \, \omega(k-1) + \omega(k)] \, [\Theta_2 \, \mathbf{c}(k-1-l) \, \omega(k-1-l) + \omega(k-l)]\}$
$= 0$ for $l > 0$.

The crosscovariance function is
$r_{ec}(l) = E\{e(k|M) \, \mathbf{c}(k-l)\}$
$= E\{[\Theta_2 \, \mathbf{c}(k-1) \, \omega(k-1) + \omega(k)] \, \mathbf{c}(k-l)\}$
$= 0$ for all l.

Hence, testing the sample estimates of these functions yields nothing, irrespective of sample length.

Locally Most Powerful test:

Let the alternative hypothesis be
H_1: $M \in \mathcal{M}_b(2) = \{\theta_1, \theta_2 | \mathbf{d}(k) = \theta_1 \, \mathbf{c}(k) + \theta_2 \, \mathbf{c}(k{-}1) \, \omega(k{-}1) + \omega(k)\}$.

The residual is
$\mathbf{e}(k|\theta_1,\theta_2) = \mathbf{d}(k) - \theta_1 \, \mathbf{c}(k) - \theta_2 \, \mathbf{c}(k{-}1) \, \mathbf{e}(k{-}1|\theta_1,\theta_2)$.

Let ∇_i be the partial differentiation operator with respect to θ_i.
Then
$\nabla_1\mathbf{e}(k) = -\mathbf{c}(k)$
$\nabla_2\mathbf{e}(k) = -\mathbf{c}(k{-}1) \, \mathbf{e}(k{-}1) - \theta_2 \, \mathbf{c}(k{-}1) \, \nabla_2\mathbf{e}(k{-}1)$
where the arguments θ_1,θ_2 in $\mathbf{e}$ have been dropped for convenience.

The Likelihood loss function and its derivatives are

$$Q(\theta_1,\theta_2) = 0.5 \, \frac{1}{N} \sum_{k=1}^{N} \mathbf{e}(k)^2$$

$$\nabla_1 Q(\theta_1,\theta_2) = \frac{1}{N} \sum_{k=1}^{N} \mathbf{e}(k) \, \nabla_1\mathbf{e}(k) = -\frac{1}{N} \sum_{k=1}^{N} \mathbf{e}(k) \, \mathbf{c}(k)$$

$$\nabla_2 Q(\theta_1,\theta_2) = \frac{1}{N} \sum_{k=1}^{N} \mathbf{e}(k) \, \nabla_2\mathbf{e}(k)$$

$$= -\frac{1}{N} \sum_{k=1}^{N} \mathbf{e}(k) \, [\mathbf{c}(k{-}1) \, \mathbf{e}(k{-}1) + \theta_2 \, \mathbf{c}(k{-}1) \, \nabla_2\mathbf{e}(k{-}1)]$$

But
$\mathbf{e}(k|\Theta_1,0) = \mathbf{d}(k) - \Theta_1 \, \mathbf{c}(k) = \Theta_2 \, \mathbf{c}(k{-}1) \, \omega(k{-}1) + \omega(k)$,
$\nabla_2\mathbf{e}(k|\Theta_1,0) = -\mathbf{c}(k{-}1) \, \mathbf{e}(k{-}1|\Theta_1,0)$
$\qquad\qquad\quad = -\mathbf{c}(k{-}1) \, \omega(k{-}1) - \Theta_2 \, \mathbf{c}(k{-}1) \, \mathbf{c}(k{-}2) \, \omega(k{-}2)$.

Hence,

$$\nabla_1 Q(\Theta_1,0) = -\frac{1}{N} \sum_{k=1}^{N} \mathbf{e}(k) \, \mathbf{c}(k)$$

$$= -\frac{1}{N} \sum_{k=1}^{N} [\Theta_2 \, \mathbf{c}(k{-}1) \, \omega(k{-}1) + \omega(k)] \, \mathbf{c}(k) \to 0 \text{ (as before)}.$$

But

$$\nabla_2 Q(\Theta_1,0) = -\frac{1}{N} \sum_{k=1}^{N} \mathbf{e}(k) \, \mathbf{c}(k{-}1) \, \mathbf{e}(k{-}1)$$

$$= - \frac{1}{N} \sum_{k=1}^{N} [\Theta_2 \, \mathbf{c}(k-1) \, \omega(k-1) + \omega(k)]$$

$$[\mathbf{c}(k-1) \, \omega(k-1) + \Theta_2 \, \mathbf{c}(k-1) \, \mathbf{c}(k-2) \, \omega(k-2)]$$

$$\rightarrow - \Theta_2$$

Hence, the test quantity $N \, \nabla Q(\Theta_1, 0) \, [\nabla^T \nabla Q(\Theta_1, 0)]^{-1} \, \nabla^T Q(\Theta_1, 0)$ tends to infinity, so the nonlinearity is detected when N becomes large enough.

The result holds also when the alternative hypothesis is not correct: Assume that the nonlinear object is actually
$$\mathbf{d}(k) = \Theta_1 \, \mathbf{c}(k) + g[\mathbf{c}(k-1), \omega(k-1)] + \omega(k)$$

$$\nabla_1 Q(\Theta_1, 0) = - \frac{1}{N} \sum_{k=1}^{N} \{g[\mathbf{c}_{k-1}, \omega_{k-1}] + \omega(k)\} \, \mathbf{c}(k)$$

$$\nabla_2 Q(\Theta_1, 0)$$

$$= - \frac{1}{N} \sum_{k=1}^{N} \{g[\mathbf{c}_{k-1}, \omega_{k-1}] + \omega(k)\} \, \mathbf{c}(k-1) \{g[\mathbf{c}_{k-2}, \omega_{k-2}] + \omega(k-1)\}$$

Since all crosscorrelations do not vanish, the nonlinearity is eventually detected, when N becomes large enough.

The unconditional test of Billings and Voon:

The third test quantity is

$$r_{eec}(l) = \frac{1}{N} \sum_{k=1}^{N} \mathbf{e}(k) \, \mathbf{e}(k-l) \, \mathbf{c}(k-l)$$

$$= \frac{1}{N} \sum_{k=1}^{N} [\Theta_2 \, \mathbf{c}(k-1) \, \omega(k-1) + \omega(k)]$$

$$[\Theta_2 \, \mathbf{c}(k-1-l) \, \omega(k-1-l) + \omega(k-\tau)] \, \mathbf{c}(k-l)$$

This tends to Θ_2 for $l = 1$, so the nonlinearity $\Theta_2 \, \mathbf{c}(k-1) \, \omega(k-1)$ is detected.

The deficiencies of second–order correlation tests can be explained intuitively as follows: It is well known that with a linear model and gaussian noise, the minimum–variance predictor is linear (Åström, 1970). Alternatively, one can postulate a linear predictor, and the same predictor will be minimum–variance even for non–gaussian noise.

However, the minimum–variance predictor for the object in the example is *nonlinear*:

$$\hat{d}(k|k-1) = \Theta_1 \, \mathbf{c}(k) + \Theta_2 \, \mathbf{c}(k-1) \, \omega(k-1) \tag{8.41}$$

where

$$\omega(k-1) = \mathbf{d}(k-1) - \Theta_1 \, \mathbf{c}(k-1) + \Theta_2 \, \mathbf{c}(k-2) \, \omega(k-2) \tag{8.42}$$

(provided the latter is a stable equation).

The best *linear* predictor is

$$\hat{d}(k|k-1) = \Theta_1 \, \mathbf{c}(k) \tag{8.43}$$

since the prediction error $e(k) = \Theta_2 \, \mathbf{c}(k-1) \, \omega(k-1) + \omega(k)$ is uncorrelated.

Hence, one has to detect that the best predictor is not linear, and that therefore it will not be enough to have a linear model. This means that one will have to detect that the residual sequence is not gaussian, and that is why one has to test a third moment.

Use both kinds of falsification

The examples illustrate the fact that, like in many other engineering problems, there is no unique rule to tell the designer what tool to use for model falsification. Unconditional falsification is blunt but safe, while conditional falsification may be sharp but double–edged. However, in this particular case one does not have to choose! One should preferably use both tools (since falsification is a one–sided affair); the conditional approach will falsify efficiently, if the alternative hypotheses are right, and if they are not, the unconditional approach will falsify also the alternative hypotheses.

Is there a structure with 'no apriori information'

It is illuminating to regard the conditional falsification as a family of tests that exploit apriori information as well as data, while unconditional falsification exploits only data. Since there is an optimal test for each structure, one might ask whether the converse is also true, in particular, whether also an unconditional test is optimal for *some* structure (= apriori information). In other words, is there a structure that may be said to carry no apriori information, and what does it look like?

─────────────── **Example 8.9** ───────────────

Consider the 'moving–average' model

$$\mathbf{d}(k) = \lambda \sum_{\tau=0}^{n} \Theta_\tau \, \omega(k-\tau)$$

where $\Theta_0 = 1$, and evaluate the reject criteria for the unconditional autocorrelation and the conditional LMP tests. The innovations are computed recursively from

$$\mathbf{w}(k) = \lambda^{-1} \, \mathbf{d}(k) - \sum_{\tau=1}^{n} \theta_\tau \, \mathbf{w}(k-\tau).$$

The n autocorrelations are

$$\xi = \frac{1}{N} \sum_{k=1}^{N} Z(k)^T \, \mathbf{w}(k)$$

where $Z(k) = [\mathbf{w}(k-1) \; \dots \; \mathbf{w}(k-n)]$

$$R_\xi \rightarrow \frac{1}{N}\frac{1}{N} \sum_{k=1}^{N} Z(k)^T Z(k)$$

and the reject criterion is

$$\xi^T R_\xi^{-1} \, \xi > \chi^2_n (1-\gamma)$$

To set up the LMP test evaluate the Likelihood function:

$$Q(\theta|n,\mathbf{d}_N) = constant + \frac{1}{2N} \sum_{k=1}^{N} \mathbf{w}(k)^2$$

$$\nabla_\theta Q(\theta|n,\mathbf{d}_N) = \frac{1}{N} \sum_{k=1}^{N} \nabla_\theta \mathbf{w}(k) \, \mathbf{w}(k) = \frac{1}{N} \sum_{k=1}^{N} \nabla_\theta \mathbf{w}(k) \, \mathbf{w}(k)$$

But

$$\nabla_\theta \mathbf{w}(k) = - \nabla_\theta \sum_{\tau=1}^{n} \theta_\tau \, \mathbf{w}(k-\tau) = [\mathbf{w}(k-1) \; \dots \; \mathbf{w}(k-n)] = Z(k)$$

$$Q_\theta(\theta) = \nabla_\theta Q(\theta|n,\mathbf{d}_N) = - \sum_{k=1}^{N} Z(k) \, \mathbf{w}(k) = - N \, \xi^T$$

$$Q_{\theta\theta}(\dot{\theta}) = \nabla_{\theta\theta}Q(\theta|n,\mathbf{d}_N)$$

$$= \frac{1}{N}\sum_{k=1}^{N}[\nabla_{\theta}\mathbf{w}(k)]^T\,\nabla_{\theta}\mathbf{w}(k) + \frac{1}{N}\sum_{k=1}^{N}\mathbf{w}(k)\,\nabla_{\theta\theta}\mathbf{w}(k)$$

$$\rightarrow \frac{1}{N}\sum_{k=1}^{N}[\nabla_{\theta}\mathbf{w}(k)]^T\,\nabla_{\theta}\mathbf{w}(k) = \frac{1}{N}\sum_{k=1}^{N}Z(k)^T\,Z(k) = N\,R_{\xi}$$

The LMP reject criterion is

$$N\,Q_{\theta}(\theta)\,Q_{\theta\theta}(\theta)^{-1}\,Q_{\theta}(\theta)^T < \chi^2_n(1-\gamma)$$

But the left member is equal to that of the unconditional 'autocorrelation' test. Hence the apriori information that the structure is a moving–average process does not make the conditional test more efficient. This particular structure may therefore be regarded as void of apriori information. The 'moving average' is the 'blackest' box there is.

---

9 Structure identification

The sequential identification procedure discussed in section 4.4 prescribes that one should carry out a sequence of falsifications of expanding structures $\mathcal{M}(n)$. In the case one has a 'validatable' problem, the stopping rule for the outer loop (the expansion) determines a *sufficient* degree of complexity for the purpose (see chapter 7 on "Validation techniques"). That may well be a structure that is simpler than one needed to describe the object as well as the data would allow. The purpose may not require as detailed a description as identification would actually be able to provide.

In the 'non—validatable' case however, one does not have a stopping rule for the outer loop; either the purpose is not well defined, or one will not be able to compute the stopping criterion $\Pr\{M \in \mathcal{G}|\mathbf{c},\mathbf{d}\} > \gamma$. In that case one will have to let the outer loop go on until the structure is not falsified by data, *i.e.* until the data source or object is modelled as well as data allows. This means that it is the experiment facilities instead that set the limit to how complex the model will be. The procedure will produce the simplest model that data does not falsify.

Hence, in neither of the cases would the task of selecting a suitable complexity n be a problem in itself. It seems to be solved in principle by applying the 'validation' (when relevant) and 'falsification' techniques in chapters 7 and 8. In many cases the complexity is tied to an 'order' of the model, for instance the number of state variables, and the problem is generally referred to as 'order determination'.

Generally, one of the two key problems in parametric system identification (the other is 'parameter estimation') is 'structure identification', *i.e.* selecting a model class $\mathcal{M}_\nu$ from a given (by 'modelling') set of disjoint classes, indexed by non—negative possibly vector—valued structure indices ν. Loosely, 'order determination' determines complexity n, while 'structure identification' also determines the structure index ν. In many cases the 'stucture identification' problem is solved immediately by 'order determination', in particular when a simple nesting is feasible apriori, but also when the nestimg is only partial. Hence a substantial part of this chapter will consider the sequential order determination problem.

However, there are still some obstacles: The repeated application of falsification techniques has its own problems, the notion that the experiment properties and not the object alone decide the model complexity may be philosophically unsatisfying to the designer, and the latter may have other information on the object (than purpose and data) that motivate a different point of view. The probably most serious obstacle in practice to sequential order determination is that when the structure index v is vector–valued, one may not be able to provide the required nesting of the feasible structures a priori. This makes the statistical testing as well as other approaches to order determination considerably more difficult. Hence the problems of both 'order determination' and 'structure identification', under these or other names, are still subjects of much research and literature (Söderström, 1987; Vansteenkiste and Spriet, 1987), and motivates a separate chapter. Coupled to these problems is that of 'model simplification', also appearing in the literature as a separate problem (Hernandez, 1987; Lamda, 1987).

> **Remark:** A general alternative to using statistical tests for determining the order of linear models is to exploit the fact that, theoretically, too high orders yield singular values of the Hankel matrices built from correlation functions of data. This leads to the problem of testing singularity of a matrix, and more precisely, to determining the rank of the Hankel matrix. These methods are generally believed to have poor efficiency, since conventional methods test singularity by computing some condition number. However, there are better rank–testing methods, which yield as efficient results as statistical tests (Fuchs, 1987).

9.1   Using a biassed Likelihood

Behind the various structure–identification techniques may or may not be a concept of 'true order' of the object. In any case, and for the sake of argument it is helpful to regard the problem of determining the structure index v in a parametric structure as a parameter estimation problem in the model $(\mathcal{A}, v, \theta)$, only that v is an integer vector, while θ is a real vector (Caines, 1986). This exploits the interpretation of the biassed Likelihood function given in section 5.1 on "Parametrization". With $M = (\mathcal{A}, v, \theta)$ the likelihood becomes

$$L(\mathcal{A}, v, \theta | c, d) = p[c, d | \mathcal{A}, v, \theta] \, p[\mathcal{A}, v, \theta]$$
$$= p[c, d | \mathcal{A}, v, \theta] \ p[\theta | \mathcal{A}, v] \, p[v | \mathcal{A}] \, p[\mathcal{A}] \tag{9.1}$$

where all factors may carry apriori information.

The bias is another set of design parameters

The last three factors specify any apriori preference or 'bias' one might have for particular θ, v, and $\mathcal{A}$ respectively. If one knows something in advance about the likely complexity, or has some limit to the structure one wants, one can express this by assigning values to $p[v|\mathcal{A}]$. It will then be possible to estimate v and θ simultaneously by searching for the maximum of the Likelihood function. If one would have alternative structures $\mathcal{A}'$ it would also be possible to weight the likelihood by their corresponding apriori probabilities. Even if they would be weighted equal (possibly because there would be no ground for a different choice), it is possible to determine the most likely structure, simply by determining which one has the highest maximum likelihood.

When estimating θ only, given v, one can do without an apriori probability $p[θ|\mathcal{A},v]$, since the Likelihood function will still have its maximum for finite θ, and if the data sequence is long, the position of the maximum will usually not change much with the factor. It will also be reasonable to write $p[v|\mathcal{A}] = p[v_n|\mathcal{A},n]\, p[n|\mathcal{A}]$, $v_n \in \mathcal{N}_n$, and weight all structure indices v_n equal that have the same complexity n. However, when estimating v and θ simultaneously, one cannot do without the factor $p[n|\mathcal{A}]$. The reason is that the dimension of θ increases with n, and this means that without a factor $p[n|\mathcal{A}]$ damping large n, the maximum with respect to θ will increase indefinitely with increasing n. The more real parameters that take part in the maximization, the larger is the reachable maximum.

> **Remark:** If the structure $\mathcal{M}(n)$ will be defined for an arbitrarily large n, the unbiassed likelihood will in fact go on increasing with n, until one has got as many real parameters as there are data points in $(\mathbf{c},\mathbf{d})$. This would correspond to a perfect fit between model and data (technically one could achieve that by taking all the primitive random variables ω in an algorithmic model $(c,d) \leftarrow A[\omega]$ to be the parameters θ) but would yield nothing but a 'data description'. Still, such models are not necessarily useless, if they are able to compress the data sequence effectively. The latter depends on the parametrization of A. Models for other purposes will need some attenuation of the likelihood for large n, in order to distinguish between how much is lasting information θ and how much is random ω.

Remark: Weighting factors assigned apriori to render estimation of v feasible are sometimes called 'subjective probabilities', since they may be used to specify not only what one knows apriori, but also how one wants the solution to be. If one believes it will suffice with a simple model, then one should build this knowledge into the Likelihood function, and thus inform the identification algorithm. It cannot know that from data alone, since noisy data is complex in itself.

This yields a 'subjective' method for determining structure index:

Structure identification by weighting of the Likelihood

Search for the maximum with respect to v and θ of the biassed Likelihood function

$\log L(v,\theta|\mathcal{A},\mathbf{c},\mathbf{d}) = \log L(\theta|\mathcal{A},v,\mathbf{c},\mathbf{d}) + \log p[v|\mathcal{A}]$,

where $\log p[v|\mathcal{A}]$ is any bias assigned objectively or subjectively, or else any design parameter used by the model maker to affect the structure index or the complexity of the model.

The method of 'subjective weighting' gives the designer much freedom to influence the complexity. However, much freedom to choose also means much responsibility to specify, and one might want some further support from theory as a guideline for the choice of weighting factors.

An independent principle is needed to determine complexity.

It is evident that if one wants to be able to eliminate models that are too complex, one cannot achieve that by analysing the (unbiassed) Likelihood function. Since the Likelihood function holds all the information in data and apriori knowledge, and if the latter leaves an uncertainty about the right model complexity, one needs more information or more specification to be able to falsify what one would like to falsify, namely too complex models. One needs an independent principle to be able to decide upon n.

9.2 Sequential falsification

The method of determining the complexity by sequential falsification, as outlined in the beginning of this chapter, may be regarded as an application of the independent principle of 'parsimony', which says that "if several models describe data, the simplest one is preferred". The method is based on the interpretation that "a model of a given complexity n describes data, if it is not falsified with a given high confidence". Hence by the principle of parsimony it is preferred before all more complex models.

In fact, he principle is inherent in the basic procedure of refinement and falsification. It tries simple structures before the complex ones, and hence accepts the simplest that is adequate. According to 'the scientist's rule' (Section 4.4) the outer loop of the procedure will be

Procedure for sequential determination of complexity:

Initialize: $n \leftarrow 0$

Repeat until stopping rule

Refine model structure: $n \leftarrow n+1$

Test whether data falsifies the structure:

Compute the ML estimate: $\hat{M} \leftarrow \arg \min Q(M|\mathcal{M},n,\mathbf{c},\mathbf{d})$

Set alternative model structure: $n \mapsto n'$

If $B(\hat{M}|\mathcal{M},n,n',\mathbf{c},\mathbf{d}) > \gamma$, then indicate *falsified* else *unfalsified*

If *unfalsified*, then accept $\hat{M}$ and stop

The procedure and its properties will depend on the type of statistical test used for the falsification, *i.e.* on the method used for selecting the alternative structure n' and computing the acceptance domain $\mathcal{B}(\gamma|\mathcal{M},n,n',\mathbf{c},\mathbf{d})$. Three general classes of tests were discussed in chapter 8 on "Falsification techniques" and they imply three different procedures.

9.2.1 Sequence of conditional tests

The 'conditional' tests were derived in section 8.4 for parametric structures and any complexity n' of the alternative structure. Remember that in statistical tests one has always to prescribe an alternative to the null hypothesis (that n is adequate), and this alternative affects the efficiency of the test profoundly. The tests will have the highest efficiencies for the least general alternatives, *i.e.* when $n' = n + 1$. This yields the first sequential method:

Structure identification by LMP–test

Identification procedure with sequential determination of complexity by the Locally Most Powerful test:

Initialize: $n \leftarrow 0$

Repeat until stopping rule

Refine model structure: $n \leftarrow n+1$

Test whether data falsifies the structure:

Compute the ML estimate:

$(\hat{v}_n, \hat{\theta}_n) \leftarrow \arg \min\{Q(v,\theta|n,\mathbf{c}_N,\mathbf{d}_N)|v \in \mathcal{N}_n, \theta \in \mathbf{R}\}$

If $B(\hat{v}_n, \hat{\theta}_n|n,n+1,\mathbf{c}_N,\mathbf{d}_N) > \gamma$,

then indicate *falsified* else *unfalsified*

If *unfalsified*, then accept $(\hat{v}_n, \hat{\theta}_n)$ and stop

The reject confidence is computed from
$B(v_n,\theta_n|n,n+1,\mathbf{c}_N,\mathbf{d}_N) = \text{Chi\_square}[2N\ \Delta Q, r]$
where $r = \dim(\theta_{n+1}) - \dim(\theta_n)$ and

$$\Delta Q = \max\{\tfrac{1}{2}\ Q_\theta(\theta_n, 0|v_{n+1})\ Q_{\theta\theta}(\theta_n, 0|v_{n+1})^{-1}\ Q_\theta(\theta_n, 0|v_{n+1})^T$$
$$|v_{n+1} \in \mathcal{N}_{n+1}, v_{n+1} > v_n\}$$

The relation between two consecutive confidence regions for two consecutive complexity numbers is illustrated in *Fig. 51* (non–composite structures).

9.2.2 Sequence of Likelihood–Ratio tests

The procedure would apply also to other types of statistical tests, but is most suited to the case when one can compute the acceptance domain $\mathcal{B}(\gamma|n,n+1,\mathbf{c},\mathbf{d})$ directly, *i.e.* in the conditional tests. The Likelihood–Ratio tests, however, (see section 8.5) define and compute the falsification rule in a more indirect way; the acceptance domain requires the

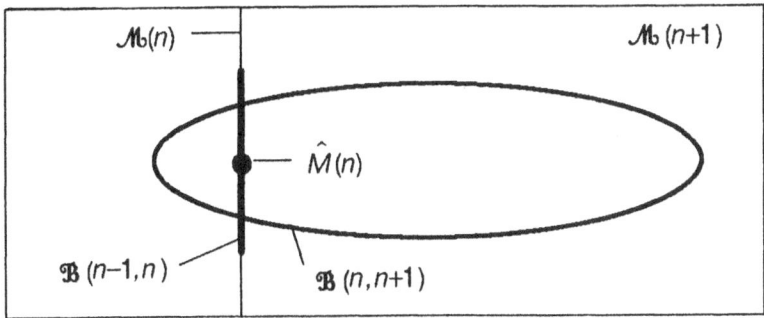

Fig. 51: Optimal confidence regions for two consecutive complexity numbers.

determination of $ML\{v,\theta|n+1,\mathbf{c},\mathbf{d}\}$ as well as $ML\{v,\theta|n,\mathbf{c},\mathbf{d}\}$. This means that it is better to rewrite the identification procedure as follows:

Structure identification by LR–test

Identification procedure with sequential determination of complexity by the Likelihood–Ratio test:

Initialize: $n \leftarrow 0$

Repeat until stopping rule

Refine model structure: $n \leftarrow n+1$

Test whether data falsifies the structure:

Compute the ML estimate:

$(\hat{v}_n, \hat{\theta}_n) \leftarrow \arg \min\{Q(v,\theta|n,\mathbf{c}_N,\mathbf{d}_N)|v \in \mathcal{N}_n, \theta \in \mathbf{R}\}$

If $B(\hat{v}_n, \hat{\theta}_n|n-1,n,\mathbf{c}_N,\mathbf{d}_N) > \gamma$,

then indicate $\mathcal{M}(n-1)$ *falsified* else *unfalsified*

If *unfalsified*, then accept $(\hat{v}_{n-1}, \hat{\theta}_{n-1})$ and stop

The reject confidence is computed from
$B(v_n, \theta_n|n-1,n,\mathbf{c}_N,\mathbf{d}_N) = \text{Chi\_square}[2\,N\,\Delta Q, r]$
where $r = \dim(\theta_n) - \dim(\theta_{n-1})$
and $\Delta Q = \min\{Q(v,\theta|n-1,\mathbf{c}_N,\mathbf{d}_N)|v \in \mathcal{N}_{n-1}, v < v_n, \theta \in \mathbf{R}\}$
$$- Q(v_n, \theta_n|n,\mathbf{c}_N,\mathbf{d}_N)$$

This includes the conventional method of determining complexity by repeated use of the χ^2–tests or F–tests (Leontaritis and Billings, 1987).

The scheme avoids computing the ML–estimates twice, once for the tested model structure and once for the alternative, but it means that one has to compute one unnecessarily complex model $(\hat{\nu}_{n+1}, \hat{\theta}_{n+1})$ in order to test $(\hat{\nu}_n, \hat{\theta}_n)$. This is the disadvantage of 'overfitting' inherent in LR–tests.

> **Remark:** Notice that with partial nesting the LMP and LR procedures may both stop prematurely and at different structures, depending on the particular nesting (see also section 9.6 on "Model structure selection").

Efficient but not safe

Both the procedures for LMP and LR tests ignore the requirement that the complexity index $n+1$ be 'sufficient'. This means that the sequence may stop prematurely, by *not* falsifying a model structure $\mathcal{M}(n)$ that is not sufficient (see Examples 8.4 and 8.7). The reason for this is that if $\mathcal{M}(n+1)$ is not sufficiently much better than $\mathcal{M}(n)$, the test will still prefer the simpler alternative $\mathcal{M}(n)$. Hence, this approach to sequential falsification is not *safe*. The result depends on how the sequence of expanding structures $\{\mathcal{M}(1), \mathcal{M}(2), \ldots, \mathcal{M}(\bar{n})\}$ has been designed. The procedure will give a false result (without warning), if the structure does not expand sufficiently with each step n, or expands in the wrong direction. The procedure is *efficient*, in the sense that the probability of rejecting a wrong model is maximized (if $\mathcal{M}$ expands enough), and also in the sense that it reduces the computing (since it is easier to compute $\mathcal{B}(\gamma|n,n+1)$ than $\mathcal{B}(\gamma|n,\bar{n})$). But that costs safety.

> **Remark:** *Fig. 52* illustrates a simple case of composite structure. Notice that the test stops prematurely, if the structures have the unsuitable simple nesting $(0,0),(1,0),(1,1)$. It stops at the correct structure with the better simple nesting $(0,0),(0,1),(1,1)$, and also with the partial nesting $(0,0),\{(1,0),(0,1)\},(1,1)$.

The safety of both the conditional tests can be improved by replacing $\mathcal{B}(\gamma|n,n+1)$ with $\mathcal{B}(\gamma|n,n')$, $n' > n+1$, even up to $n' = \bar{n}$. But again, this costs efficiency (reduced 'power') as well as computing.

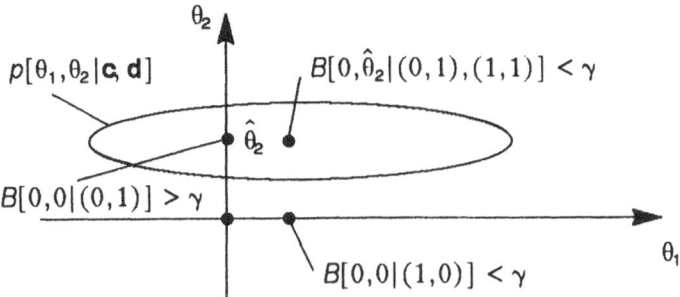

Fig. 52: Illustrating falsification of model structure when the alternative structure is composite. Structure (0,0) is falsified, if the alternative is {(1,0),(0,1)}, but not if it is (1,0). Structure (0,1) is accepted (correctly) with the alternative (1,1).

9.2.3 Sequence of unconditional tests

In case one would have doubts about whether the structure $\mathcal{M}(n+1)$ is sufficient, one will have to resort to unconditional tests. Exploiting for instance 'Unconditional Falsification Rule 3' in section 8.2 yields the procedure

---

Structure identification by unconditional test

Identification procedure with sequential determination of complexity by unconditional test:

[Initialize: $n \leftarrow 0$

Repeat until stopping rule

[Refine model structure: $n \leftarrow n+1$

Test whether data falsifies the structure:

[Compute any optimal estimate:

$(\hat{\nu},\hat{\theta}) \leftarrow \arg \min\{Q(\nu,\theta|n,\mathbf{c}_N,\mathbf{d}_N)|\nu \in \mathcal{N}_n, \theta \in \mathbf{R}\}$

[If $B(\hat{\nu},\hat{\theta}|n,\mathbf{c}_N,\mathbf{d}_N) > \gamma$, then indicate *falsified* else *unfalsified*

[If *unfalsified*, then accept $(\hat{\nu},\hat{\theta})$ and stop

The reject confidence is computed from

$$B(\hat{v},\hat{\theta}|n,\mathbf{c}_N,\mathbf{d}_N) = \max[B_1,B_2]$$

where

$$B_1 = \text{Chi\_square}[\tfrac{1}{2} N \| \tfrac{1}{N} \sum_{k=1}^{N} \mathbf{w}(k|\hat{v},\hat{\theta})^2 - 1\|^2, \dim(\mathbf{w})]$$

$$B_1 = \text{Chi\_square}[\xi^T R_\xi^{-1} \xi, \dim(\xi) - \dim(\hat{\theta})]$$

$$\xi = \tfrac{1}{N} \sum_{k=1}^{N} \mathbf{Z}(k|\hat{v},\hat{\theta})^T \mathbf{w}(k|\hat{v},\hat{\theta}), \text{ with } E\{\xi|\hat{v},\hat{\theta}\} = 0$$

$$R_\xi = \tfrac{1}{N}\tfrac{1}{N} \sum_{k=1}^{N} \mathbf{Z}(k|\hat{v},\hat{\theta})^T \mathbf{Z}(k|\hat{v},\hat{\theta})$$

A good way to resolve the choice between efficiency (conditional tests) and safety (unconditional tests) is to use both tests. If one of the tests falsifies, then the structure should be considered as falsified, and the search for a better structure should continue. The conditional tests are good at revealing errors in the model, when there is a better one in the alternative structure, and the unconditional tests are able to reveal when the latter condition is not true.

9.3 Philosophy revisited:
Equivalence *vs* goodness

Needless to say, the general approach to the identification problem used in this book ('the scientist's rule' and, optionally, 'the engineer's rule') is not the only one. Other principles are advocated and used successfully by other schools of identification. The following is a survey of various ideas that are used to determine complexity, and in general to approach the identification problem with unspecified structure index.

As mentioned, there are some objections to determining complexity by statistical tests, the most common is that the method contains a 'subjective' element, namely the apriori risk. The objection may be refuted as being unimportant in practice, but other objections may carry more weight. Fundamentally, the question of what complexity to select for the final model brings out into the open the problem of defining what is to be understood by 'the identification problem'.

The first definition of the identification problem is probably that of Zadeh (1962): "Identification is the determination, on the basis of input and output, of a system within a specified class of systems, to which the system under test is equivalent".

A key issue in this definition is the concept of 'equivalence' or agreement between the mathematical model and the physical object. Since the responses of the two never agree in practice, this leaves room for different definitions of 'equivalence', more or less difficult to accept from a philosophical or practical point of view.

There are two common approaches to this problem:
1) Assume that there is at least one model 'equivalent' to the physical object, and try to find one.
2) Reject the idea of an 'equivalent' model, and look for the 'best' one instead.

In the first case one has to define 'equivalence', and, in practice, a measure of 'approximation' plus a threshold for what is 'sufficiently good approximation' (Gaines, 1977; Willems, 1986). In the second case, one has to define a measure of 'goodness' to maximize (Caines, 1986; Ljung, 1978, 1987a). Ljung (1978) reviews a number of common ways to specify loss functions.

> **Remark:** A third approach is the following: If the model is to be used for making some decisions, for instance control, do not try to find a single model. Instead, aim at a *set* of models, with or without probabilities attach to each model, and base the decision on the set (Peterka, 1981). This is particularly reasonable for the purpose of adaptive control, where it is often neither necessary nor possible to obtain a unique model that can be regarded as 'equivalent' in some well-defined sense.

> **Remark:** A fourth way out is to accept only exact agreement between model and data, and instead 'model' the data sequence by coding it so efficiently that the code can be used for decisions (Rissanen, 1983; Rissanen and Langdon, 1981). For an application of this idea see Example 1.2.

> **Remark:** There are also approaches to model making that originate in other 'cultures' than systems and control theory, and therefore use other concepts for defining the problem, for instance (Nurmi, 1978; Schneeweiss, 1987). Mehra (1979) surveys some unconventional approaches to identification, structure selection, and input

design. In particular, the Group Method of Data Handling (GMDH) has attracted much interest. No attempt has been made here to include these methods, or to find relations to system identification based on stochastic dynamic models.

Unsatisfying but useful

It is often recognized that there can be no exact model for most physical objects, but also that it is still reasonable to base much of the identification *theory* on the fiction of a 'true model' (or an 'equivalent model'), since the concept is "useful and operational" (Gevers and Bastin, 1982). In the background there is obviously some idea of 'continuity': If the 'true model' is 'close' to the object, in some unspecified meaning, then the effect of the best decision based on the model should also be 'close' to that of the best decision for the actual object.

In particular, this motivates the use of *deterministic* modelling, which assumes that the object is also deterministic and of a form that can be modelled exactly. This allows defining 'equivalence' as exact agreement between given characteristics of model and object, for instance their responses, as well as in a number of other ways pertinent to the purpose of the modelling (Kubrusly, 1984). Even if this is clearly an unrealistic assumption in most cases, the theory is still useful for answering questions of 'identifiability', that is, for investigating what kind of information about the object that might be possible to get out of a given experiment. Hence, deterministic theory is still possible to use, and it has the important advantage that the assumption of 'exactness' allows derivation of powerful results. The question of 'approximation' does appear eventually, but not until one has arrived at a final model and has to decide whether it is acceptable for the purpose.

'Equivalence' is a matter of interpreting the innovations

It would still seem that assuming equivalence in deterministic models would be unrealistic in most practical cases. Quite often objects are affected substantially by unknown input, even if the dynamics themselves would be possible to describe quite exactly. This is a reason for using a stochastic description. And that makes it impossible to base a definition of 'equivalence' on exact agreement between the responses of model and object.

However, even in this case the concept of equivalence between deterministic models may allow sufficient freedom to be useful. Guidorzi (1981) defines a deterministic state–vector model to be 'equivalent' to the true object, if for every initial state of the object there is an initial state of the model which generates the same response as the object for every possible stimulus. Hence, if the object is $d_K = A[c_K,x(0)]$, then there should exist $x(0)$ such that $A[c_K,x(0)] = A[c_K,x(0)]$ for all c_K. The state $x(0)$ must also be observable. This allows one to model disturbances as affecting the initial state, which is not an unrealistic approximation, if the model is to be valid only for a short application interval. This suits the purpose of feedback control.

If one would want to have equivalence for a longer application interval in the presence of random disturbance, one would have to increase the dimension of $x(0)$, in practice up to the same order of magnitude as the number of data points in the response (Willems, 1986). It is therefore a short step to replace the unknown $x(0)$ by the likewise unknown innovations sequence w_K, to obtain the external algorithmic model $d_K = A[c_K,w_K]$. The observability condition is still satisfied, since there is a one-to-one correspondence between d_K and w_K (section 5.5.1 on "External models").

The consequence of trying this is that almost any A would then be 'equivalent', unless one would place some further restrictions on w_K. These restrictions depend on how one will interpret the w_K sequence. Two ways are common:

● *Deterministic modelling*: A limited part of w is carrying information, and is interpreted as 'state' — it affects control decisions. The remaining part is interpreted as 'error', not to affect decisions. This yields a natural definition of approximation: Two models are 'approximately equivalent' if states are the same *and* errors are 'small' according to some given measure.

● *Stochastic modelling*: Both the 'state' and the 'error' parts of w are stochastic variables. The errors are not necessarily small; instead it is required for 'equivalence' that the distributions of ω and w are the same. The relation $d = A[c,w]$ defining the innovations yields $w = W[c,d] = W[c,A(c,\omega)]$. Now, if w has the same distribution as ω, then A and A are equivalent in the sence used in this book (section 6.1): From (5.6),

$$p[d|c,A] = \int \delta[d - A(c,\omega)]\, p[\omega]\, d\omega$$
$$= \int \delta[d - A(c,w)]'\, p[w]\, dw = p[d|c,A]$$

In both cases equivalence is defined only for external models. What differs is whether one chooses to regard the innovations so small that they can be neglected, or so regular that they can be described by a distribution. Notice that the latter need not be known completely; it may depend on unknown parameters.

> **Remark:** Falsification does not need Source Equivalence to reject a model that does not agree with data, and neither does conditional falsification. In fact, they *test* Source Equivalence (which therefore becomes a 'hypothesis'). Neither do the sequential order determination procedures based on those methods (section 9.2) need Source Equivalence to arrive at a best model in $\mathcal{M}_b(n)$. One has only to define the alternative, which is another hypothesis. But to *define* reasonable hypotheses for the alternatives one needs Source Equivalence. The assumption is needed only to motivate why one would possibly be satisfied with a 'best' model. Obviously, the alternative needs assumptions about the true source to be meaningful.

> **Remark:** The *validation* methods in chapter 7 do assume something about the 'true source' in order to yield correct probabilities of purposivity, but neither they require Source Equivalence in theory. They assume Purpose Equivalence instead, which is a different condition on the data source (see section 4.5 on "Conditions for Bayesian validation"), and possibly easier to accept. It says simply that the difference between the best model and the true source must be negligible for the purpose, but not necessarily negligible for describing the data distribution.

> **Remark:** In principle, identifications *algorithms* do not need assumptions about reality (since they work with models and data), while *specifications* do — they are the designer's link to reality. Hence, in all cases it is the designer's specifications of what is a 'best' model that decides the outcome, and it is immaterial *how* he/she arrived at these specifications (whether by conceiving an equivalent model or not). The distinction between an 'equivalent' and a 'best' model is therefore not clear in practice, and not even in principle. One might argue that this is too academic a point to argue about, but again one might not.

A criterion function is needed in any case

The alternative of regarding identification as an optimization problem is also an early established idea (Åström and Eykhoff, 1971). In principle, the design of a criterion should be based on the purpose of the model, but since that problem is not solved for general cases, other criteria are used instead (Söderström, 1980).

Even in the first approach (defining 'equivalence') a criterion is often used as a mathematical tool: Eykhoff (1974) defines 'equivalence' by means of a loss function comparing the response data from object and model: Two models are 'equivalent', if they have the same losses. If, as is reasonable, the loss is minimum for the 'true' model (= the object), then 'object–equivalent' models are those that minimize the loss, provided the minimum is an admissible model. However, the last sub-clause is most important — the crucial step is taken when one decides whether or not to assume that there is an 'object–equivalent' model in the set of admissible models.

Hence, most general identification methods (originating in the 'automatic control culture') do rest on criteria functions $V(M|\mathbf{c},\mathbf{d})$. They may be expressed as 'loss' or 'gain' functions, but evaluate to what extent any model M agrees with any data set $(\mathbf{c},\mathbf{d})$, or else how 'good' the model is in some prespecified meaning, that can be expressed apriori in terms of model and data.

The differences come from different specifications

The differences between identification methods originate in the different answers given to the following general questions:

● What set may M belong to?
● How to design V?
● How to define a 'best' model $\hat{M}$, given V?
● How to search for $\hat{M}$?

The answer to the first question is the result of the 'modelling' step. It mainly affects the *techniques* used to find $\hat{M}$, and not so much the *principles* used to define the identification problem. Since the possible model structures are many, so are the solution techniques, and the latter occupy most of the identification literature. They are not the main subject in this book.

In contrast to this, there are only few general answers to the two middle questions. They give rise to different 'schools' of identification, and the ideas will be outlined in the next sections. Also the fourth question has few answers, which will also be outlined in the sequel, except in cases where M is a parametric model with a given structure, for which case there are numerous search strategies.

9.4   Designing the criterion:
Description *vs* purpose

The most common answer to the question of designing the criterion function is that it should be designed with the purpose in mind. Strictly speaking, this is not an answer, unless it is also stated how this should be done, but some guidelines have been suggested for linear models and quadratic criteria. Even some optimization is feasible (see below). In some cases the relation to the purpose is obvious. Generally, the heuristic approach to designing the criterion is reasonable, when one wants to have a model whose response simulates the data, or a model that predicts well (for forecasting or control), since then it is fairly easy to write down a reasonable criterion. This case may also have natural design parameters, whose values have an obvious relation to the purpose. An example is the prediction range, determined by the unavoidable delay of the intended control.

A common choice is to set $V(M|c_N,d_N) = h[e_N]$, where e_N is the residual sequence obtained by inverting the model relation $(c_N,d_N) \leftarrow A(e_N)$ (section 5.5.2). Shaping the h–function is a powerful tool for emphasizing many of the properties one wants the resulting model to have. In many model structures also the Maximum Likelihood criterion can be formulated in this way, $V(M|c_N,d_N) = L(M|c_N,d_N)$. The main stream with this philosophy, the 'residuals approach', is represented by the 'Prediction Error Methods' (Ljung, 1987a).

Purposes other than forecasting cause difficulties.

The heuristic approach has the advantage that one does not have to make the controversial assumption that there is a 'true' or even an 'equivalent' model in the widest model structure. It has the disadvantage that one must specify a criterion that expresses the purpose in terms of model and data only.

For some purposes, like prediction, this should be no problem in practice. In general, however, the purpose is more easily expressed in terms of the *object* (or 'true model'), since a purposive model of the object must unavoidably depend on at least some properties of the object. An obvious case is when one wants the physical parameters of the object and the model to agree (within a prespecified accuracy), but Example 4.2 illustrates that the same difficulty may occur even when the purpose is feedback control, and one would not be interested a priori in a 'correct' model. It is generally not a trivial matter to transform a given function of model and object into one of model and data.

> **Remark:** Zypkin (1987) determines an 'optimal' criterion function as that which minimizes the expected value of the squared parameter error. If that principle is applied to the validity function defined by (7.22), but with an arbitrary weighting matrix D, then this implies that the optimum weighting is $D = R_\theta^{-1}$. Maximizing the resulting validity function $V(\theta|\mathbf{c}_N,\mathbf{d}_N) \triangleq$ Chi_square$(2N\epsilon,\mathbf{r},\rho^2)$ will mean minimizing $\rho^2 = (\theta - \hat{\theta})^T R_\theta^{-1} (\theta - \hat{\theta})$. Hence, the optimal criterion will be minimized by the ML estimate.

Optimize design parameters

Ljung's approach to the problem of getting the purpose into the criterion function can be described like this: Let the criterion $V(M|\mathbf{c},\mathbf{d},\mathcal{H})$ depend on a number of design parameters $\mathcal{H}$, which may be residual filters, weighting functions, and even properties of the stimulus, for instance $\mathcal{H} \triangleq (h,\mathbf{X})$. Let $D(\mathbf{M},\hat{M})$ be the criterion one wants to minimize, with respect to the intended use of the model $\hat{M}$ that results from minimizing $V(M|\mathbf{c},\mathbf{d},\mathcal{H})$. Hence, $\hat{M}$ depends on $(\mathbf{c},\mathbf{d},\mathcal{H})$. Define the optimization criterion $J(\mathcal{H},\mathbf{M}) = E\{D[\hat{M}(\mathbf{c},\mathbf{d},\mathcal{H}),\mathbf{M}]\}$, and minimize $J(\mathcal{H},\mathbf{M})$ with respect to $\mathcal{H}$. This seemingly roundabout way yields the desired criterion, but generally the optimal design depends on the unknown equivalent model $\mathbf{M}$.

If would be conceivable to solve this problem by an iterative procedure, successively replacing $\mathbf{M}$ by the $\hat{M}$ obtained as a result of the previous iteration. That is the rule, when the design is done with respect to the experiment $\mathbf{X}$ (Goodwin and Payne, 1977). This means that the optimal $\mathcal{H}$ will be determined by an implicit relation, which aggravates the problems of analysis.

Ljung (1986a, 1986b, 1987a) develops this idea for the case that the model and true object are both linear and the D–function is quadratic in the deviation between model and object. He considers the problem of optimizing design parameters, such as prefiltering and weighting of the prediction errors, for some special purposes (simulation, prediction, minimum–variance control design). In some important cases this leads to explicit design rules, based on the predefined weighting, motivated by the purpose, of the model errors for different frequencies. Gevers and Ljung (1986) also design the optimal spectrum of the experiment stimulus in this way.

The theory requires the assumption of a 'true' model.

Use Bayes' rule

If the purpose is formulated by means of the criterion $G(\mathbf{M}, M) > 0$, where $\mathbf{M}$ is any Purpose–Equivalent model (that, is any model for which G is positive satisfies the purpose), then it is possible to transform this criterion depending on the true object into one that depends on data: Define the gain function as the probability that M satisfies the purpose. Then,

$$V(M|\mathbf{c},\mathbf{d}) = \Pr\{G(\mathbf{M}, M) > 0|\mathbf{c},\mathbf{d}\}$$

$$= \int \mathrm{Ind}\{G(\mathbf{M}, M) > 0\} \; dP[\mathbf{M}|\mathbf{c},\mathbf{d}], \qquad (9.2)$$

and the probability can be evaluated from Bayes' rule, as is done in section 7.1 on "Validating parametric models". That is the approach taken in this book.

The price one has to pay is to assume that there is at least a Purpose–Equivalent model. Otherwise one cannot define the probability from Bayes' rule. However, notice again, that the Purpose–Equivalent model need not have to be Source Equivalent, and that one need not actually compute the Purpose–Equivalent model. This should remove some of the natural hesitation to involving the 'true source' in the criterion. It remains that the difference must be negligible between the effect on the purpose of using the best model in the structure one chooses, and that of using the true source (if this were possible).

Use Popper's rule

When there is no obvious purpose, one may adopt the 'scientist's rule'. In this case one is simply interested in inferring as much from data as possible. If, further, one accepts Popper's principle of "choosing the model that passes the most severe test" (Section 4.3 on "Fitting"), then the criterion is $V(M|\mathbf{c},\mathbf{d}) = B(M|\mathbf{c},\mathbf{d})$, where B defines the set $\mathfrak{B}$ of unfalsified models for different levels γ of approximation, $\mathfrak{B}(\gamma|\mathbf{c},\mathbf{d}) = \{M|B(M|\mathbf{c},\mathbf{d}) \leq \gamma\}$, as discussed in section 4.2 on "Model structure, data descriptions, and purposive models". If one further applies the principles of optimal statistical tests in Chapter 8 on "Falsification techniques", then $V(M|\mathbf{c},\mathbf{d}) = L(M|\mathbf{c},\mathbf{d})$, where L is the Likelihood function.

Hence, the Likelihood function is a measure of to what extent the model agrees with data, even when it cannot be expressed as a function only of the residuals, but it is not necessarily the best choice for all purposes. Since, it exhausts the data, it may yield models that are more complex than necessary for the purpose.

In conclusion, one has to choose between wanting a *description* of the source or satisfying a *purpose*. In the first case one must conceive *equivalence* for the description to make sense. In the second case one must define a measure of *goodness*. The 'residuals' approach chooses the latter. It has difficulties of satisfying other than special purposes. Poppers's approach chooses the former. Bayes' approach needs another assumption about the data source (than equivalence), and makes it possible to take more general purposes into account, but the techniques are less well developed.

Generally, there is a problem whenever the natural criterion for the fitting differs from that suiting the purpose (Gevers and Bastin, 1982).

9.5   Defining the optimal order:

Accuracy *vs* complexity

Usually, criteria functions $V(M|\mathbf{c},\mathbf{d})$, when minimized over sets $\mathcal{M}(n)$ of different degrees n of complexity, decrease when n increases. It is simply easier to get a model agree with data, when it has more degrees of freedom for fitting. Hence, there is an obvious and basic conflict between accuracy and complexity; one wants one large and the other small, but they are in fact coupled.

> **Remark:** Fasol and Jorgl (1980) wrote an elementary tutorial on the model making problems, supported by many examples from the engineering sciences. In particular, the authors stress the basic necessity to compromize between accuracy and complexity for a given purpose. In some cases the 'compromise' can be made 'optimal', if one has some reasonable principle to base the compromise on.

Remark: Gaines (1977) bases the definition of 'identification' on the two concepts of 'complexity' and 'approximation', specified in such a way that approximation always increases with decreasing complexity. This defines a set of *admissible* models, but not a single 'best' model. An association with the concepts in this book is obtained with 'complexity' as the number of parameters and 'approximation' measured by one of the B or V functions. Notice that it is the independent principle defining weights on complexity (section 9.1) that defines a single 'best' model.

Remark: Willems (1986) introduces a measure of 'misfit' as a function of model and data, which corresponds to the distance function $B(M|\mathbf{c},\mathbf{d})$, and also a 'complexity measure' associated with each model M. He then defines 'optimality' in two ways: 1) Fixing complexity, choose the model that minimizes misfit. 2) Fixing maximum tolerance, choose the model that minimizes complexity. Apparently, the approach is less obvious in the case of stochastic models. The treatment is abstract mathematics.

Remark: Goodwin and Salgado (1989) define 'undermodelling' by a frequency function of the error between object and model. The crucial assumption is that there be a known statistical distribution for the error of a 'nominal' model. Other models yield a bias function due to undermodelling, in addition to the statistical error. Specifying a tolerance for the undermodelling determines the minimum complexity.

Remark: In all cases the controversy is between the two demands of parsimony and flexibility. Ljung and Glover (1981) state that "The question of how to reach a sensible compromise is the heart of the identification problem".

When one has to choose between two conflicting quantities V and n, one can either limit one and optimize the other, or else invoke some independent principle to compromize between the two. The alternatives lead to what are sometimes called 'subjective' and 'objective' methods, but will here be called 'classical' and 'modern' methods respectively, all on grounds that will be discussed below.

To formulate the difference let V be a loss function defined for the expanding set of parametric structures used in this book, $M_n \in \mathcal{M}(n)$, where $\mathcal{M}(n') \subset \mathcal{M}(n)$ for $n' < n$. The scalar integer n is the complexity indicator, most easily interpreted as the number of free parameters θ. Let $\hat{M}_n$ minimize $V(M|\mathbf{c},\mathbf{d})$ over the set $\mathcal{M}(n)$. Then $V(\hat{M}_n|\mathbf{c},\mathbf{d})$ decreases with increasing n.

'Subjectivity' comes from limiting one of the unknowns

The simplest alternative, limiting n, obviously does not solve the problem, since this means fixing n to the preset value, which is usually unknown, at least in 'black–box' cases. The alternative of limiting $V(\hat{M}_n|\mathbf{c},\mathbf{d})$ and minimizing n is better, since one may have a knowledge of what accuracy the purpose would require.

However, the limiting can also be made in a more sophisticated way: It is to invoke the principle of 'parsimony', to keep complexity low. Formulate the criterion in such a way that $V(\hat{M}_n|\mathbf{c},\mathbf{d})$ is the risk that the falsification procedure will yield a more complex model than needed to describe data. That risk can be computed using the techniques in Chapter 8. Then it is possible to define the 'best' model as the simplest one that agrees with data (with a given risk α), that is, by $\hat{n} = \min\{n|V(\hat{M}_n|\mathbf{c},\mathbf{d}) < \alpha\}$. The procedure for determining complexity becomes one of those in section 9.2.

As pointed out, a disadvantage of the repeated falsification scheme, is that it requires a value for the preset risk value α. That may not actually be a problem in practice, since the value is between zero and one, one should generally make it small, (0,01 say), the result is not very sensitive to the risk value, and the cost of making a wrong decision is usually not very large (one might get a slightly overparametrized model).

However, the requirement of a partially nested model set may be a limitation; the tests need an alternative hypothesis to every model tested, and this means that a wider set must always be defined.

The 'objective' methods try and evade these limitations by invoking some independent principles. Their 'objectivity' does of course depend on the principles (which may have different degrees of 'subjectivity'). Stoica *et al* (1986) use the more neutral attribute 'modern', since the approach is no doubt of a later date. In agreement with this, I prefer to call the 'subjective' methods 'classical'. (For one thing, there is no value attached to these attributes, unless, of course, one believes that 'modern' is always 'better'.) The 'modern' methods are outlined in the following subsections.

9.5.1 Approximating the expected loss

The given criterion $V(M|\mathbf{c},\mathbf{d})$ generally creates 'data descriptions'. In particular, the prediction–error criterion $V(M|\mathbf{c}_N,\mathbf{d}_N) = \sum_k \mathbf{e}(k|M)^2$ rewards models that predict the *data sample* well. Most often one wants a criterion that rewards models that predict *other* data samples coming from the same source. It would therefore be reasonable to minimize $E\{V(M|\mathbf{c},\mathbf{d})\}$, where the expectation is taken over the possible data sets generated by the true source. The obvious obstacle to this is that the expected value depends on the unknown object.

The dilemma is resolved in principle, if one can make an independent experiment to evaluate the loss. The ideal test is of course to evaluate the loss from data obtained by stimulating the object in the same way as when the model will be applied, but even other experiment will do, if repeated and independent.

> **Remark:** Statistical inference from multiple data samples has been studied by Goodrich and Caines (1979b) for linear systems. Baram (1980b) uses multiple independent data samples to validate linear models, possibly of lower order than needed to describe an equivalent model. Repeated experiments do of course increase the credibility of the identification results — they make it possible to test the condition of Reproducibility (section 3.2.2). Söderström and Kumamaru (1985) devised a method for falsification of linear, multivariable, transfer–function models, when several data samples are available. They exploited the fact that it is possible to compute the distribution of Kullback's Discrimination Index (KDI) under the null hypothesis. The method can be used to detect changes in the object behaviour.

Basing the evaluation of the loss on independent samples is usually indicated by the prefix 'cross'. If, as is generally the case, one has only one data sample available for both estimation and validation, then it remains to try and estimate the expected value from the single data set. Two main ideas are the following:

Evaluate the expectation analytically

In order to illustrate the first method of approximating the expected value let the criterion be $V(n,\theta|\mathbf{c}_N,\mathbf{d}_N) = -\log L(\theta|n,\mathbf{c}_N,\mathbf{d}_N)/N$, where $L(\theta|n,\mathbf{c}_N,\mathbf{d}_N) = p[\mathbf{c}_N,\mathbf{d}_N|n,\theta]$ is the (unbiassed) Likelihood function, and the structure is parametric. Assume that V is a smooth function of

θ, and can be expanded in a Taylor series around Θ, which minimizes $\overline{V}(n,\theta) \triangleq E\{V(n,\theta|\mathbf{c}_N,\mathbf{d}_N)\}$. If there is an equivalent model in $\mathcal{M}(n)$, then (n,Θ) corresponds to such a model according to the Lemma of the Maximum Likelihood (section 6.2.1). Let $\hat{\theta}$ minimize $V(n,\theta|\mathbf{c}_N,\mathbf{d}_N)$. It generally deviates from Θ, but assume that the estimate is consistent, so it tends to Θ, when $N \to \infty$. Let ∇ be the gradient operator (row vector) with respect to θ. Then, $\nabla^T\nabla$ is the second-order differentiation operator (matrix). The derivation follows Söderström and Stoica (1989):

Suppress $\mathbf{c}_N,\mathbf{d}_N$ for convenience, and develop V in a Taylor series:

$$E\{V(n,\hat{\theta})\} = E\{V(n,\Theta) + \nabla V(n,\Theta) \ (\hat{\theta} - \Theta) + \mathcal{O}\|\hat{\theta} - \Theta\|^2\} \tag{9.3}$$

Since $\hat{\theta}$ is the ML estimate of Θ, it holds for long samples (see section 8.3.1, eqn (8.18))

$$\hat{\theta} - \Theta = -[E\nabla^T\nabla V(n,\Theta)]^{-1} \ \nabla^T V(n,\Theta) + o(N^{-1/2}) \tag{9.4}$$

Apply the following lemma (Kendall and Stuart, 1967)

$$E\{\nabla^T\nabla \log L(\Theta)\} = - E\{\nabla^T\log L(\Theta) \ \nabla\log L(\Theta)\} \tag{9.5}$$

Proof:

Since $\int p[\mathbf{c},\mathbf{d}|n,\theta] \ d(\mathbf{c},\mathbf{d}) = 1$ for all θ, and $L(\theta) = p[\mathbf{c},\mathbf{d}|n,\theta]$, it follows that

$$\int \nabla L(\theta) \ d(\mathbf{c},\mathbf{d}) = \nabla \int L(\theta) \ d(\mathbf{c},\mathbf{d}) = 0,$$

$$\int \nabla^T\nabla L(\theta) \ d(\mathbf{c},\mathbf{d}) = \nabla^T\nabla \int L(\theta) \ d(\mathbf{c},\mathbf{d}) = 0.$$

Hence,

$$E\{\nabla \log L(\Theta)\} = E\{L(\Theta)^{-1} \ \nabla L(\Theta)\}$$

$$= \int L(\Theta)^{-1} \ \nabla L(\Theta) \ p[\mathbf{c},\mathbf{d}|n,\Theta] \ d(\mathbf{c},\mathbf{d})$$

$$= \int \nabla L(\Theta) \ d(\mathbf{c},\mathbf{d}) = 0.$$

$$E\{\nabla^T\nabla \log L(\Theta)\} = E\{\nabla^T[L(\Theta)^{-1} \ \nabla L(\Theta)]\}$$

$$= E\{-L(\Theta)^{-2} \ \nabla^T L(\Theta) \ \nabla L(\Theta) + L(\Theta)^{-1} \ \nabla^T\nabla L(\Theta)\}.$$

The last term is

$$E\{L(\Theta)^{-1} \ \nabla^T\nabla L(\Theta)\} = \int L(\Theta)^{-1} \ \nabla^T\nabla L(\Theta) \ p[\mathbf{c},\mathbf{d}|n,\Theta] \ d(\mathbf{c},\mathbf{d})$$

$$= \int \nabla^T\nabla L(\Theta) \ d(\mathbf{c},\mathbf{d}) = 0.$$

Hence,

$$E\{\nabla^T\nabla\log L(\Theta)\} = E\{-L(\Theta)^{-2} \nabla^T L(\Theta) \nabla L(\Theta)\}$$
$$= - E\{\nabla^T\log L(\Theta) \nabla\log L(\Theta)\}. \qquad \blacksquare$$

It follows from (9.5) that

$$E\{\nabla^T\nabla V(n,\Theta)\} = N E\{\nabla^T V(n,\Theta) \nabla V(n,\Theta)\} \qquad (9.6)$$

Inserting this into (9.4) yields

$$\hat{\theta} - \Theta = -N^{-1} E\{\nabla^T V(n,\Theta) \nabla V(n,\Theta)\}^{-1} \nabla^T V(n,\Theta) + o(N^{-1/2}) \qquad (9.7)$$

Inserting this into (9.3) yields

$$E\{V(n,\hat{\theta})\}$$

$$= E\{V(n,\Theta) - N^{-1} \nabla V(n,\Theta) E\{\nabla^T V(n,\Theta) \nabla V(n,\Theta)\}^{-1} \nabla^T V(n,\Theta)$$
$$+ o(N^{-1})\}$$

$$= E\{V(n,\Theta)\} - N^{-1} \mathrm{Tr}[E\{\nabla^T V(n,\Theta) \nabla V(n,\Theta)\}^{-1}$$
$$E\{\nabla^T V(n,\Theta) \nabla V(n,\Theta)\}] + o(N^{-1})\}$$

$$= E\{V(n,\Theta)\} - \dim(\Theta)/N + o(N^{-1})\} \qquad (9.8)$$

Hence,

$$\min_{\theta} E\{V(n,\theta)\} = E\{V(n,\Theta)\} = E\{V(n,\hat{\theta}) + \dim(\theta)/N + o(N^{-1})\}$$

$$= E\{\min_{\theta} V(n,\theta|c_N,d_N) + \dim(\theta)/N + o(N^{-1})\} \qquad (9.9)$$

and

$$\min_{n,\theta} E\{V(n,\theta)\} = E\{\min_{n,\theta} [V(n,\theta|c_N,d_N) + \dim(\theta)/N + o(N^{-1})]\} \ (9.10)$$

This means that if $\dim(\theta)/N$ is added to the loss function, $V(n,\theta|c_N,d_N)$ the minimum value will be an unbiassed estimate of the wanted $E\{V(n,\Theta)\}$. The term compensates the bias obtained by evaluating the loss on the same data as was used for determining the estimate. It corresponds to the weighted Likelihood function $\exp[-\dim(\theta)]$ $L(\theta|n,c_N,d_N)$.

Remark: A different reasoning, but leading to the same solution for long data samples is the following: Since the θ–parameter is undetermined when testing the class $\mathcal{M}b(n)$, it seems reasonable to use the expected value of the log likelihood conditional on the given data. But it has been shown (Kashyap and Ramachandra Rao, 1976) that for long samples

$$E\{\log p[\mathbf{c}_N,\mathbf{d}_N|n,\Theta]|\mathbf{c}_N,\mathbf{d}_N\} \approx \max_{\theta} \log L(\theta|n,\mathbf{c}_N,\mathbf{d}_N) - \dim(\theta).$$

Hence, maximizing this function of n, called 'the likelihood of the class' yields the same result as maximizing $\log L(\theta|n,\mathbf{c}_N,\mathbf{d}_N) - \dim(\theta)$ with respect to both n and θ.

Remark: The criterion was invented by Akaike (1974), and is known as AIC when used in connection with the prediction–error models. The loss function is then AIC $\triangleq N \log [\det R(\theta)] + 2 \dim(\theta)$. For a further discussion of this and related criteria see section 9.5.2.

Evaluate the sample average

The second way to estimate the expected value of the criterion function is to take a *sample average* based on data *independent* of those used to estimate the model parameters (but not obtained from an independent experiment). When one has only one data sample available for both estimation and validation, it is often recommended that one cut the sample in half, and use each half for each purpose. Then it is obviously reasonable to switch the two halves, and repeat the validation.

The following improvement was suggested by Stoica *et al.* (1986) for criteria of the form $V = \dfrac{1}{N}\sum_{k=1}^{N} h[e(k)]$ (see also Janssen *et al*, 1988): Divide the sample of length N into several segments of length K. Base the evaluation on the r:th segment I_r, and estimate the parameters based on the remainder $I - I_r$ of the sample. Repeat for all k and sum the losses. Then

$$C_1 = \sum_r \sum_{k \in I_r} h[e(k|\hat{\theta}_r)],$$

$$\hat{\theta}_r = \arg \min_{\theta} \sum_{k \in I-I_r} h[e(k|\theta)] \tag{9.11}$$

should approximate the expected value of V.

The prespecified value of K is a useful design parameter. It reflects the length of the 'application phase', that is, the length of time the model should be valid (see section 1.3 on "The purpose"). Since $K \ll N$, the criterion (9.11) suits short prediction horizons, as in the case of adaptive control.

For long application phases cases Stoica *et al* (1986) suggest the alternative criterion

$$C_2 = \sum_r \sum_{k \in I - I_r} h[e(k|\hat{\theta}_r)],$$

$$\hat{\theta}_r = \arg \min_\theta \sum_{k \in I_r} h[e(k|\theta)] \tag{9.12}$$

where the evaluation intervals are now the much longer $I - I_r$.

> **Remark:** Analysis of these criteria reveals that, asymptotically, they too have the form $N\, V(\theta) + k \dim(\theta)$. The criterion is called GAIC (for Generalized AIC), and should be familiar by now, except for the factor k. That factor is a design parameter and should reflect the intended use of the model (Stoica *et al*, 1986). AIC sets it to $k = 2$.

In summary: The 'classical', 'subjective' methods of repeated testing require a (partially) nested structure and a preset risk. The second condition is not a serious limitation; the first might be. The 'modern', 'objective' methods can theoretically do without both limitations. However, they are developed for less general structures, so far.

9.5.2   Other 'modern' principles for determining complexity

The 'modern' methods of order determination have motivated much research, since Akaike (1974) first presented his AIC criterion. A number of other principles have been tried, yielding modified Likelihood functions with weights on complexity. A few of those will be outlined briefly:

Akaike (1974), Rissanen (1976), and others have suggested that one base a general–purpose loss function on information–theoretical concepts. The following principle by Jaynes (1957) applies to statistical inference: "The minimally prejudiced assignment of probabilities is that which maximizes the entropy subject to the given information about the situation". An optimization principle is therefore that "the parameters in a model which determine the value of the maximum entropy should be assigned values which minimize the maximum entropy" (Rissanen, 1976). Another formulation is that one should "maximize ones ignorance of the object, subject to the observed constraints" (Ljung and

Glover, 1981). These and similar formulations of the Maximum Entropy principle still leave room for different ways of defining 'entropy', and result in different loss functions.

Remark: Akaike (1974) arrived at his AIC criterion by applying the Maximum Entropy principle, and defining entropy as the negative of KDI $= E\{\log L(\mathbf{M}|\mathbf{c},\mathbf{d})/L(M|\mathbf{c},\mathbf{d})|\mathbf{M}\}$ (Kullback's Information Criterion). Notice that this is equivalent to minimizing the expected loss function, when the latter is the negative log Likelihood function. In view of the derivation of the complexity–dependent loss function (9.10), this explains why the AIC criterion is of the same form.

Remark: Akaike (1976) argues that if identification is based on information concepts (the Likelihood function is a consistent estimate of its expected value, *i.e.* Kullback's criterion), then there is no need for assuming that there is a Source–Equivalent model is the structure that defines the Likelihood function. The Likelihood function makes a reasonable criterion anyhow. But independent of that, the model designer may simply observe that no matter what reasonable principle one prefers to invoke in order to arrive at a purpose–independent criterion, the result seems to be the negative log Likelihood function, possibly with a term dim(θ) added as penalty for high model complexity.

Remark: By also introducing structural assumptions with the Maximum Entropy principle Rissanen (1976) derives a loss function that includes a measure of parameter uncertainty, in addition to the two measuring prediction error (the negative log likelihood) and complexity (number of parameters). A consequence of this is that equivalent structures may have different losses, depending on their parametrizations. This does in fact introduce a new aspect on the loss function design, that is not covered by the Likelihood function with or without weights on complexity: As the purpose is usually parameter dependent, it enters into the loss function in this way.

Because of the ambiguity in defining entropy and/or some difficulty of interpreting the concept in practice, more concrete principles have also been suggested:

Prefer the model that is most purposive

The simplest and most well known principle of the purpose–oriented kind is to minimize the prediction error of the model. This is obviously well suited to the purpose of designing feedback control of the object. The principle leads to maximizing

$$\log L(v,\theta|\mathbf{c},\mathbf{d}) - \tfrac{1}{2} N \log\{[1 + \dim(\theta)/N]/[1 - \dim(\theta)/N]\}$$

with respect to v and θ, where N is the length of the data sample (Söderström and Stoica, 1989). It can therefore be interpreted as a weighted ML criterion, with weights $p[v|\mathcal{A}]$ determined by the principle and decreasing with increasing complexity. For long data samples, the weights will tend to $\exp[-\dim(\theta)]$.

> **Remark:** The criterion is known as FPE (Final Prediction Error), when used in connection with the MPE (Minimum Prediction Error) criterion for estimation of the real parameters θ (Akaike, 1974). The loss function is then
>
> $$\text{FPE} \triangleq [\det R(\theta)] \; [1 + \dim(\theta)/N]/[1 - \dim(\theta)/N],$$
>
> where $R(\theta) = \dfrac{1}{N}\sum_{k=1}^{N} \mathbf{e}(k|\theta)\,\mathbf{e}(k|\theta)^{T}$,
>
> and $\mathbf{e}(k|\theta)$ are the one–step prediction errors.

Rissanen (1978) proposed the principle of 'shortest data description'. The resulting models have the shortest average length (in bits) of all possible descriptions of the given data sample. Hence, the criterion is suited to cases where the purpose requires only a 'data description' (see section 5.8 on "Classification of models by purpose"). The effect of adopting any of the principles is a weight of approximately the same size $\exp[-\dim(\theta)]$ in front of the unbiassed likelihood.

Prefer the simplest model

The principle of parsimony is also applicable here. One can apply this principle in a number of ways: One way is to use 'subjective probabilities' with more weight on low–complexity models, another one is to use sequential falsification, both as described above.

The principle of parsimony also yields the model with the highest statistical accuracy in addition to the simplest one (Stoica and Söderström, 1982). In the light of this, it can also be interpreted as preferring the most purposive model, where the purpose is to obtain an accurate model. Applied to the prediction problem this leads again to the weights $p[v|\mathcal{A}] = \exp[-\dim(\theta)]$.

> **Remark:** Perel'man (1984) surveys other criteria that dependend on model complexity.

All order–determination algorithms do similar operations

The different principles all lead to some loss function to minimize, but in practice they operate in the same way as sequential hypothesis testing or with weights on complexity. They require the fitting of parameters for a number of tentative complexity indices n, and they compare losses with bias in favour of models with low n.

The main feature of those methods is that they compute the weights, or biasses, or risk levels automatically, instead of demanding the values apriori from the operator. On that basis they are sometimes called 'objective' and the methods based on tests 'subjective' (Unbehauen and Göhring, 1974). An argument against subjective methods is that there is no satisfactory way of determining apriori risk. Objective methods do eliminate this conceptual difficulty, but are in fact also based on more or less subjectively adopted principles (Leontaritis and Billings, 1987).

> **Remark:** The similarity of the results of different principles is illustrated well in a method of deriving a family of risk–dependent criteria pointed out by Leontaritis and Billings (1987): By applying a sequence of Likelihood–Ratio tests (Section 8.5) to candidates of different complexities, and taking into account the condition that there should be no conflict between the corresponding test results, independent of the particular sequence of tests, it is possible to conclude the following: The model selected should minimize a criterion of the form $L(\theta|v,c,d) \exp[-n\,\lambda/2]$, where λ depends on a prespecified risk that the selected model will have one parameter too many. Again, this criterion has the familiar form. In particular, the value of $\lambda = 2$ corresponds to the AIC criterion. Conversely, AIC corresponds to a risk value of 15.6% for obtaining one parameter too many.

> **Remark:** It has been observed experimentally that both the FPE and the AIC criteria tend to overparametrize the models (Kashyap, 1980). This is easy to see from the fact that they are approximately equivalent to a statistical test with 15.6% risk of getting a too complex model (Söderström and Stoica, 1989). In fact it has been shown that in order to obtain a criterion that yields asymptotically unbiassed estimates for the complexity indicator, one has to modify the weights so that they depend also on the sample length, namely as $\exp[-k(N)\,\dim(\theta)]$, where $k(N) \rightarrow \infty$ and $k(N)/N \rightarrow 0$ (Kashyap, 1980). A common value is $k(N) = 2\log(N)$ (Stoica *et al*, 1987).

Remark: Kashyap (1980) also gives a practical case where the AIC criterion *underparametrizes* the model: Using AR models for speech synthesis, and AIC for order determination, usually results in orders between 2 and 5. But the synthesized speech does not sound good, until the order is between 8 and 14. This illustrates the importance of having the purpose in the criterion function — the purpose is not to predict the speech signal one sampling interval.

Remark: The following are two good references for clarifying and unifying various methods for order determination: Stoica *et al* (1986) for 'modern', and Leontaritis and Billings (1987) for 'classical' statistical methods.

In spite of their 'subjectivity' I still prefer the classical methods of sequential testing from an engineering point of view. Even if the statistical test has its conceptual difficulties, it is still a very practical device for determining complexity n. The point is that one has often an intuitive understanding of what is meant by 'risk'. And that is precisely the kind of link between theory and practice that one needs to apply identification theory, and which I want to stress the importance of in this book. In practice there is usually no problem to assign a value to the risk, since the tests are not sensitive to reasonable risk levels (such as 0.1 or 0.01 or 0.001). In fact, the freedom to choose the risk value (or the subjective biasses) is an *asset*, since it provides a means to control the complexity of the final model. And it is easy to understand how these design parameters work (low risk reduces complexity), which is a valuable quality in an engineering design method.

9.5.3 The Bayesian approach

An alternative to the problem of deciding on an optimal structure index is again offered by Bayes' idea (Peterka and Karny, 1979). It is to refrain from pointing out a particular structure index ν, and compute instead the aposteriori probability $p[\nu|\mathcal{A},\mathbf{c},\mathbf{d}]$. The rationale for this is that any subsequent decision will need only this distribution, in addition to the real–parameter distribution $p[\theta|\mathbf{c},\mathbf{d},\mathcal{A},\nu]$, the model $p[c,d|\mathcal{A},\nu,\theta]$, and the utility function of the purpose.

The problem of the approach is of course that one will have to keep track of a large number of models with different structures and complexity, in order to let all affect the decision according to their different probabilities. This is avoided, if instead one selects one of the models as 'representative'. However, this is clearly an approximation, and any type of estimation destroys information.

The following derivation of $p[v|\mathcal{A},c_N,d_N]$ (Peterka and Karny, 1979) illustrates the technique of exercising with Bayes' rule:

$$p[v|c_N,d_N] = p[c_N,d_N|v]\ p[v]/\sum_v p[c_N,d_N|v]\ p[v] \qquad (9.13)$$

Since the controller does not depend on v ('Natural condition of control' implies an Isolated Experimenter; see section 3.5 on "Experiments on dynamic objects"):

$$p[c_N,d_N|v] = \prod_{k=1}^{N} p[d(k)|c_{k-1},d_{k-1},v]\ p[c(k)|c_{k-1},d_k,v]$$

$$= \prod_{k=1}^{N} p[d(k)|c_{k-1},d_{k-1},v]\ p[c(k)|c_{k-1},d_k] \qquad (9.14)$$

Insert this into (9.13):

$$p[v|c_N,d_N]$$

$$= \prod_{k=1}^{N} p[d(k)|c_{k-1},d_{k-1},v]\ p[v]/\sum_v \prod_{k=1}^{N} p[d(k)|c_{k-1},d_{k-1},v]\ p[v] \qquad (9.15)$$

But

$$p[d(k)|c_{k-1},d_{k-1},v] = \int \prod_{k=1}^{N} p[d(k)|c_{k-1},d_{k-1},\theta,v]\ p[\theta|v]\ d\theta \qquad (9.16)$$

Bayes' rule again:

$$p[\theta|c_{k-1},d_{k-1},v] = \prod_{k'=1}^{k-1} p[d(k')|c_{k'-1},d_{k'-1},\theta,v]\ p[\theta|v]/\lambda(v|k-1) \qquad (9.17)$$

where

$$\lambda(v|k) = \int \prod_{k'=1}^{k} p[d(k')|c_{k'-1},d_{k'-1},\theta,v]\ p[\theta|v]\ d\theta \qquad (9.18)$$

Insert (9.17) into (9.16). Then

$$p[d(k)|c_{k-1},d_{k-1},v] = \lambda(v|k)/\lambda(v|k-1) \qquad (9.19)$$

Insert (9.15). Then

$$p[v|c_N,d_N] = \lambda(v|N)\ p[v]/\sum_\mu \lambda(\mu|N)\ p[\mu] \qquad (9.20)$$

which yields

$$p[\nu|c_N, d_N]/p[\mu|c_N, d_N] = \lambda(\nu|N)\, p[\nu]/\lambda(\mu|N)\, p[\mu] \qquad (9.21)$$

This determines the aposteriori probabilities of any structure ν, provided one can evaluate the integral in (9.18). Notice also that the apriori probabilities $p[\nu]$ are important.

In case one would want the result for prediction, the predictor would be

$$p[d(k)|c_{k-1}, d_{k-1}] = \sum_\nu p[d(k)|c_{k-1}, d_{k-1}, \nu]\, p[\nu|c_N, d_N]$$
$$= \sum_\nu \lambda(\nu|k)\, p[\nu]/\sum_\nu \lambda(\nu|k-1)\, p[\nu] \qquad (9.22)$$

This demonstrates that one will not need actually to estimate the structure index ν. Neither is there a need for any particular nesting. However, the main problem in practice (besides the exhaustive search over ν) is to carry out the integration in (9.18), when the number of parameters is not small.

9.6   Model structure selection

In addition to 'order determination' and 'structure identification' the literature on 'system identification' contains contributions on topics like 'system classification', 'model structure determination/selection/testing', 'model reduction/simplification', 'pole–zero cancellation', and others using similar keywords. Their meanings are usually defined only for special cases, and often differently (when at all). A review of the concepts and of their relations in general cases would therefore be cumbersome to make, and probably also to read. The following will instead review the various identification *problems* as defined in this book, and then attach suitable labels to them.

As pointed out earlier, one can achieve 'structure identification' (selecting a class $\mathcal{M}_\nu$) in two ways: 1) by defining a loss function that depends on ν and searching for the minimum loss, or 2) by defining an indicator for a 'good' model structure, and stop the search at the first structure that passes the test.

The first approach regards also the class selection problem as an estimation problem, and finds a solution by fitting both integer and real parameters ν and θ. In addition to the problems of *postulating* a suitable loss function, taking into account both complexity and accuracy, one has to determine a *search strategy*. In particular, determining the sequence in which to search for the integer vector v will be a difficult problem.

The second approach is to base the selection on a specified risk that one accept a model with one parameter too many (Leontaritis and Billings, 1987). It was adopted in chapter 4 on "The identification problem". It implies that the sequence in which to evaluate losses or test for good models must be given. This is implicit in the 'principle of parsimony' — that simpler model structures are to be tried first. However, what a 'simpler' model structure would mean must still be *postulated*, as well as a strategy for searching within structures of equal complexity. All together, the second approach leads to problems of the same magnitude as finding a search strategy for integer variables needed in the first approach.

> **Remark:** Statistical tests can also be used to discriminate between several alternatives (Leontaritis and Billings, 1987). This increases the possibilities of using statistical tests in structure identification.

It is the number of alternatives that decides the problem magnitude

The difficulty of the structure identification problem clearly increases by a magnitude, if the admissible structures are so many that deciding in what order to try them becomes crucial for the applicability of either an order–testing or a loss–minimization scheme. This is not uncommon with composite structures, since they are often combinations of several structural elements, and trying all combinations is usually out of the question. Hence, even if there is little reason in practice for distinguishing between search and sequential testing, there is still good reason for distinguishing between structure identification methods, depending on whether or not they consider the testing order or search strategy.

The methods described so far do not. Label them *'order determination'* methods. They are obviously applicable in cases where it is possible to specify a simple nesting, so that the structure index v is determined uniquely by the complexity number n (which is determined by search in an obvious order). They are also applicable when only a partial nesting is available, so that the tested structures are composite, but the alternatives with given complexity are few enough to render the minimization of losses over the composite structure practically feasible. Also in this case the structure index is determined, when the complexity number has been determined by sequential testing.

The label *'model structure selection'* will be used for methods that do prescribe rules for search strategy or testing sequence. This means that 'structure selection' will require a (usually recursive) procedure for determining, in each particular case, *which* structures, out of the multitude of possible structures, that should be put to test or evaluated in some other way.

This book does not treat the problem of nesting the given structure indices, and hence not the general 'structure selection' problem. The nesting is assumed to be the result of previous 'modelling' activity, and hence not part of the identification task. However, a few hints on where to look for solutions are given below.

> **Remark:** The text book of Kashyap and Ramachandra Rao (1976) approaches the 'structure identification' problem in several ways, and suggests rules for determining the structure of linear polynomial operator models (AR, ARMA, ARIMA, etc.), but does not consider the problem of search strategy or the sequencing problem. The book suggests minimization of various prediction–error criteria or sequences of Likelihood–Ratio tests. It also contains detailed techniques for multivariable linear systems.

> **Remark:** How difficult the structure identification problem is in practice depends very much on the possibilities to set up reasonable structures. To facilitate the task one can either aim at having few alternatives, and use ones apriori information about the object to eliminate most of the possible combinations (the 'grey–box' approach). Or else one can choose structures of a form that facilitates the fitting and selection procedure (the 'black–box' approach). Billings (1980) and Billings and Leontaritis (1982) have suggested such structures. However, even this may quickly get out of hand, and this makes it the more important to make use of apriori information to guide the search. This is a strong point in favour of the 'grey–box' approach.

> **Remark:** The Bayesian approach (section 9.5.3) needs no nesting, but again there is a problem of the same magnitude as that of searching for a vector–valued v: one has to decide for what structures to compute the aposteriori probabilities, unless one can afford to do it for all.

There are few solutions

The structure selection problem in the case of many possible models structures of the same complexity, is generally difficult and not much explored (Natke and Samirowski, 1988). Attempts that promise some success use 'heuristics' or else originate in 'unconventional' ways of formulating the problem (Mehra, 1979).

However, a few solutions of the structure selection problem without a postulated sequencing of the structures have been proposed. They are restricted to special structures: Karny (1983) has presented an algorithm to determine the order in which to test alternative structures. It applies to linear polynomial operator models with several unknown orders.

The paper by van Overbeek and Ljung (1982) presents an algorithm for searching over composite structures, *i.e.* going between different parametrizations $\mathcal{M}_\nu \subset \mathcal{M}(n)$ of the same complexity. It uses the idea of switching structure (parametrization), when the searching over the current parameter set becomes illconditioned. It applies to 'black–box' linear time–invariant multivariable state–vector models of the 'innovations' type, *i.e.* 'external' models (see section 5.5.1).

Two heuristic search principles are the SFI and SBE methods (Leontaritis and Billings, 1987), the first one searches 'forward' from the simplest to the most complex model, the second 'backwards' from the most complex to the simplest. Both reduce the maximum number of structures investigated to $\bar{n}(\bar{n}+1)/2 + 1$ from a total of $2^{\bar{n}}$ for exhaustive search, but they do not guarantee that the minimum will always be reached in that way. The SBE is a model reduction scheme, and can therefore be used in conjunction with sequential falsification, based on the convenient alternative of simple nesting.

The restriction of 'linearity–in–the–parameters' simplifies the nonlinear structure identification problem in a crucial way (as it does the fitting problem). In the general case the innovations should be possible to express as linear combinations of known nonlinear functions of data. It might be difficult to motivate this restriction in practice. However, one can often expand NARMAX–type of models for external model structures to achieve linearity in the parameters (Chen, Billings, and Luo, 1989). Desrochers and Mohseni (1984) devised a selection algorithm that searches for the best (in the least–squares sense) linear combination of n nonlinear elements in only n steps. To achieve this the designer must specify apriori that the model structures be searched according to a particular pattern, namely along a 'tree structure': Each structure (combination) is a node in the given 'tree', where each 'branch' corresponds to a possible development to higher complexity, *i.e.* involving one more element. Hence, again the substantial reduction in computing is obtained at the price of substantial apriori specification. Kortmann and Unbehauen (1988) treated the nonlinear 'black–box' case and suggested two methods for deciding which of the many possible coefficients of nonlinear function elements that should be non–zero and thus included in the model structure.

Billings, Korenberg, and Chen (1988) and Chen, Billings, and Lou (1989) developed selection methods under similar assumptions. The solutions apply to nonlinear 'black–box' cases, where the conceivable parameters are many.

The latter reference also contains a lucid survey and discussion of the relations between structure selection methods that assume linearity in the parameters. In essence, the non–zero parameters are picked recursively, in each step as that which contributes most to the loss reduction — a quite natural strategy, also called 'the Stepwise Inclusion Method' (Perel'man, 1984). Since usually only few, n say, of the many possible $\bar{n}$ parameters contribute significantly, the number of tests will be approximately $n \, \bar{n}$, and far from the number of possible combinations.

The 'Group Method of Data Handling' uses a heuristic approach and applies to more general systems. It has therefore attracted much interest (*e.g.* Farlow, 1986). It was first suggested by Ivakhnenko (1970).

Nakamori (1989) developed comprehensive interactive modelling software for large systems, employing several structure selection techniques.

Pole–zero cancellation (*e.g.* Söderström, 1975), and in general all model reduction schemes, are other ways of selecting model structures in linear cases. The problem of order reduction is usually to find models of lower order, that either minimize some criterion for the deterioration in accuracy, or else preserve some preselected parameters of the full-order model. Also those results depend indirectly on the purpose; the reduced model depends very much on what properties of the model the designer wants to preserve, and this depends on the purpose. For instance, de Willemagne and Skelton (1987) designed a method taking into account the frequency range in which the model is to operate. It applies to linear, multivariable state–vector models.

In summary, there are two alternatives, either expanding or reducing the structures. Both have their disadvantages: Expansion requires apriori sequencing (*i.e.* apriori information). Reduction requires the fitting and evaluation of models of too high complexity. The choice depends on what one knows apriori. In any case, combinations may help to alleviate the drawbacks.

9.7   Terminology revisited

The following summarizes the key concepts used in this book with the purpose of defining the various problems in mathematical terms.

● The *experiment system* or *data source* is the system that generates data: $(\mathbf{X},\mathbf{S}) \mapsto (\mathbf{c},\mathbf{d})$.

● A *model* is another (algorithmic) system that generates data taking values in the same space: $M \mapsto (c,d)$.

> Example: A rational transfer function driven by an uncorrelated gaussian sequence with zero mean and unit variance. Call the latter sequence 'standard white noise' for simplicity.

● A *model structure* $\mathcal{M}$ is the set of all models M that agree with apriori assumptions and hypotheses.

> Example: All rational transfer functions driven by standard white noise.

● *Experiment design* is the specification of a sequence of experimenters $\{\mathbf{X}(m)|m=1,\ldots,\bar{m}\}$.

> Example: $\mathbf{X}(m)$ generates a sequence of standard white noise as control signal $\{\mathbf{c}(k)|k=1,\ldots,100\ m\}$. The integer m is the 'extension index' of the experiment.

● *Modelling* is the specification of an indexed, expanding sequence of model sets $\{\mathcal{M}(n)|n=1,\ldots,\bar{n}\}$, such that $\mathcal{M}(n) \supset \mathcal{M}(n-1)$. The 'widest model set' is $\bar{\mathcal{M}} \triangleq \mathcal{M}(\bar{n})$.

> Example: $\mathcal{M}(n)$ is all transfer functions with numerator and denominator of degree up to n (including those with common zeroes) and driven by standard white noise.

● *Identification* means picking one model M, given the sequence $\{\mathcal{M}(n)\}$ and the data provided by the experiment sequence $\{\mathbf{X}(m)\}$. Thus, identification includes the extension of the experiment (within the given sequence) needed to establish a satisfactory model.

> Example: Determine numerator and denominator coefficients as well as the smallest order needed to satisfy the purpose of the

model design. Determine also the simplest experiment needed to establish the model.

The necessary 'parametrization' requires more concepts:

● A *parametric model* is the triple $M = (\mathcal{A}, \nu, \theta)$, where $\mathcal{A}$ is the 'form' of the model, the nonnegative, possibly vector–valued integer ν is the 'structure index', and the real θ is the 'parameter vector'.

> **Example:** $\mathcal{A}$ is all rational transfer functions, driven by standard white noise, ν is the two polynomial orders, and θ is the array of coefficients, the first coefficient of the denominator excluded.

● A *parametric (simple) model structure* is a doublet $\mathcal{M}_\nu = (\mathcal{A}, \nu)$. With a slight abuse of notation it also means a model set such that $M = (\mathcal{A}, \nu, \theta) \in \mathcal{M}_\nu$. A 'composite model structure' is a set of model structures: $(\mathcal{A}, \mathcal{N}) = \{\mathcal{M}_\nu | \nu \in \mathcal{N}\}$.

● *Partial sequencing* (of model structures) means partitioning and ordering the set of structure indices ν in the 'widest structure' $\overline{\mathcal{M}}$ into a sequence $\mathcal{N} = \{\mathcal{N}_1, \mathcal{N}_2, \ldots, \mathcal{N}_{\bar{n}}\}$, such that all model structures in $(\mathcal{A}, \mathcal{N}_n)$ have the same complexity, for instance the same number r_n of parameters. The integer n is the 'complexity number'.

> **Example:** If $\mathcal{A}$ is all rational transfer functions driven by standard white noise, then a reasonable sequencing $\mathcal{N}_n = \{\nu_1, \nu_2 | \nu_1 + \nu_2 = n - 1\}$. All members in $\mathcal{N}_n$ have $n-1$ parameters.

● Two parametric structures $(\mathcal{A}, \nu)$ and $(\mathcal{A}, \nu')$ are *nested* if for every θ there is some θ' such that $p[c, d | \mathcal{A}, \nu, \theta] = p[c, d | \mathcal{A}, \nu', \theta']$ with $\nu' > \nu$. That is, $\mathcal{M}_\nu \subset \mathcal{M}_{\nu'}$.

● *Partial nesting* of a sequenced set of structures $(\mathcal{A}, \mathcal{N})$ means that for every $\nu_n \in \mathcal{N}_n$, $n < \bar{n}$, there is some $\nu_{n+1} \in \mathcal{N}_{n+1}$ such that $(\mathcal{A}, \nu_n)$ and $(\mathcal{A}, \nu_{n+1})$ are nested, and for every $\nu_n \in \mathcal{N}_n$, $n > 1$, there is some $\nu_{n-1} \in \mathcal{N}_{n-1}$ such that $(\mathcal{A}, \nu_{n-1})$ and $(\mathcal{A}, \nu_n)$ are nested.

● *Simple nesting* means that $(\mathcal{A}, \mathcal{N})$ is nested and each index set $\mathcal{N}_n$ contains only one structure index ν_n, *i.e.* every structure is non–composite.

> **Example:** If $\overline{\mathcal{M}}$ is all rational transfer functions driven by standard white noise, then the following nestings are simple:
> $\mathcal{N} = \{(0,0), (1,1), (2,2), \ldots, (\bar{n}, \bar{n})\}$
> and

$$\mathcal{N} = \{(0,0),(1,0),(1,1),(2,1),\ldots,(\bar{n},\bar{n})\},$$

but not for instance:

$$\mathcal{N} = \{(0,0),[(1,0),(0,1)],[(2,0),(1,1),(0,2)],\ldots,(\bar{n},\bar{n})\}.$$

● The *parametric modelling* task will be to specify $(\mathcal{M},\mathcal{N})$. This yields the expanding sequence by

$$\mathcal{M}(n) = \mathcal{M}(n-1) \cup (\mathcal{A},\mathcal{N}_n), \quad \mathcal{M}(-1) = \emptyset,$$

$$(\mathcal{A},\mathcal{N}_n) \triangleq \bigcup_{\nu \in \mathcal{N}_n} \mathcal{M}_\nu, \quad \mathcal{M}_\nu \triangleq (\mathcal{A},\nu).$$

● The *parametric identification* task means determining (ν,θ) given $(\mathcal{A},\mathcal{N})$.

● *Structure identification* means determining ν given $(\mathcal{A},\mathcal{N})$.

● *Model structure selection* means determining ν and $\mathcal{N}$ given $\mathcal{A}$.

● *Model order reduction* means finding an adequate model in $(\mathcal{A},\nu')$, $\nu' < \nu$, when one in $(\mathcal{A},\nu)$ is known. The problem arises when the structure index is vector–valued and has been determined by a method requiring simple nesting, but simple nesting cannot include all feasible structures $(\mathcal{A},\nu)$.

> **Example:** Let $\bar{\mathcal{M}}$ be all rational transfer functions driven by standard white noise, let the postulated ordering be $\mathcal{N} = \{(0,0),(1,0),(1,1),\ (2,1),\ldots,(\bar{n},\bar{n})\}$, and let the structure index $(0,1)$ be adequate. Then $n = 3$, the first model to be validated has the index $\nu_3 = (1,1)$, but there is a valid model with index $(0,1) < (1,1)$ not included in $\mathcal{M}(3)$.

> **Remark:** The still more difficult problem of 'model structure determination' involves also defining and indexing the widest model class $\bar{\mathcal{M}}$ of tentative models. It is evident that this problem cannot be solved only by using mathematics.

10   A unified design procedure

The purpose of this chapter is to compile the results in previous chapters into a general procedure for model design, including both modelling and identification. The procedure provides a partial algorithmic specification, and thus sets a 'frame' for the structuring of proper software for model design. The complexity and detail of actual program specifications then depend on how many and how general structures $\mathcal{M}$ and $\mathcal{G}$ the program will provide for and how many validation, falsification, and fitting techniques that are implemented. Its usefulness will also depend much on the man–machine interface, *i.e.* input and display facilities, and the computer's ability to put the right questions, give the right information, and provide 'help' facilities.

Computer–aided design of models involves three 'actors':
● *The experiment system*: This is the system of experimenter and object that produces stimulus **c** (according to **X**) and response data **d** (according to **S**). The 'experimenter' may be a human being or a machine, but must be separable from the object.
● *The model designer*: This is the person who provides apriori specifications and assumptions/hypotheses about the object and the experimenter, drawing from the two sources of 'apriori information on the object' and 'engineering sense'.
● *The computer*: It executes well–defined tasks, operating on primary data and intermediate results.
They do their tasks in an *interactive* procedure of model design.

In essence, the task of the model designer is of two kinds:
● Giving a sequence of 'execute' commands to the computer to apply selected identification tools.
● Entering apriori information expressed as real and integer parameter values and model forms.

The first task means implicitly *accepting* the preconditions behind each selected tool, and upon those preconditions hinges the validity of the result. The computer may be basically unable to check some of the preconditions (the 'assumptions') and may therefore come out with a wrong result without warning, and this is the cause of the 'pitfalls' everpresent in model making. This holds in particular for assumptions concerning the circumstances under which the experiment data was obtained. Parameter values and model form are generally easier to check by the computer (by 'falsification'), and this kind of apriori information is less of a 'pitfall'.

What computers can and cannot do

As stated in the introduction, design of models involves two steps, *viz.* the 'modelling' step, *i.e.* deciding on model structure $\mathcal{M}$, purpose $\mathcal{G}$, and experimenter **X**, and the 'identification' step, *i.e.* determining a model M from given $(\mathcal{M},\mathcal{G},\mathbf{c},\mathbf{d})$. The second task is a mathematical problem, and can be handled by the computer, provided algorithms are known and the numerical problems are manageable. Otherwise, the designer may have to intervene with a helping hand also in the identification. That is a rule more than exception in complicated cases, since few identification algorithms are safe.

The modelling is not an entirely mathematical problem and is therefore a task for the designer, although the computer may provide a helping screen display also during the modelling step. The latter involves ascertaining that the *conditions* for proper model structure design and experiment are satisfied. That cannot be concluded from data, and therefore the computer cannot do the modelling alone. The best any computer can do is to display the critical assumptions. The designer must *acknowledge* and thereby take the *responsibility*. He/she has more information to work with than the computer has.

The 'identification' part of the procedure is necessarily the more detailed, simply because it is easier to give detailed specifications to the computer than to the human designer. The 'modelling' part contains mainly the conditions for proper identification. It serves to highlight those conditions and pinpoint the circumstances under which each one has to be satisfied, and thus to help the designer to stay clear of 'pitfalls'. As a minimum requirement (although not satisfied by contemporary identification programs), the computer should display them. A good (future) program should provide guidelines and possibly also computing aid to their verification.

> **Remark:** Some of the tasks of the 'model designer' may be possible to place on an 'expert system' (Sage, 1987b). Such systems have been reported by Gomide *et al* (1988), Haest *et al* (1988), and Betta and Linkens (1990).

10.1 Summary of conditions for proper identification

According to the analysis so far a number of basic conditions on the experiment system guarantee a correct result of identification. There are relations between them expressed by a number of lemmas. Of course, the designer may have difficulties in *verifying* the conditions in the particular case, and then the 'guarantee' provided by the lemmas will not be more worth than the designer's assumptions. However, the purpose of the analysis is to *separate out* what assumptions are the responsibility of the designer, as distinguished from those the computer can possibly falsify. This yields precisely the assumptions the computer should ask permission for and the designer acknowledge.

In the most common case that the purpose of identification requires at least an External Object Description the designer must acknowledge the assumptions summarized below. Since they have been expressed previously in mathematics, they will be expressed here in natural language to support the intuitive understanding. It is obviously imperative that the designer does indeed have this understanding, since he/she does have the responsibility.

● *Reproducibility*: It must be at least conceivable to repeat the experiment and obtain data with the same distribution (even if not the same data). If not, the model may not be valid at a later time.

● *Isolated Experimenter*: There must be no other channel for influence between stimulus and response than via the object or the experimenter. There are two 'pitfalls' here:
1) 'Hidden variables' causing false or 'nonsense' correlations (Bohlin, 1987a). In closed loop there may be unrecorded feedback, for instance from a human operator.
2) Reversed causality. Generally, the definition of what is object and

controller in closed loop hinges on causality. Synchronization error may ruin that.

● *Sufficient Stimulation*: The experiment must be relevant and representative for the application. A special case is 'persistently exciting' signals for linear systems. It is generally hazardous to apply models outside their tested range of validity.

● *Purpose Equivalence*: There must be a model within the assumed widest set of structures such that there would be no more purposive information to be gained from data by trying to model more properties of the data source. The latter properties must neither affect the purpose directly, nor the estimates of any properties of the model that do affect the purpose.

● *Source or Object Equivalence*: There must be a model within the assumed widest set of structures such that its responses behave statistically as those of the data source. This assumes that it is feasible at all to approximate reality sufficiently well within the framework of the designer's most general model structure. If not, the 'accept' results of tests may be misleading.

● *Consistency*: The data distribution must be possible to estimate from a long data sample. This requires two things:
1) The Likelihood function must converge, and the limiting function must depend only on the source, and not on the particular data sample from the source. This means that the pertinent information in the data must be distributed over most of the sample. Spurious transients (such as 'outliers') must not dominate, unless they can be modelled in such a way that they do not affect the convergence, for instance by 'clipping'. If there are too many unexplained phenomena in the experiment data, or only a small part of the data sample (for instance start–up transients) contains substantial information, then identification will obviously have difficulties using statistical methods.
2) The complexity of the model must not grow too fast with the time interval of its applicability. Generally, the number of parameters needed to describe a data sample by a model must be much smaller than the number of data in the sample. This does not necessarily mean stationarity or time–invariance. But the source must not unpredictably change its dynamics too often. Dynamics may be allowed to change, even frequently, but that must be possible to model within some of the structures.

Remark: A good identification program should illuminate the meanings of those conditions further (on request), for instance by giving examples of realistic cases, where the conditions are not satisfied, and the consequences thereof.

Remark: The Purpose Equivalence is probably the most difficult to interpret correctly. It does however affect only the validation. This may be a reason for choosing the 'Scientist's approach', even when the purpose is well defined (see section 4.6.2 on "Purposive models without validation").

Other sufficient conditions are feasible and some have been derived in previous chapters, partially related to each other. The choice of conditions to verify apriori, or else accept without further evidence (and at a risk), depends of course on what is the designer's knowledge about the object. The choice also depends on the how demanding the purpose of the modelling will be. However, to sort out what is needed is the computer's responsibility. A graphic summary of the most important conditions and relations is given in *Fig. 53*.

The relations between different conditions shown in the figure correspond to the following lemmas:

Lemma of the Closed Loop:
Reproducibility, Sufficient Stimulation, Isolation
⇒ *Representability, Applicability*

Lemma of Object Equivalence:
Sufficient Stimulation, Isolation, Source Equivalence
⇒ *Object Equivalence*

Lemma of the Structured Source:
Source Equivalence, Bounded Structure, Likelihood Convergence
⇒ *Consistency*

Lemma of Object Identifiability:
Sufficient Stimulation, Isolation, Object Equivalence, Consistency
⇒ *Object Identifiability*

Sufficient but not necessary

Even with the given purpose, the conditions listed above may not necessarily be the minimum set of assumptions. Ultimately, one would like to have as proper 'assumptions' only those that can never be tested, regard the others as 'hypotheses', and apply suitable test algorithms. This may be a topic for further research; so far there are not enough general al-

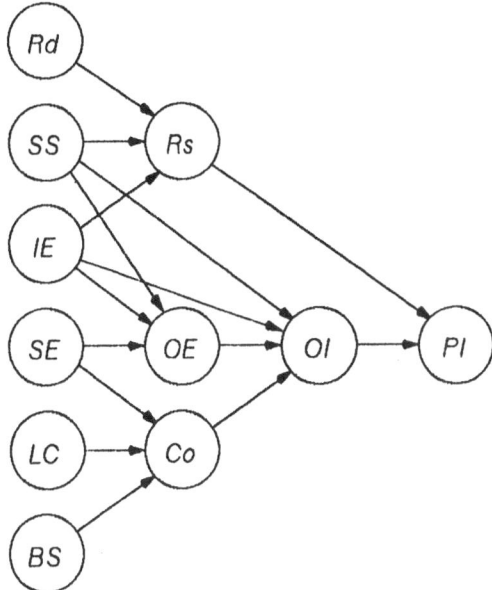

Fig. 53: Illustrating the most important conditions for proper identification of a dynamic system. The arrows indicate implications. The acronyms mean the following:

Rd   *Reproducibility*
IE   *Isolated Experimenter*
SS   *Sufficient Stimulation*
SE   *Source Equivalence*
LC   *Likelihood Convergence*
BS   *Bounded Structure*
Rs   *Representability*
OE   *Object Equivalence*
Co   *Consistency*
OI   *Object Identifiability*
PI   *Proper Identification*

gorithms for this, and neither is there any proof that the above list is necessary.

On the contrary, the analysis in section 8.1 on "Statistical tests" reveals that the condition of Source Equivalence is in fact testable, and thus to be regarded as a 'hypothesis'. However, the more efficient tests, the conditional tests, do assume Source Equivalence for their alternative hypotheses. In addition, the condition of Consistency may be partly testable, since it should be possible to investigate experimentally whether a specified Likelihood function does indeed converge when fed with given data. The second condition, however, that the limit should also be independent of the particular data sequence, and only depend on the source, is obviously not testable from a single data set. On the other hand, some conditions for Consistency, like a Bounded Structure, Time Invariance, and the existence of an Information State depend only on $\overline{\mathcal{M}}$ and are therefore testable. Hence, it may be possible to formulate other, as well as fewer conditions for proper identifications.

However again, even if it would be feasible to reduce the assumptions down to one or a few 'universal' assumptions, an important issue is still whether these assumptions would be possible to *interpret* in physical terms by a human model designer. The list of 'assumptions' is compiled to strike a compromise between 'minimality' and 'intrepretability'.

10.2 Identification procedures

Introducing a selection of the techniques developed in chapters 7 – 8 into the sequential identification procedures outlined in section 4.4 yields two general algorithms for design of nested parametric models $M = (\mathcal{A}, \nu, \theta)$. In particular, the validation rule in section 7.2.2 and the rule for 'conditional falsification of structures' in section 8.4 (the LMP test), yields the following procedure:

Identification by the engineer's rule:

Procedure for validatable identification with LMP falsification:

Initialize: $n, m \leftarrow 0$; Indicate *unvalidated* and *unfalsified*

Repeat until outer stopping rule

Refine model structure: $n \leftarrow n+1$

Repeat until inner stopping rule

If *unfalsified*, then refine the experiment and observe:

$m \leftarrow m+1$; $X(m) \mapsto (\mathbf{c}_N, \mathbf{d}_N)$

Fit a model and test whether it is valid:

Compute the MP model:

$(\hat{\nu}, \hat{\theta}) \leftarrow \arg \max \{V(\nu, \theta | n, \mathbf{c}_N, \mathbf{d}_N) | \nu \in \mathcal{N}_n, \theta \in \mathbf{R}\}$

> If $V(\hat{v},\hat{\theta}|n,\mathbf{c}_N,\mathbf{d}_N) \geq \gamma$,
> then indicate *validated* else *unvalidated*

If *validated*, then stop inner and outer loops

Test whether data falsifies the structure:

> Compute the ML estimate:
> $(\hat{v},\hat{\theta}) \leftarrow \arg\max\{Q(v,\theta|n,\mathbf{c}_N,\mathbf{d}_N)|v \in \mathcal{N}_n, \theta \in \mathbf{R}\}$

> If $B(\hat{v},\hat{\theta}|n,n+1,\mathbf{c}_N,\mathbf{d}_N) > \gamma$,
> then indicate *falsified* else *unfalsified*

If *falsified*, then stop inner loop

The corresponding procedure without a purpose (the 'scientist's rule') will be

---

Identification by the scientist's rule:

Procedure for identification with LMP falsification:

Initialize: $n,m \leftarrow 0$; Indicate *unfalsified*

Repeat until outer stopping rule

Refine model structure: $n \leftarrow n+1$

Repeat until inner stopping rule

> If *unfalsified*, then refine the experiment and observe:
> $m \leftarrow m+1$; $X(m) \mapsto (\mathbf{c}_N,\mathbf{d}_N)$

Test whether data falsifies the structure:

> Compute the ML estimate:
> $(\hat{v},\hat{\theta}) \leftarrow \arg\max\{Q(v,\theta|n,\mathbf{c}_N,\mathbf{d}_N)|v \in \mathcal{N}_n, \theta \in \mathbf{R}\}$

> If $B(\hat{v},\hat{\theta}|n,n+1,\mathbf{c}_N,\mathbf{d}_N) > \gamma$,
> then indicate *falsified* else *unfalsified*

If *falsified*, then stop inner loop

If *unfalsified*, then stop outer loop

10.3 Procedure for modelling and identification

Collecting the results in sections 10.1–2 yields a formal procedure for model design, that clarifies the various tasks and responsibilities of the designer and computer in a correct modelling session. In particular, the 'engineering approach' to modelling (modelling for a purpose), together with the 'structured approach' to falsification, may be summarized into the following formal algorithm:

Procedure for design of dynamic models

Design of parametric dynamic models with a purpose:

Modelling:

Specify type of identification problem:

Type of purpose: *purpose_type* ← **Designer**

Type of model: *menu* ← **Computer**(*purpose_type*);
model_type ← **Designer**(*menu*)

Type of Object: *menu* ← **Computer**(*model_type*);
variable_list ← **Designer**(*menu*)

Type of experiment: *menu* ← **Computer**(*model_type*);
experiment_type ← **Designer**(*menu*)

Check conditions for proper identification:

If *model_type* ≠ Data Description, do

message ← **Computer**(*'Check Reproducibility'*);
Go/Stop_command ← **Designer**(*message*)

If *model_type* ≠ Source Description, do

message ← **Computer**(*'Check Sufficient stimulation'*);
Go/Stop_command ← **Designer**(*message*)

If *experiment_type* = Manual or Spontaneous, do

message ← **Computer**(*'Check Isolated Experimenter'*);
Go/Stop_command ← **Designer**(*message*)

Indicate *Representability*

Set experiment:
$\{X(m)\,|\,m{=}1,\ldots,\bar{m}\} \leftarrow$ **Designer**(*variable_list*)

Set model structure:
$\{\mathcal{M}(n)\,|\,n{=}1,\ldots,\bar{n}\} \leftarrow$ **Designer**(*variable_list*)

Check conditions for feasibility:

If *model_type* $\neq$ *Data Description*, do

message $\leftarrow$ **Computer**(*'Check Source Equivalence'*);
Go/Stop_command $\leftarrow$ **Designer**(*message*)

message $\leftarrow$ **Computer**(*'Check Likelihood Convergence'*);
Go/Stop command $\leftarrow$ **Designer**(*message*)

message $\leftarrow$ **Computer**(*'Check Bounded Structure'*);
Go/Stop_command $\leftarrow$ **Designer**(*message*)

Indicate *Consistency*

If *model_type* $\neq$ *Source Description*,
then Indicate *Object Equivalence* and *Object Identifiability*

Indicate *Proper Identification*

Set purpose:

$\mathcal{G} \leftarrow$ **Designer**$(X,\mathcal{M})$

Check condition of Purpose Equivalence:
message $\leftarrow$ **Computer**(*'Check Purpose Equivalence'*);
Go/Stop_command $\leftarrow$ **Designer**(*message*)

Identification:

Initialize: $n,m \leftarrow 0$; Indicate *unvalidated* and *unfalsified*

Repeat until outer stopping rule

Refine model structure: $n \leftarrow n{+}1$

Repeat until inner stopping rule

If *unfalsified*, then refine the experiment and observe:
$m \leftarrow m{+}1$; $(\mathbf{c}_N,\mathbf{d}_N) \leftarrow$ **Experimenter**(X,m)

Fit a model and test whether it is valid:

Compute the MP model and its validity:
$(\hat{\nu},\hat{\theta}),V_{\!}(\hat{\nu},\hat{\theta}) \leftarrow$ **Designer/Computer**$(\mathcal{G},\mathcal{M},n,\mathbf{c}_N,\mathbf{d}_N)$

If $V(\hat{\nu},\hat{\theta}) \geq \gamma$, then indicate *validated* else *unvalidated*

If *validated*, then stop inner and outer loops

> Test whether data falsifies the structure:
>> Compute the ML estimate and reject confidence:
>> $(\hat{\nu},\hat{\theta}), B(\hat{\nu},\hat{\theta}) \leftarrow$ **Computer**$(\mathcal{M},n,\mathbf{c}_N,\mathbf{d}_N)$
>> If $B(\hat{\nu},\hat{\theta}) > \gamma$, then indicate *falsified* else *unfalsified*
> If *falsified*, then stop inner loop

The classification variables involved in the procedure have the following possible values:

> *purpose_type* $\in$ {*Transmission, Forecasting, Feedback control,*
>   *Feedforward control, System design*}
> *model_type* $\in$ {*Data Description, Source Description,*
>   *External Object Description, Internal Object Description,*
>   *Physical Object Description*}
> *experiment_type* $\in$ {*Automatic, Manual, Feedback, Spontaneous*}
> *variable_list* = {*variables*}
>   where
> *variables* $\in$ {*parameters a, stimulus c, disturbance v,*
>   *actuator output u, process output z, sensor output y,*
>   *sampler output d*}

Not all Designer choices of *model_type* and *variable_list* (**defining the object**) are meaningful. The *menus* presented by the Computer are therefore restricted by the following rules:

> *menu* = {*model_type*} $\leftarrow$ **Computer**(*purpose_type*)
>   where *model_type* must satisfy at least
>   *Transmission* $\mapsto$ *Data Description*
>   *Forecasting* $\mapsto$ *Source Description*
>   *Feedback control* $\mapsto$ *External Object Description*
>   *Feedforward control* $\mapsto$ *Internal Object Description*
>   *System design* $\mapsto$ *Physical Object Description*

> *menu* = {*variable_list*} $\leftarrow$ **Computer**(*model_type*)
>   where *variable_list* must satisfy at least
>   *Data Description* $\mapsto$ *sampler output d*
>   *Source Description* $\mapsto$ *sampler output d*
>   *External Object Description* $\mapsto$ (*stimulus c, sampler output d*)
>   *Internal Object Description* $\mapsto$ (*stimulus c, disturbance v,*
>   *actuator output u, process output z, sensor output y,*

sampler output d)
Physical Object Description ↦ (*parameters a, stimulus c,*
disturbance v, actuator output u, process output z,
sensor output y, sampler output d)

Remark: The *menu* of *experiment types* cannot be restricted apriori by the computer. What type of experiment that is needed in each case depends on the amount of apriori information in $\mathcal{M}$. In the extreme case that modelling is done from apriori knowledge alone, one need no experiment at all, independent of what would be the purpose of the modelling.

This structure of a program for model design will have to be filled with algorithms pertinent to various cases of model structures $\mathcal{M}$ and purposes $\mathcal{G}$. They are the subjects of other books (Eykhoff, 1974; Goodwin and Payne, 1977; Ljung, 1987a; Ljung and Söderström, 1983; Söderström and Stoica, 1989).

References

Remark: The list is not exhaustive, and no attempt has been made to refer to an 'original reference'. Instead, easily accessible journals are preferred. However, the literature on the foundations of modelling and identification is sparse, and seems to favour conferences as the forum of presentation. A possible explanation is that many ideas are speculative. Many of the references in this book to basic identification concepts and techniques come from the text books of Ljung (1987a) and of Söderström and Stoica (1989), whose terminology is much in agreement with this book. For the reason of easy accessibility the Systems & Control Encyclopedia by Sing has also been exploited. The reader who wants more detail is recommended to take on from there, using the excellent reference lists both in the Encyclopedia and in the text books.

Abramowitz, M. and I. A. Stegun (1964). *Handbook of Mathematical Functions*. National Bureau of Standards, Applied Mathematics Series, 55. Washington DC, USA.

Akaike, H. (1974). A new look at the statistical model identification. *IEEE Trans. Autom. Control, AC-19*, 716–723.

Akaike, H. (1976). Canonical correlation analysis of time series and the use of an information criterion. In R. K. Mehra and D. G. Lainiotis (Ed.) *System identification: Advances and case studies*. Academic Press, New York. pp. 27–96.

Anderson, B. D. O. and J. B. Moore (1971). *Linear Optimal Control*. Prentice-Hall, Englewood Cliffs, NJ.

Åström, K. J. (1970). *Introduction to Stochastic Control Theory*. Academic Press, New York.

Åström, K. J. (1972). System identification. *Proceedings of the International Symposium on Stability of Stochastic Dynamical Systems, Coventry, England*, pp. 35–55. Lecture Notes in Mathematics, Springer-Verlag, Berlin.

Åström, K. J. (1980). Maximum likelihood and prediction error methods. *Automatica, 16, 551-574.*

Åström, K.J. and P. Eykhoff (1971). System identification – a survey. *Automatica, 7, 123-162.*

Åström, K. J. and T. Söderström (1974). Uniqueness of the maximum likelihood estimates of the parameters of an ARMA model. *IEEE Trans. Autom. Control, AC-19, 769-773.*

Åström, K. J. and B. Wittenmark (1989). *Adaptive Control.* Addison-Wesley.

Baram, Y. (1980a). Nonstationary model validation from finite data records. *IEEE Trans. Autom. Control, AC-25, 10-19.*

Baram, Y. (1980b). Model validation using mismatched filters. *Int. J. Control, 32, 191-198.*

Bellman, R. and K. J. Åström (1970). On structural identifiability. *Math. Biosci., 7, 329-339.*

Betta, A. and D. A. Linkens (1990). Intelligent knowledge-based system for dynamic system identification. *IEE Proceedings, 137, 1-12.*

Billings, S. A. (1980). Identification of nonlinear systems – A survey. *IEE Proc. D, 127, 272-285.*

Billings, S. A., M. J. Korenberg, and S. Chen (1988). Identification of non-linear output-affine systems using an orthogonal least-squares algorithm. *Int. J. Control, 41, 303-344.*

Billings, S. A. and I. J. Leontaritis (1982). Parameter estimation techniques for nonlinear systems. *Proc. IFAC Symposium on Identification and System Parameter Estimation.* Washington DC, USA.

Billings, S. A. and W. S. F. Voon (1983). Structure detection and model validity tests in the identification of nonlinear systems. *IEE Proceedings, 130, 193-199.*

Blackman, R. B. and J. W. Tukey (1958). *The Measurement of Power Spectra.* Dover, New York.

Bohlin, T. (1971). On the problem of ambiguities in Maximum Likelihood identification. *Automatica, 7, 199-210.*

Bohlin, T. (1978). Maximum-power validation of models without higher-order fitting. *Automatica, 14, 137-146.*

Bohlin, T. (1986). Computer–aided grey–box identification. In C. I. Byrnes and A. Lindquist (Ed.), *Modelling, Identification and Robust Control*. Elsevier, North–Holland.

Bohlin, T. (1987a). Identification: Practical aspects. In M. G. Sing (Ed.), *Systems & Control Encyclopedia*, Vol. 4. Pergamon Press, Oxford. pp. 2301–2307.

Bohlin, T. (1987b). Evaluation of likelihood functions for nonlinear grey–box identification. *Preprints, IFAC 10th World Congress on Automatic Control*, Munich, FRG, vol 10. pp. 219–224.

Bohlin, T. (1987c). Validation of identified models. In M. G. Sing (Ed.), *Systems & Control Encyclopedia*, Vol. 7. Pergamon Press, Oxford. pp. 4996–5001.

Bose, M. K. (1987). Realization theory: Multivariate. In M. G. Sing (Ed.), *Systems & Control Encyclopedia*, Vol. 6. Pergamon Press, Oxford. pp. 3969–3972.

Caines, P. E. (1978). Stationary linear and nonlinear system identification and predictor set completeness. *IEEE Trans. Autom. Control, AC–23*, 583–594.

Caines, P. E. (1986). On the scientific method and the foundations of system identification. In
C. I. Byrnes and A. Lindquist (Ed.), *Modelling, Identification and Robust Control*. Elsevier, North–Holland.

Caines, P. E. and L. Ljung (1976). Prediction error estimators: asymptotic normality and accuracy. *Proceedings of the 1976 IEEE Conference on Decision and Control including the 15th Symposium on Adaptive Processes*, Clearwater, Fla., USA. pp. 652–658.

Caines, P. E. and S. P. Sethi (1979). Recursiveness, causality, and feedback. *IEEE Trans. Autom. Control, AC–24*, 113–115.

Chen, S., S. A. Billings, and W. Luo (1989). Orthogonal least squares methods and their applications to non–linear system identification. *Int. J. Control, 50*, 1973–1896.

Desrochers, A. and S. Mohsen (1984). On determining the structure of a non–linear system. *Int. J. Control, 40*, 923–938.

de Willemagne, C. and R. E. Skelton (1987). Model reduction using a projection formulation. *Int. J. Control, 46*, 2141–2169.

Elmqvist, H. (1980). Manipulation of continuous models based on equations to assignment statements. *Simulation of Systems '79. Proceedings of the 9th IMACS Congress*, Sorrento, Italy, 1979. pp. 15–21.

El-Sherief, H. and N. K. Sinha (1979). Identification and modeling for linear multivariable discrete–time systems: a survey. *J. Cybern.*, 9, 43–71.

Eykhoff, P. (1974). *System Identification*. Wiley, London.

Eykhoff, P. (1987). Identification: History. In M. G. Sing (Ed.), *Systems & Control Encyclopedia*, Vol. 4. Pergamon Press, Oxford. pp. 2270–2273.

Eykhoff, P. (1988). A bird's eye view on parameter estimation and system identification. *Automatisierungstechnik*, 36, 413–420 and 472–479.

Farlow, St. J. (1986). *Self–organizing methods in modelling: GMDH–type algorithms*. Marcel Decker, New York.

Fasol, K. H. and H. P. Jorgl (1980). Principles of model building and identification. *Automatica*, 16, 505–518.

Fel'dbaum, A. (1965). *Optimal Control Systems*. Academic Press.

Fowler, T. B. (1989). Application of stochastic control techniques to chaotic nonlinear systems. *IEEE Trans. Autom. Control, AC–34*, 201–205.

Fuchs, J. J. (1987). On estimating the order of an ARMA process. *Automatica*, 23, 779–782.

Gaines, B. R. (1977). System identification, approximation and complexity. *Int. J. Gen. Syst.*, 3, 145–174.

Gevers, M. and G. Bastin (1982). What does system identification have to offer? *IFAC Symposium on Identification and System Parameter Estimation*, Washington DC., USA.

Gevers, M. and L. Ljung (1986). Optimal experiment design with respect to the intended model application. *Automatica*, 22, 543–554.

Glover, K. (1987). Identification: Frequency–domain methods. In M. G. Sing (Ed.), *Systems & Control Encyclopedia*, Vol. 4. Pergamon Press, Oxford. pp. 2264–2270.

Gomide, F., W. C. Amaral, L. V. R. Arruda, G. Favier, and W. Fontanini (1988). Expert system identification: The supervisory approach. *4th IFAC Symposium on Computer Aided Design in Control Systems*. Beijing, China. pp. 1215–1220.

Goodrich, R. L. and P. E. Caines (1979a). Necessary and Sufficient Conditions for local second–order identifiability. *IEEE, AC–24*, 125.

Goodrich, R. L. and P. E. Caines (1979b). Linear system identification from nonstationary cross–sectional data. *IEEE, AC–24*, 403–411.

Goodwin, G. C. (1982). Experiment design. *Proc. IFAC Symposium on Identification and System Parameter Estimation*. Washington DC, USA 1982.

Goodwin, G. C. and R. I. Payne (1977). *Dynamic System Identification: Experiment Design and Data Analysis*. Academic Press, New York.

Goodwin, G. C. and M. E. Salgado (1989). A stochastic embedding approach for quantifying uncertainty in the estimation of restricted complexity models. *Int. J. Adaptive Control and Signal Processing, 3*, 333–356.

Goodwin, G. C. and K. S. Sin (1984). *Adaptive Filtering Prediction and Control*. Information and System Science Series, Prentice–Hall, Englewood Cliffs, NJ.

Graebe, S. (1990). Theory and Implementation of Gray Box Identification. Thesis, TRITA–REG 90/3 and 90/6, Automatic Control, Royal Institute of Technology, Stockholm, Sweden.

Gruhl, J. (1979). Model validation. *Proceedings of the International Conference on Cybernetics and Society*, Denver, CO, USA. pp. 536–541.

Guidorzi, R. P. (1981). Invariants and canonical forms for systems structural and parametric identification. *Automatica, 17*, 117–133.

Gustavsson, I., L. Ljung, and T. Söderström (1977). Identification of processes in closed loop – identifiability and accuracy aspects. *Automatica, 13*, 59–75.

Haest, H., G. Bastin, M. Gevers, and V. Wertz (1988). An expert work station for system identification. *4th IFAC Symposium on Computer Aided Design in Control Systems*. Beijing, China. pp. 1355–1360.

Hernandez, D. B. (1987). Model structure simplification. In M. G. Sing (Ed.), *Systems & Control Encyclopedia*, Vol. 5. Pergamon Press, Oxford. pp. 3079–3082.

Ivakhnenko, A. G. (1970). Heuristic self–organization in problems of engineering cybernetics. *Automatica*, *6*, 207–219.

Jamshidi, M. and C. J. Hergel (1985). *Computer–aided control systems engineering*. Elsvier, Amsterdam.

Janssen, P., P. Stoica, T. Söderström, and P. Eykhoff (1988). Cross-validation ideas in model structure selection for multivariable systems. *4th IFAC Symposium on Computer Aided Design in Control Systems*. Beijing, China. pp. 671–675.

Jaynes, E. T. (1957). Information theory and statistical mechanics. *Phys. Rev.*, *106*, 520–630.

Kashyap, R. L. (1980). Inconsistency of the AIC rule for estimating the order of autoregressive models. *IEEE–AC*, *25*, 996–998.

Kashyap, R. L. and A. Ramachandra Rao (1976). *Dynamic Stochastic Models from Empirical Data*. Academic Press, New York.

Karny, M. (1983). Algorithms for determining the model structure of a controlled system. *Kybernetika*, *19*, 164–178.

Kendall, M. G. and A. Stuart (1967). *The Advanced Theory of Statistics*, Vol. 2. Griffin, London.

Kendall, M. G. and A. Stuart (1969). *The Advanced Theory of Statistics*, Vol. 1. Griffin, London.

Kortmann, M. and H. Unbehauen (1988). Two algorithms for model structure determination of nonlinear dynamic systems with applications to industrial processes. *Identification and System Parameter Estiomation 1988. Selected papers from the Eighth IFAC/IFORS Symposium, Beijing, China*. Vol. 2, pp. 640–656.

Kramer, S. C. and H. W. Sorenson (1988a). Bayesian parameter estimation. *IEEE Trans. Autom. Control*, *33*, 217–222.

Kramer, S. C. and H. W. Sorenson (1988b). Recursive Bayesian estimation using piece-wise constant approximations. *4th IFAC Symposium on Computer Aided Design in Control Systems*. Beijing, China. pp. 789–801.

Kubrusly, C. S. (1984). On identifiability of linear dynamical systems. *Mat. Appl. & Comput.*, *3*, 219–256.

Lamda, S. S. (1987). Model simplification. In M. G. Sing (Ed.), *Systems & Control Encyclopedia*, Vol. 5. Pergamon Press, Oxford. pp. 3071–3077.

Lehmann, E. L. (1959). *Testing Statistical Hypotheses*. Wiley, London.

Leontaritis, I. J. and S. A. Billings (1987). Model selection and validation methods for non–linear systems. *Int. J. Control*, *45*, 311–341.

Leontaritis, I. J. and S. A. Billings (1985). Input–output parametric models for non–linear systems. *Int. J. Control*, *41*, 303–344.

Ljung, L. (1978). Convergence analysis of parametric identification methods. *IEEE Trans. Autom. Control*, *AC–23*, 770–783.

Ljung, L. (1979). Asymptotic behaviour of the extended kalman filter as a parameter estimator for linear systems. *IEEE Trans. Autom. Control*, *AC–24*, 36–50.

Ljung, L. (1982a). Model validation. *Identification and System Parameter Estimation. Proceedings of the Sixth IFAC Symposium, Washington DC, USA*, pp. 73–75.

Ljung, L. (1982b). Aspects on the system identification problem. *Signal Processing*, *4*, 445–456.

Ljung, L. (1986a). Building models for a specified purpose using system identification. *Simulation of Control Systems. Selected Papers from the IFAC Symposium, Vienna, Austria*. pp. 1–5.

Ljung, L. (1986b). System identification, model simplification and the design objective. *Proceedings of the 25th IEEE Conference on Decision and Control, Athens, Greece*, pp. 783–788.

Ljung, L. (1987a). *System Identification. Theory for the User*. Prentice-Hall, Englewood Cliffs, USA.

Ljung, L. (1987b). Identification: Maximum likelihood method. In M. G. Sing (Ed.), *Systems & Control Encyclopedia*, Vol. 4. Pergamon Press, Oxford. pp. 2284–2287.

Ljung, L. (1987c). Identification: Basic problems. In M. G. Sing (Ed.), *Systems & Control Encyclopedia*, Vol. 4. Pergamon Press, Oxford. pp. 2239–2245.

Ljung, L. and P. E. Caines (1979). Asymptotic normality of prediction error estimators for approximate system models. *Stochastics, 3*, 29–46.

Ljung, L. and K. Glover (1981). Frequency domain versus time domain methods in system identification. *Automatica, 17*, 71–86.

Ljung, L. and T. Söderström (1983). *Theory and Practice of Recursive Identification*. MIT Press, Cambridge, Mass.

Lofgren, L. (1987). Complexity of systems. In M. G. Sing (Ed.), *Systems & Control Encyclopedia*, Vol. 1. Pergamon Press, Oxford. pp. 704–709.

Madanski, A. (1959). The fitting of straight lines when both variables are subject to error. *J. Amer. Statist. Assoc., 54*, 173.

Marcus, S. I. (1987). Stochastic integrals and stochastic calculus. In M. G. Sing (Ed.), *Systems & Control Encyclopedia*, Vol. 4. Pergamon Press, Oxford. pp. 4631–4637.

Mehra, R. K. (1979). Nonlinear system identification: Selected survey and recent trends. *Fifth IFAC Symposium on Identification and System Parameter Estimation*. Darmstadt, FRG.

Monaco, S. and D. Normand–Cyrot (1987). Input–output maps of nonlinear discrete–time systems. In M. G. Sing (Ed.), *Systems & Control Encyclopedia*, Vol. 4. Pergamon Press, Oxford. pp. 2523–2527.

Nakamori, Y. (1989). Development and application of an interactive modeling support system. *Automatica, 25*, 185–206.

Natke, H. G. and M. Samirowski (1988). On methods of structure identification for a class of nonlinear mechanical systems. *Identification and System Parameter Estimation. Selected papers from the Eighth IFAC/IFORS Symposium, Beijing, China*. Vol. 2, pp. 637–642.

Nguyen, V. V. and E. F. Wood (1982). Review and unification of linear identifiability concepts. *SIAM Rev., 24*, 34–51.

Niedzwiecki, H. (1990). Recursive functional series modeling estimators for identification of the time–varying plants – More bad than good news? *IEEE Trans. Autom. Control, AC–35*, 610–616.

Nurmi, H. (1978). On strategies of cybernetic model–building. *Kybernetes, 7*, 13–18.

Perel'man, I. I. (1984). Methodology of selecting the model structure for plant identification. *Autom. & Remote Control, 44*, 1389–1408.

Peterka, V. (1981). Bayesian system identification. *Automatica, 17*, 41–53.

Peterka, V. and M. Karny (1979). Bayesian system classification. *Proc. 5th IFAC Symposium on Identification and System Parameter Estimation*, Darmstadt, FRG.

Porat, B. (1987). Stationary time series and their spectra. In M. G. Sing (Ed.), *Systems & Control Encyclopedia*, Vol. 7. Pergamon Press, Oxford. pp. 4556–4566.

Rada, R. (1987). Artificial intelligence. In M. G. Sing (Ed.), *Systems & Control Encyclopedia*, Vol. 1. Pergamon Press, Oxford. pp. 305–307.

Reckhow, K. H. (1987). Validation of simulation models: Philosophy and statistical methods of confirmation. In M. G. Sing (Ed.), *Systems & Control Encyclopedia*, Vol. 7. Pergamon Press, Oxford. pp. 5011–5015.

Rissanen, J. (1976). Minimax entropy estimation of models for vector processes. In R. K. Mehra and D. G. Lainiotis (Ed.), *System identification: Advances and case studies*. Academic Press, New York. pp. 27–97.

Rissanen, J. (1978). Modelling by shortest data description. *Automatica, 14*, 465–471.

Rissanen, J. (1983). A universal data compression system. *IEEE Trans. Inf. Theory, IT–29*, 656–664.

Rissanen, J. and G. Langdon (1981). Universal modeling and coding. *IEEE Trans. Inf. Theory, IT–27*, 12–23.

Sage, A. P. (1987a). Bayes' rule. In M. G. Sing (Ed.), *Systems & Control Encyclopedia*, Vol. 1. Pergamon Press, Oxford. p. 414.

Sage, A. P. (1987b). Expert systems. In M. G. Sing (Ed.), *Systems & Control Encyclopedia*, Vol. 2. Pergamon Press, Oxford. pp. 1569–1573.

Schneeweiss, C. (1987). On a formalisation of the process of quantitative model building. *Eur. J. Oper. Res., 29*, 24–41.

Söderström, T. (1975). Test of pole–zero cancellation in estimated models. *Automatica, 11*, 537–41.

Söderström, T. (1977). On model structure testing in system identification. *Int. J. Control, 26*, 1–18.

Söderström, T. (1981). On a method for model structure selection in system identification. *Automatica, 17*, 387–388.

Söderström, T. (1987). Identification: Model structure determination. In M. G. Sing (Ed.), *Systems & Control Encyclopedia*, Vol. 4. Pergamon Press, Oxford. pp. 2287–2292.

Söderström, T. and K. Kumamaru (1985). Some model validation criteria based on a Kullback discrimination index. *Proceedings of the 24th IEEE Conference on Decision and Control*, Fort Lauderdale, FL, USA, vol 1, pp. 219–224.

Söderström, T., L. Ljung, and I. Gustavsson (1978). A theoretical analysis of recursive identification methods. *Automatica, 14*, 231–244.

Söderström, T. and P. Stoica (1989). *System Identification*. Prentice Hall International, London.

Stoica, P. and T. Söderström (1982). On the parsimony principle. *Int. J. Control, 36*, 409–418.

Stoica, P., P. Eykhoff, P. Janssen, and T. Söderström (1986). Model-structure selection by cross–validation. *Int. J. Control, 43*, 1841–1878.

Strejc, V. (1981). Trends in identification. *Automatica, 17*, 7–21.

Unbehauen, H. and B. Göhring (1974). Tests for determining model order in parameter estimation. *Automatica, 10*, 233–244.

Vadja, S. (1983). Structural identifiability of dynamical systems. *Int. J. Systems Sci., 14*, 1229–1247.

Vadja, S. and H. Rabitz (1989). State isomorphism approach to global identifiability of nonlinear systems. *IEEE Trans. Autom. Control, AC–34*, 220–223.

Wahlberg, B. (1988). On the identification of continuous time dynamical systems. *Identification and System Parameter Estimation 1988. Selected papers from the Eighth IFAC/IFORS Symposium, Beijing, China*. Vol. 1, pp. 435–440.

van den Boom, A. J. W. (1988). CADACS developments in Europe. *4th IFAC Symposium on Computer Aided Design in Control Systems*. Beijing, China. pp. 65–73.

van Overbeek, A. J. M. and L. Ljung (1982). On–line structure selection for multivariable state–space models. *Automatica, 18*, 529–543.

Vansteenkiste, G. C. and J. A. Spriet (1987). Identification: Model structure determination. In M. G. Sing (Ed.), *Systems & Control Encyclopedia*, Vol. 8. Pergamon Press, Oxford. pp. 4690–4692.

Wall, K. D. and J. H. Westcott (1974). Macroeconomic modelling for control. *IEEE, AC–19*, 862–873.

Wellstead, P. E. (1981). Non–parametric methods of system identification. *Automatica, 17*, 55–69.

Wilks, S. S. (1962). *Mathematical Statistics*. Wiley, London.

Willems, J. C. (1986). Modelling, approximation and complexity of linear systems. In C. I. Byrnes and A. Lindquist (Ed.), *Modelling, Identification and Robust Control*. Elsevier, North–Holland.

Willems, J. C. (1987). Modelling linear systems. *Preprints, IFAC 10th World Congress on Automatic Control*, Munich, FRG, vol 9. pp. 1–10.

Young, P., J. Naughton, C. Neethling, and S. Shellswell (1973). Macro–economic modelling – a case study. *3rd IFAC Symposium on Identification and System Parameter Estimation*, the Hague, Netherlands.

Zadeh, L. A. (1962). From circuit theory to system theory. *Proc. IRE, 50*, 856–865.

Zypkin, J. S. (1987). *Grundlagender informationellen Theorie der Identifikation*. VEB Verlag, Berlin.

Glossary of notations

See also section 9.7 "Terminology revisited"

General symbols and notations

$\{x\}$	A set of elements x
$\{x\|C\}$	A set of elements x satisfying condition C
$\|$	'Conditional' symbol
$x(t)$	Instantaneous value of x at time t
$x(k)$	Sampled value #k of x
x_t	History of x up to time t
x_k	Record of length k of sampled value
$\text{Ind}\{\mathcal{E}\}$	Indicator function for the event $\mathcal{E}$
$\text{Pr}\{\mathcal{E}\}$	Probability of the event $\mathcal{E}$
$P[x] \triangleq \text{Pr}\{\mathbf{x} \leq x\}$:	Cumulative probability distribution function of $\mathbf{x}$ at the point x
$p[x]$	Probability density of $\mathbf{x}$ at the point x
$p[x,y]$	Simultaneous probability density of $\mathbf{x}$ and $\mathbf{y}$ at the point (x,y)
$p[x\|y]$	Conditional probability density of $\mathbf{x}$ given that $\mathbf{y} = y$
$E\{\mathbf{x}\}$	Expected value of $\mathbf{x}$
$E\{\mathbf{x}\|\mathbf{y}\}$	Expected value of $\mathbf{x}$ given $\mathbf{y}$
$\text{Cov}\{\mathbf{x},\mathbf{y}\}$	Covariance matrix of $\mathbf{x}$ and $\mathbf{y}$
$ML\{\mathbf{M}\|\mathcal{M},\mathbf{d}\}$:	The Maximum Likelihood estimate of $\mathbf{M}$ given structure $\mathcal{M}$ and data $\mathbf{d}$
ϕ	Gaussian density function
Φ	Gaussian cumulative distribution function
$\text{Chi\_square}(\chi^2,r)$:	Chi–square distribution function with r degrees of freedom
$\text{Chi\_square}(\chi^2,r,\rho^2)$:	Non–central chi–square distribution function with r degrees of freedom and 'offset' ρ

$\chi_r^2(\alpha)$ The value of the chi–square distribution function for r degrees of freedom and probability $1-\alpha$

R   Real space (of any dimension)

$f(\bullet)$ A function with unspecified argument

δ Dirac's or Kronecker's delta function

$\delta\theta$ The variation of θ

(x,y) A pair of elements

$(x,\bullet)$ A set of pairs with one element unspecified

$a \rightarrow b$ a tends to b (convergence symbol)

$a \leftarrow b$ b replaces a (assignment symbol)

$a \mapsto b$ b is a function of a (mapping symbol)

$\hat{a}$ Estimate of a

$\nabla_\theta = \mathrm{grad}_\theta$: Gradient operator = row vector of partial differentiation operators with respect to components in θ

$\nabla_{\theta\theta} \triangleq \nabla_\theta^T \nabla_\theta$: The second–order gradient operator matrix

Bold characters generally mean actual values (known or unknown), such as data and 'true' object and parameter values

Sets and variables

a Physical parameters

A Model algorithm (including parameter values)

$\mathcal{A}$ Model algorithm (excluding parameter values)

A^c, A^d Model algorithms for experimenter and object

$(\mathcal{A}, \mathcal{N}_n)$ Composite structure of complexity n

$(\mathcal{A}, \nu, \theta)$ Parametric model

$a_\theta = \mathrm{grad}_\theta \mathcal{P}(\nu,0)$: Scaling matrix

$B(M|\mathbf{c},\mathbf{d})$ 'Distance' between model and data

$B(\nu,\theta|n,\mathbf{c},\mathbf{d})$: Confidence function for rejecting parametric models $(\mathcal{A},\nu,\theta)$ of maximum complexity n

$B(\nu,\theta|\nu',\mathbf{c},\mathbf{d})$: Confidence function for rejecting parametric models $(\mathcal{A},\nu,\theta)$ within the structure $(\mathcal{A},\nu')$, $\nu' > \nu$

$B(\theta|\nu',\mathbf{c},\mathbf{d}) = B(\nu,\theta|\nu',\mathbf{c},\mathbf{d})$

$B(\hat{\nu},\hat{\theta}|n,n',\mathbf{c},\mathbf{d})$: Confidence function for rejecting parameteric ML models $(\hat{\nu},\hat{\theta})$ of complexity n within structures of complexity n'.

$\mathcal{B}(\mathbf{c},\mathbf{d})$ The set of unfalsified models

$\mathcal{B}(\gamma|\mathbf{c},\mathbf{d})$ The acceptance domain = The set of models not falsified with confidence γ

$\mathcal{B}(\gamma|n,\mathbf{c},\mathbf{d})$: Acceptance domain for models of complexity n

$\mathcal{B}(\gamma|\nu,\mathbf{c},\mathbf{d})$: Acceptance domain for parametric models with structure index ν

$\mathcal{B}(\gamma|n,n',\mathbf{c},\mathbf{d})$: Acceptance domain for structures $\mathcal{M}(n)$ within $\mathcal{M}(n')$

$\mathcal{B}(\gamma|\nu',\nu,\mathbf{c},\mathbf{d})$: Acceptance domain for parametric structures $(\mathcal{A},\nu')$ within the alternative structure $(\mathcal{A},\nu)$, $\nu > \nu'$

$\mathcal{B}_\infty \triangleq \mathcal{B}(\mathbf{c}_\infty,\mathbf{d}_\infty)$: The set of unfalsified models for infinite sample length

c Control variables

$\mathbf{c}$ The stimulus

$(\mathbf{c},\mathbf{d})$ The data

$c_N \triangleq \{c(k)|k=1,\ldots,N\}$: A stimulus/control sequence

C Actuator model

d Response variables

$\mathbf{d}$ The response

D Sampler model

$d_N \triangleq \{d(k)|k=1,\ldots,N\}$: A response sequence

$\mathcal{D}(\gamma|M)$ The complement of the 'critical domain' for data generated by model M

$e(k) \triangleq [\nabla_d \mathbf{w}^d(k|M^d)]^{-1}\, \mathbf{w}^d(k)$: Residual

E The 'error' model

F Internal plant model

F^ν, F^z, F^y Internal models for environment, process proper, and sensor

$G(\mathbf{M},M)$ Tolerance function for purposive models M

$\mathcal{G}(\mathbf{M})$ The set of purposive models

$\mathfrak{J}$ Identifier or state–transition function

$\mathfrak{J}(c,d) \triangleq |\det \operatorname{grad}_{cd} W(c,d)|$: Jacobian of W

$\mathcal{J}^d(c,d) \triangleq |\det \operatorname{grad}_d W^d(c,d)|$: Jacobian of W^d

H_0, H_1	'Null' and 'alternative' hypotheses
k	Sampling counter
K	Length of application interval
$L(M|\mathbf{c},\mathbf{d})$	Likelihood of M given $(\mathbf{c},\mathbf{d})$
m	Experiment number
$\bar{m}$	Maximum extension of experiment
M	Model of the data source
$\hat{M}$	Optimal model
$\mathbf{M}$	Source–Equivalent or Purpose–Equivalent model
$\mathcal{M}$	The set of all tentative models = the model structure
$\mathcal{M}(n)$	Model structure of complexity n

$\bar{\mathcal{M}} = \mathcal{M}(\bar{n})$: Widest model structure

M^c, M^d	Experimenter and object models
$\mathbf{M}^c, \mathbf{M}^d$	Experimenter– and Object–Equivalent models

$\mathcal{M}_\nu \triangleq (\mathcal{A},\nu)$: Simple parametric structure

$\mathcal{M}(\nu)$	All structures $\mathcal{M}_{\nu'}$, $\nu' \leq \nu$
n	Structure number or complexity indicator
$\mathbf{n}$	Minimum sufficient complexity
$\bar{n}$	Maximum admissible complexity
N	Length of data sample

$\mathcal{N} = \{\mathcal{N}_1, \mathcal{N}_2, \dots, \mathcal{N}_{\bar{n}}\}$: Partially ordered set of structure indices

$\mathcal{N}_n$	Set of structure indices of complexity n
$\hat{p}$	Sample density of data $(\mathbf{c},\mathbf{d})$
$p[c,d|M]$	Probabilistic source model

$p[c(k),d(k)|c_{k-1},d_{k-1},k,M]$: Probabilistic dynamic source model

$\mathcal{P}$	Parameter map

$Q(M|\mathbf{c}_N,\mathbf{d}_N) \triangleq -\log L(M|\mathbf{c}_N,\mathbf{d}_N)/N$

$$= -\frac{1}{N}\sum_{k=1}^{N} \log p[\mathbf{c}(k),\mathbf{d}(k)|\mathbf{c}_{k-1},\mathbf{d}_{k-1},k,M]$$

= Likelihood loss function

$\bar{Q}_N(M|\mathbf{X},\mathbf{S}) \triangleq E\{Q(M|\mathbf{c}_N,\mathbf{d}_N)|\mathbf{X},\mathbf{S}\}$: Expected Likelihood loss

$\bar{Q}(M\vert\mathbf{X},\mathbf{S})$	Asymptotic value of the expected Likelihood loss
$\mathfrak{R}(\mathfrak{C})$	Range of model validity
R_θ	Covariance matrix of the limiting aposteriori distribution of θ
R_ξ	Covariance matrix of parameter–free statistic
S	Tentative object
$\mathbf{S}$	The 'true' object
$\mathcal{G}$	Set of Source–Equivalent systems
$\mathcal{G}^d$	Set of Object–Equivalent systems
t	Time variable
T	Application interval
$\mathbf{T}$	The experiment interval
t_k	Sampling time #k
u	Exogenous plant input
$V(M\vert\mathbf{c},\mathbf{d})$	Validation criterion or loss function
$\mathcal{V}(\gamma\vert\mathbf{c},\mathbf{d})$	The set of validated models with confidence γ
$\mathbf{w}(k\vert M) \triangleq W[\mathbf{c}_k,\mathbf{d}_k,k\vert M]$	: Innovation
$\mathbf{w}^c,\mathbf{w}^d$	Innovations for experimenter and object models
$W(c,d)$	The inverse of the external source model $(c,d) = A(\omega)$
$W^d(c,d)$	The inverse of the external object model $d = A^d(c,\omega^d)$
x	State vector
X	Experimenter/controller
$\mathbf{X}$	The 'true' experimenter
$X(m)$	Experiment of extension m
$\bar{X} = X(\bar{m})$	Most extensive experiment
$\mathfrak{C}$	Set of admissible controllers
$(\mathbf{X},\mathbf{S})$	The data source
$(\mathbf{X},\mathbf{S},\mathbf{T})$	The experiment
y	Sensor output
z	Dependent plant output
$\mathbf{Z}(k)$	Weights in cross–correlation tests
α	Risk (of the first kind)

ϵ	Margin for acceptable approximation
γ	Threshold value/Confidence
Γ_θ	Choleski factor of covariance of aposteriori parameter distribution
θ	Free parameters
ν	Structure index
ν_n	Structure index of complexity n
(ν^M, θ^M)	Parametric model to be validated or falsified
$(\hat{\nu}, \hat{\theta})$	Maximum Likelihood estimates
(ν_n, θ_n)	Parametric model of complexity n
ξ	Parameter–free statistic
ω	'Hidden variable' or 'noise'
ω^c, ω^d	Experimenter and object noise

Subject index